Supercritical Fluid Processing of Food and Biomaterials

Supercritical Fluid Processing of Food and Biomaterials

Edited by

S. S. H. RIZVI
Professor of Food Engineering
Institute of Food Science
Cornell University
Ithaca

SPRINGER-SCIENCE+BUSINESS MEDIA, B.V.

First edition 1994

© 1994 Springer-Science+Business Media Dordrecht
Originally published by Chapman & Hall in 1994
Softcover reprint of the hardcover 1st edition 1994

Typeset in 10/12pt Times by ROM-Data Corporation, Falmouth

ISBN 978-1-4613-5907-4   ISBN 978-1-4615-2169-3 (eBook)
DOI 10.1007/978-1-4615-2169-3

A catalogue record for this book is available from the British Library

Library of Congress Catalog Card Number: 93–74990

∞ Printed on acid-free text paper, manufactured in accordance with
ANSI/NISO Z39.48-1992 (Permanence of Paper)

Preface

The need for understanding the fundamentals of supercritical fluid processing and their applications to ever-widening ranges of materials and conditions continues to expand. There has been much interest in the use of supercritical fluids as solvents in bioprocessing of food and related materials. Admittedly, a few successful applications of supercritical fluids could be cited but these are minuscule in comparison with the potential applications as yet undeveloped and unexploited.

This volume is based on the papers presented at the symposium on *Supercritical fluid processing of biomaterials: Basics of process design and applications* organized during the 8th World Congress of Food Science and Technology held in Toronto, Sept. 29–Oct. 4, 1991. The coverage represents the breadth of interest in this field around the world. I am indeed indebted to the authors who so willingly brought their work to the symposium and provided revised manuscripts of their papers for publication. I would also like to acknowledge the assistance of Professor M. LeMaguer of the University of Guelph for co-chairing the symposium.

The organization and successful completion of the symposium and the production of this volume is due to the assistance of the Technical Program Committee of the Congress and the cooperation of many people. I express my appreciation to them all.

S. S. H. Rizvi

Series foreword

The 8th World Congress of Food Science and Technology, held in Toronto, Canada, in 1991 attracted 1400 delegates representing 76 countries and all five continents. By a special arrangement made by the organisers, many participants from developing countries were able to attend. The congress was therefore a most important international assembly and probably the most representative food science and technology event in that respect ever held. There were over 400 poster presentations in the scientific programme and a high degree of excellence was achieved. As in previous congresses, much of the work reported covered recent research and this will since have been published elsewhere in the scientific literature.

In addition to presentations by individual researchers, a further major part of the scientific programme consisted of invited papers, presented as plenary lectures by some of the leading figures in international food science and technology. They addressed many of the key food issues of the day including advances in food science knowledge and its application in food processing technology. Important aspects of consumer interest and of the environment in terms of a sustainable food industry were also thoroughly covered. The role of food science and technology in helping to bring about progress in the food industries of developing countries was highlighted.

This book is part of a series arising from the congress and including full bibliographical details. The series editors are Professor M. A. Tung of the Technical University of Nova Scotia, Canada; and Dr G. E. Timbers of Agriculture Canada, Ottawa, Canada. The book presents some of the most significant ideas which will carry food science and technology through the nineties and into the new millenium. It is therefore essential reading for anyone interested in the subject, including specialists, students and general readers. IUFoST is extremely grateful to the organisers from the Canadian Institute of Food Science and Technology for putting together a first class scientific programme and we welcome the publication of this book as a permanent record of the keynote papers presented at the congress.

Dr D. E. Hood
(President, International Union of Food Science & Technology)

Contributors

A. G. Arreola   Food Science and Human Nutrition Department, FSB 341, University of Florida, IFAS, Gainesville, FL 32611-0163, USA.

M. E. Bailey   Department of Food Science and Human Nutrition, 249 Agricultural Engineering, University of Missouri, Columbia, MO 65211, USA.

M. Balaban   Food Science and Human Nutrition Department, FSB 341, University of Florida, IFAS, Gainesville, FL 32611-0163, USA.

A. Bertucco   Istituto di Impianti Chimici, University of Padova, Via Marzolo, Padova 9-35131, Italy.

A. R. Bhaskar   Institute of Food Science, Cornell University, Ithaca, NY 14853, USA.

L. P. Buckley   Atomic Energy of Canada Ltd., Research Company, Chalk River Nuclear Laboratories, Chalk River, Ontario K0J 1J0, Canada.

D. P. Byskal   Atomic Energy of Canada Ltd., Research Company, Whiteshell Laboratories, Pinawa, Manitoba R0E 1L0, Canada.

A. Castera   ITERG—French Institute for Fats and Oils, Rue Monge—Parc Industriel, F33600 Pessac, France.

R. R. Chao   Department of Food Science and Human Nutrition, 249 Agricultural Engineering, University of Missouri, Columbia, MO65211, USA.

C. B. Chidambara Raj   Institute of Food Science, Cornell University, Ithaca, NY 14853, USA.

J. A. Cornell   Department of Statistics, University of Florida, Gainesville, FL 32611, USA.

J. P. Ecalard	Even, Ploudaniel, France.
D. A. Evans	Norac Technologies Inc., Edmonton, Alberta T6E 5V2, Canada.
I. Flament	Firmenich S.A., Scientific Research Division, PO Box 239, CH-1211 Geneva 8, Switzerland.
G. Frakman	Norac Technologies Inc., Edmonton, Alberta T6E 5V2, Canada.
G. B. Guarise	Istituto di Impianti Chimici, University of Padova, Via Marzolo, Padova 9-35131, Italy.
K. Hatakeda	Government Industrial Research Institute, Tohoku 2-1, Nigatake-4-chome, Miyagino-ku, Sendai 983, Japan.
J.-H. Hsu	Department of Chemical Engineering, National Tsing Hua University, Hsinchu, Taiwan 30043, Republic of China.
H. Huang	Department of Food Science, Rutgers—The State University, New Brunswick, NJ 08903, USA.
K.-Y. Hwang	Separation Process Laboratory, Division of Chemical Engineering, KIST, PO Box 131-650, Cheongryang, Seoul 130-650, Korea.
Y. Ikushima	Government Industrial Research Institute, Tohoku 2-1, Nigatake-4-chome, Miyagino-ku, Sendai 983, Japan.
S. Ito	Government Industrial Research Institute, Tohoku 2-1, Nigatake-4-chome, Miyagino-ku, Sendai 983, Japan.
U. Keller	Firmenich S.A., Scientific Research Division, PO Box 239, CH-1211 Geneva 8, Switzerland.
J.-D. Kim	Separation Process Laboratory, Division of Chemical Engineering, KIST, PO Box 131-650, Cheongryang, Seoul 130-650, Korea.
Ž. Knez	Faculty of Technical Sciences, Department of Chemical Engineering, Smetanova u. 17, 62000 Maribor, Slovenia.
I. Krmelj	Faculty of Technical Sciences, Department of Chemical Engineering, Smetanova u. 17, 62000 Maribor, Slovenia.

B.-C. Lee   Separation Process Laboratory, Division of Chemical Engineering, KIST, PO Box 131-650, Cheongryang, Seoul 130-650, Korea.

Y. Y. Lee   Separation Process Laboratory, Division of Chemical Engineering, KIST, PO Box 131-650, Cheongryang, Seoul 130-650, Korea.

W. Majewski   Separex, Chemin des Blanches Terres, BP9, 54250 Champigneulles, France.

B. Manohor   Central Food Technological Research Institute, Mysore 570013, Karnataka, India.

M. R. Marshall   Food Science and Human Nutrition Department, FSB 341, University of Florida, IFAS, Gainesville, FL 32611-0163, USA.

P. Mengal   Separex, Chemin des Blanches Terres, BP9, 54250 Champigneulles, France.

V. K. Mishra   College of Home Sciences, Gujarat Agricultural University, Sardar, Krushinagar 385506, Dist. Banaskanta, North Gujarat, India.

S. J. Mulvaney   Department of Food Science, Stocking Hall, Cornell University, Ithaca NY 14853, USA.

K. Nakamura   Department of Biological and Chemical Engineering, Gunma University, Kiryu, Gunma 333376, Japan.

U. Nguyen   Norac Technologies Inc., Edmonton, Alberta T6E 5V2, Canada.

B. Ooraikul   Department of Food Science and Nutrition, 308 Home Economics Building, University of Alberta, Edmonton, Alberta T6G 2M8, Canada.

P. Pallado   Istituto di Impianti Chimici, University of Padova, Via Marzolo, Padova 9-35131, Italy.

C. A. Passey   St. Hyacinthe Food Research and Development Centre, 3600 Casavant Blvd. West, St. Hyacinthe, Quebec J2S 8E3, Canada.

A. J. Peplow   Food Science and Human Nutrition Department, FSB 341, University of Florida, IFAS, Gainesville, FL 32611-0163, USA.

M. Perrut   Separex, Chemin des Blanches Terres, BP 9, 54250 Champigneulles, France.

F. Posel   Faculty of Technical Sciences, Department of
                            Chemical Engineering, Smetanova u. 17, 62000
                            Maribor, Yugoslavia.

S. S. H. Rizvi   108 Stocking Hall, Cornell University, Ithaca, NY
                            14850-7210, USA.

N. Saito   Government Industrial Research Institute, Tohoku
                            2-1, Nigatake-4-chome, Miyagino-ku, Sendai 983,
                            Japan.

K. U. Sankar   Central Food Technological Research Institute,
                            Mysore 570013, Karnataka, India.

C.-S. Tan   Department of Chemical Engineering, National
                            Tsing Hua University, Hsinchu, Taiwan 30043,
                            Republic of China.

F. Temelli   Department of Food, Science and Nutrition, 308
                            Home Economics Building, University of Alberta,
                            Edmonton, Alberta T6G 2M8, Canada.

S. W. Vance   477 Bartolo Street, Pittsburgh, PA 15243, USA.

S. Vijayan   Atomic Energy of Canada Ltd., Research Com-
                            pany, Chalk River Nuclear Laboratories, Chalk
                            River, Ontario K0J 1J0, Canada.

C. I. Wei   Food Science and Human Nutrition Department,
                            FSB 341, University of Florida, IFAS, Gainesville,
                            FL 32611-0163, USA.

L. Wünsche   Firmenich S.A., Scientific Research Division, PO
                            Box 239, CH-1211 Geneva 8, Switzerland.

Z. R. Yu   Institute of Food Science, Cornell University,
                            Ithaca, NY 14853, USA.

Contents

7   Selecting a pump for supercritical fluid service   93

S.W. VANCE

8   Natural antioxidants produced by supercritical extraction   103

U. NGUYEN, D.A. EVANS and G. FRAKMAN

9   Separation of ethanol/water solution with supercritical CO_2 in the presence of a membrane   114

J.-H. HSU and C.-S. TAN

19 *In situ* monitoring of selective extraction of a mixture of higher fatty acids with supercritical carbon dioxide 244

Y. IKUSHIMA, N. SAITO, K. HATAKEDA and S. ITO

1 Fundamentals of processing with supercritical fluids

S. S. H. RIZVI, Z. R. YU, A. R. BHASKAR and
C. B. CHIDAMBARA RAJ

Abstract

The solvent power of dense gases for less volatile lipophilic materials is a subject of significant practical interest. Phase equilibrium data and solubility are the basis of successful engineering design of supercritical processing plants. Appreciable mass transfer resistances are frequently encountered in the extraction of natural products requiring high solvent circulation rates resulting in high recompression costs which may make the process unattractive. Investigations on liquid feed materials are relatively less common than that of solid feeds. Scale-up rules for continuous processing of biomaterials are not very well understood and further study is warranted. The fundamentals of processing with supercritical fluids and related issues are reviewed in this article.

1.1 Introduction

Separation processes play a crucial role in biomaterial processing. A gas, when compressed isothermally to pressures more than its critical pressure, exhibits enhanced solvent power in the vicinity of its critical temperature (Diepen and Scheffer, 1948). Such fluids are called supercritical fluids (SCF) and their corresponding thermodynamic state is illustrated in Figure 1.1. A supercritical fluid exhibits desirable transport properties that enhance its adaptability as a solvent for liquid extraction processes. This is indicated in Figure 1.2 (Debenedetti, 1986); note that the density of a SCF is closer to that of liquids and its viscosity is low comparable to that of gases. High density of SCF contributes to high diffusivity equivalent to that of liquids; hence the faster dissolution of solute particles in SCF has contributed to the increasing usage of SCF as solvents for extraction purposes. Although numerous such supercritical fluids can be adapted as solvents for liquid extraction processes (Schneider *et al.*, 1980), carbon dioxide is by far the most extensively used solvent due to its non-toxic, inert and non-flammable nature while remaining an inexpensive

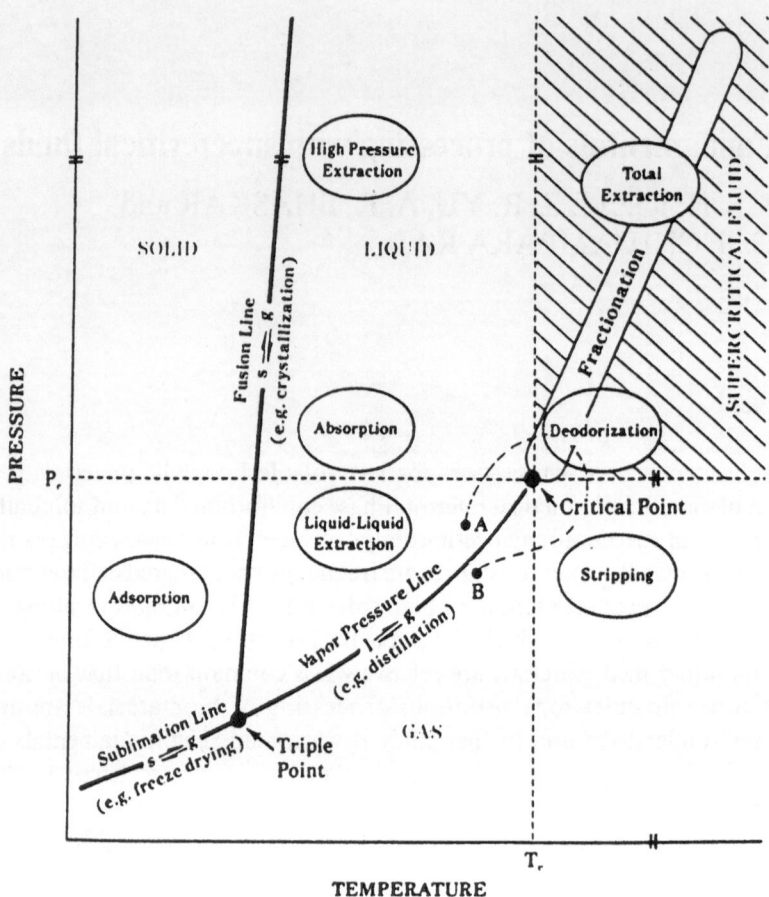

Figure 1.1 Phase diagram of a pure material and the thermodynamic state of various separation processes (Rizvi, 1987).

and environmentally acceptable substance. This fact is easily perceived from the information presented in Table 1.1, which is adapted from a list of 73 pure supercritical fluids arranged in the order of ascending critical temperature up to 719 K by Rainwater (1991). Products of biological origin are often thermally labile, lipophilic, non-volatile and required to be kept and processed around room temperature. Carbon dioxide (CO_2) has a critical temperature of 31°C which makes it a particularly attractive medium for the extraction of biological materials. In Table 1.1, fluids other than CO_2 showing critical temperature in its vicinity are often difficult to handle and to obtain in a pure form, may be toxic or give rise to explosive mixtures or ecologically prohibited or highly reactive chemicals. For such logical reasons, supercritical fluid extraction (SFE) using CO_2 has emerged as an attractive unit operation for the processing of food and biological materials. However, poor thermodynamic

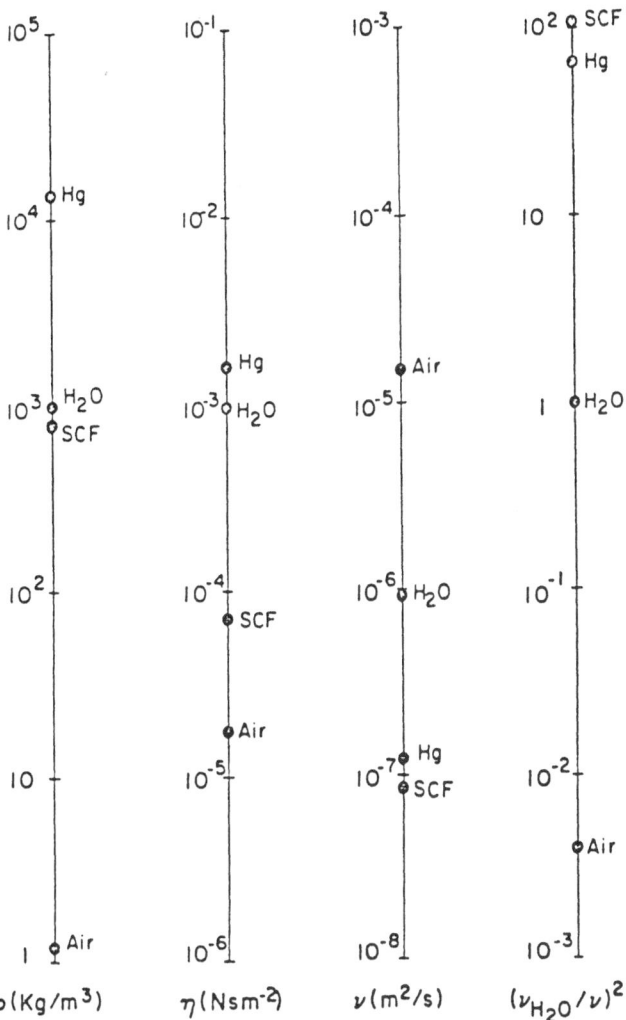

Figure 1.2 Comparison of physical properties of air, water, mercury (at 298 K and 1 bar) and CO_2 (SCF at 150 bar and 310 K) (Debenedetti, 1986).

descriptions of a supercritical solvent–solute mixture, high capital costs for a supercritical solvent extraction process as compared to an ordinary liquid extraction process at atmospheric conditions and an almost thorough absence of engineering data to facilitate scale-up and design are the prime factors that have contributed to the limited acceptance of SFE processes on a commercial scale. A sound engineering design requires reliable data on the transport of any given biological material into the supercritical carbon dioxide (SC-CO_2). This includes study of the thermodynamic properties such as fluid-liquid or fluid-solid equilibrium data for the biomaterial of interest in the neighborhood of the temperatures and pressures where processing is technically and

Table 1.1   Critical points of gases (Rainwater, 1991)

Fluid	Critical temperature (K)	Critical pressure (MPa)
Helium 3	3.311	0.115
Helium 4	5.188	0.227
Hydrogen	33.25	1.297
Neon	44.40	2.6545
Nitrogen	126.24	3.398
Carbon monoxide	132.85	3.494
Argon	150.66	4.860
Oxygen	154.58	5.043
Methane	190.55	4.595
Krypton	209.46	5.49
Carbon tetrafluoride	227.6	3.74
Ethylene	282.35	5.040
Xenon	289.7	5.87
Chlorotrifluoromethane	302.0	3.92
Carbon dioxide	304.17	7.386
Ethane	305.34	4.871
Acetylene	308.70	6.247
Nitrous oxide	309.15	7.285
Sulfur hexafluoride	318.82	3.765
Hydrogen chloride	324.55	8.263
Bromotrifluoromethane	340.08	3.956
Propylene	365.05	4.600
Chlorodifluoromethane	369.27	4.967
Propane	369.85	4.247
Hydrogen sulfide	373.40	8.963

economically attractive. A reasonable measurement of these parameters for the biomaterial + SC-CO_2 mixture in the vicinity of the pressure and temperature of interest, any mass transfer rate limitations from the bulk material into the supercritical solvent and the selectivity for the desired chemical species in comparison with the rest of the solubles in the biomaterial will also aid the design process and help to make cost-saving revisions of the design. In the following sections, we review the methods of measurement, correlation and utility of each of these parameters in an economical process framework. In addition, this paper describes the use of SC-CO_2 as a medium for enzymatic reactions.

1.2   Phase equilibrium and solubility

1.2.1   Measurement

Several techniques have been developed to measure the solubility of a solute in supercritical fluids. Precise measurements pertaining to the temperature and pressure dependence of solubility are indispensable; variation of solubility with density must also be ascertained. A consistent and accurate

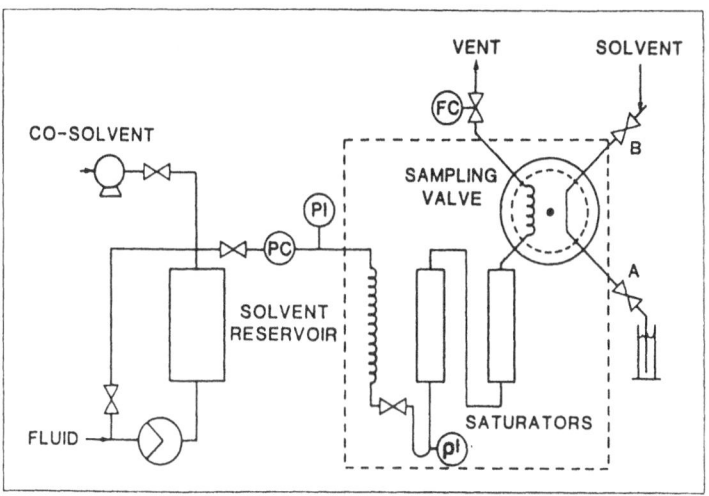

Figure 1.3 Dynamic flow apparatus for solubility measurements (Wong and Johnston, 1986).

thermodynamic phase equilibrium model is required to correlate the experimental solubility data and to predict the latter at different conditions of pressure and temperature for the purpose of process design. Methods of solubility measurement fall into four general categories, namely dynamic or flow methods, static or equilibrium methods, chromatographic methods and spectroscopic methods. The merits of each of these techniques have been compared by Bruno (1991). The dynamic or flow apparatus for solubility measurement is quite popular due to its simplicity. One version of this apparatus from the study of Wong and Johnston (1986) is shown in Figure 1.3. The solvent (with a co-solvent, if desired) flows into a packed extraction vessel containing the solute at a given flow rate. It is crucial that the solute and the solvent reach equilibrium at the experimental pressure and temperature before the solvent exits the flow cell. The compressed fluid mixture is then sampled and allowed to expand. The solute is then measured by gas chromatography, spectrophotometry, spectrofluorimetry or any other suitable analytical technique. The solubility thus measured must be independent of flow rate. Another alternative that is commonly used to obtain supercritical fluid-liquid equilibrium data is the static recirculation method. The components of interest are circulated through or agitated in a high pressure cell to establish equilibrium between the fluid and liquid phases. A static recirculation method was developed by Yu *et al.* (1992a) and it was used to measure phase equilibria of binary, ternary and multicomponent systems, as shown in Figure 1.4.

1.2.2 Theory

1.2.2.1 Solubility parameter. The solvent strength of a compressed gas (super-critical fluid) can be characterized by the Hildebrand solubility parameter

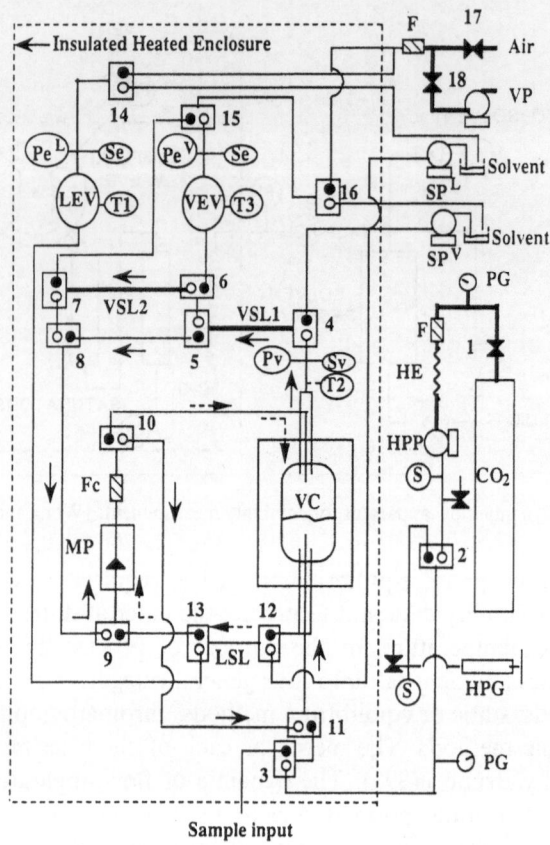

Figure 1.4 Schematic diagram of fluid-liquid equilibrium apparatus (Yu *et al.*, 1992).

which roughly measures the power of the solvent to dissolve various substances (Giddings *et al.*, 1969; Johnston, 1984). By applying van der Waals equation of state to dense gases (although deviations are known to occur), Giddings *et al.* (1968) showed that the solubility parameter, δ, can be related to the cohesive energy density and thus the density of the gas as given by

$$\delta = 1.25 \, P_c^{1/2} \, [\rho / \rho_{liq}] \tag{1.1}$$

where P_c is the critical pressure, ρ is the gas density and ρ_{liq} is the liquid density. At low pressures, the density of a gas is rather small and at near-critical conditions the density increases rapidly to that of the liquid. Hence, as seen from the expression above, the solubility parameter varies from zero at normal pressures increasing rapidly as the critical pressure is approached (Figure 1.5). Such a rapid change in solvent strength provides the basis for this powerful solvent extraction technique; the biomaterial is solubilized in the supercritical fluid at the pressure and temperature conditions of its highest solubility and then a small change in temperature or pressure or both will bring a substantial

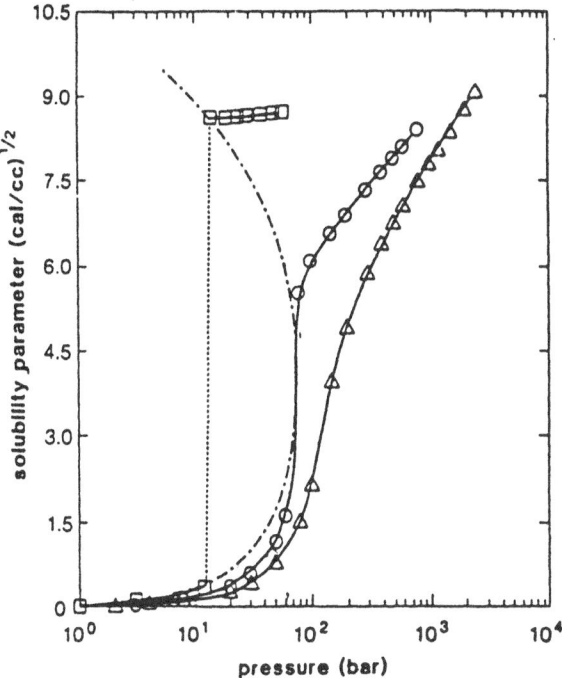

Figure 1.5 Hildebrand solubility parameter for CO_2 (Johnston, 1989). $\square$, – 30°C; $\bigcirc$; 31°C; $\triangle$, 70°C.

decrease in the solvent power of CO_2 and precipitate a single solute or fractionate a group of solutes exhibiting similar physico-chemical properties in sequential pressure and/or temperature manipulations.

1.2.2.2 Solubility as a function of density of pure CO_2. Supercritical extraction is based on the observation that near its critical temperature, the density of a pure supercritical fluid is highly sensitive to pressure. As a first approximation, since the solvent power of a fluid generally increases with density, slight changes in pressure at near-critical conditions can drastically affect solubilities. A density-based correlation model for solubility prediction is calculated from experimental data with statistical analysis. An empirical equation for determination of the solubilities of solids and liquids in dense gases was described by Chrastil (1982) and then improved by del Valle and Aguilera (1988). Chrastil proposed an empirical equation to calculate the solubility from the solvent density and corresponding operation temperature. The equation relates the solubility of the solute (c, g/l) to the solvent density (ρ, g/ml) and the temperature (T, °C):

$$\ln c = c_0 + \frac{c_1}{T} + c_2 \ln \rho \qquad (1.2)$$

del Valle and Aguilera presented a modified equation from Chrastil's model:

$$\ln c = c_0 + \frac{c_1}{T} + \frac{c_{11}}{T^2} + c_2 \ln \rho \tag{1.3}$$

where the c_0, c_1, c_{11}, c_2 are regressed constants in the equations. Using this method, it is easier to calculate solubility in the limited pressure and temperature ranges of measurement.

1.2.2.3 Equation of state. Accurate quantitative description of supercritical fluid mixture solubility behavior is essential to the process design and evaluation of extraction of natural materials. Cubic equations of state are popular for this purpose due to their simplicity. The van der Waals equation, the Soave modification of the Redlich–Kwong equation (1972), and the Peng–Robinson equation (1976) have been used for this purpose. As an example, consider the Peng–Robinson equation of state,

$$P = \frac{RT}{v-b} - \frac{a}{v(v+b) + b(v-b)} \tag{1.4}$$

where P is the pressure, R the gas constant, T the temperature, v the molar volume, a the attraction parameter and b the van der Waals co-volume. To extend the range of state from pure fluids to mixtures, the P–v–T relationship in the Peng–Robinson equation is displayed as

$$P = \frac{RT}{v-b_m} - \frac{a_m}{v(v+b_m) + b(v-b_m)} \tag{1.5}$$

where

$$a_m = \sum_{i=1}^{N} \sum_{j=1}^{N} X_i X_j a_{ij} \tag{1.6}$$

$$b_m = \sum_{i=1}^{N} X_i b_i \tag{1.7}$$

In equations (1.6) and (1.7), a_m is the mixture attraction parameter, a_{ij} is the binary attraction parameter between components i and j, b_m is the mixture van der Waals co-volume and X is the mole fraction in the liquid or fluid phase. The calculated mole fraction in liquid phase (x) and mole fraction in fluid phase (y) in the equation can be obtained from the fluid-liquid phase equilibria,

$$f_k^L (P, T, x_k) = f_k^F (P, T, y_k) \tag{1.8}$$

$$x_k \, \varphi_k^L = y_k \, \varphi_k^F \tag{1.9}$$

and material balance

$$x_1 + x_2 + \cdots + x_k = 1 \quad \text{and} \quad y_1 + y_2 + \cdots + y_k = 1 \qquad (1.10)$$

where f is the fugacity, φ the fugacity coefficient, L the liquid phase, V the fluid phase, and k the component identification. In the fluid-solid phase equilibria, the mole fraction of solute in the fluid phase can be obtained from the following thermodynamic relationship (Prausnitz et al., 1986).

$$y_2 \; \varphi_2^F \; P = P_2^S \; \varphi_2^S \; \exp\left(\int_{P_2^S}^{P} \frac{v_2^S}{RT} \, dP\right) \qquad (1.11)$$

Since the solid molar volume (v_2^s) is approximately constant with pressure, the fugacity coefficient (φ_2^s) is nearly equal to 1. Therefore, the mole fraction of solute in SC-CO_2 (y_2) in equation (1.11) can be integrated and rewritten as follows:

$$y_2 = \frac{P_2^S \exp\left(\dfrac{Pv_2^S}{RT}\right)}{\varphi_2^F P} \qquad (1.12)$$

As shown in equation (1.6), a_{ij} is applied to correlate the experimental data. The most commonly used mixing rules are presented by van der Waals. The van der Waals mixing rule uses one interaction parameter (k_{ij}) as shown below.

$$a_{ij} = \sqrt{a_i \, a_j}(1 - k_{ij}) \qquad (1.13)$$

The Panagiotopoulos and Reid (1986) mixing rule utilizes two interaction parameters (k_{ij} and k_{ji}) for a_{ij} to improve the accuracy of equations of state

$$a_{ij} = \sqrt{a_i \, a_j} \, ((1 - k_{ij}) + (k_{ij} + k_{ji}) \, X_i) \qquad (1.14)$$

The binary interaction parameters in equation (1.13) or (1.14) are regressed from the experimental data. An example of phase equilibria of methyl oleate in SC-CO_2 was correlated with the Peng–Robinson equation using the Panagiotopoulos and Reid mixing rule as shown in Figure 1.6 (Yu et al., 1992a).

A number of suggestions have evolved to improve the performance of the mixing rules (Huron and Vidal, 1979). However, many researchers including Mansoori and Ely (1985), Ely (1986), Mart et al. (1986) and Serbanovic and Djordjevic (1987) have observed that all the equation of state models incorporating mixing rules for supercritical fluid mixtures are very sensitive to the interaction parameters. There is some theoretical basis to the cubic equations of state; but these adjustable parameters have no theoretical basis. Often the success of a particular model to correlate phase equilibria depends upon the number of adjustable parameters introduced (Ekart et al., 1991); hence a model is not necessarily superior because it gives better representation of a given system than other models.

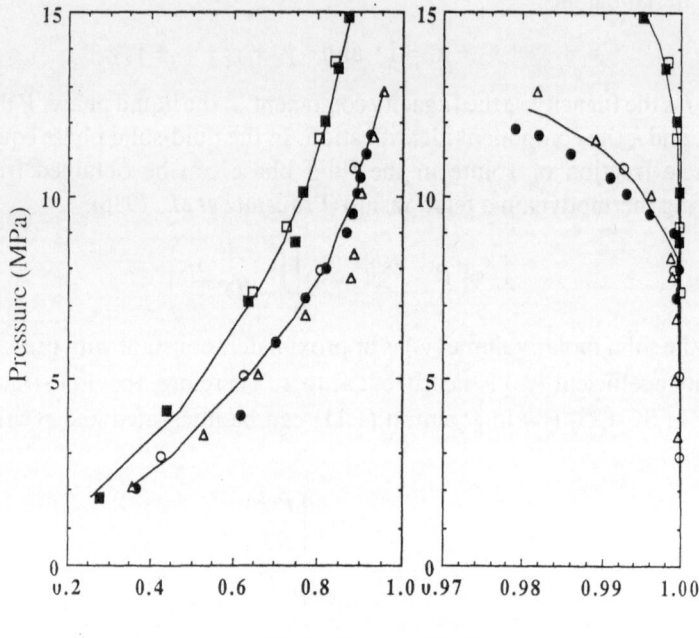

Mole fraction of CO$_2$

Figure 1.6   Measured and calculated fluid-liquid equilibrium data for the SC-CO$_2$ + methyl oleate system (Yu *et al.*, 1992a). ○ ⊔, 313.15 and 333.15 K, ● ■, Inomata *et al.* (1989) at 313.15 and 333.15 K; △, Gunawan (1990) at 313.15 K; ——, Peng–Robinson equation of state with Panagiotopoulos and Reid mixing rule at 313.15 and 333.15 K.

1.2.2.4   Enhancement factor.   Another term that is frequently used to quantify the increase in solubility in supercritical fluids is the enhancement factor (*E*); it is the ratio of actual solubility to the solubility in an ideal gas ($y_i = P_2^s/P$) as defined by equation (1.15).

$$E = \frac{y_2}{y_i} = \frac{\exp\left(\dfrac{P v_2^s}{RT}\right)}{\varphi_2} \qquad (1.15)$$

Large enhancement factors of the order of 10^5–10^7 are frequently observed due to small values of φ_2. As an example, the enhancement factor for the solubility of naphthalene in SC-CO$_2$ is shown in Figure 1.7. Small amounts of co-solvents (or entrainers) have been shown to increase significantly the solubility of certain substances in supercritical fluids. It may be attributed to certain orientation forces or specific solute/co-solvent interactions such as hydrogen bond formation. Wong and Johnston (1986) have measured the co-solvent-induced solubility enhancements for sterols. For example, in a 3.5 mol% methanol/SC-CO$_2$ mixture, the solubility of cholesterol was 7.2 times greater than its solubility in pure SC-CO$_2$ at the same conditions of pressure and temperature.

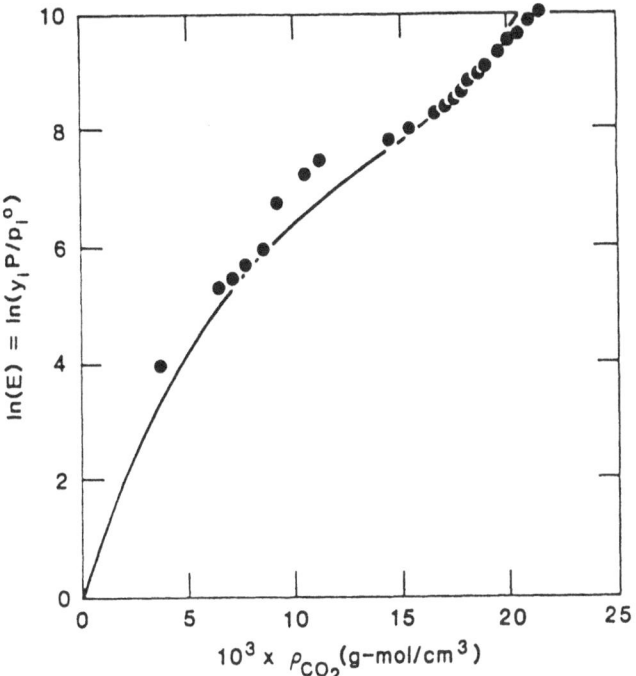

Figure 1.7 Enhancement factor for the solubility of naphthalene (Cochran *et al.*, 1987).

1.2.2.5 Distribution coefficient and selectivity. The distribution coefficient $(K = y/x)$ is the key variable that determines the distribution of a biomaterial between the liquid (or raffinate) (x) and fluid (or extract) phases (y). For a multicomponent system, the distribution coefficient for component i (K_i) is defined as

$$K_i = \frac{y_M}{x_M} \frac{C_i^F}{C_i^L} \tag{1.16}$$

where y_M and x_M are the mass fractions of mixture (M) in the fluid (F) and liquid (L) phases, respectively, and C_i^F and C_i^L are the concentrations of component i in the mixture in that phase. As an example, the distribution of cholesterol in the fluid and liquid phases at different pressures and temperatures is shown in Figure 1.8. The results show that the K values of cholesterol increase as the pressure increases at constant temperature, but decrease as the temperature increases at constant pressures below 30 MPa.

In order to understand the separation of component i from component j, selectivity is used as an index. For example, for the SC-CO_2 + triglycerides + cholesterol system, the selectivity is defined as

$$\beta_{cholesterol/triglycerides} = \frac{K_{cholesterol}}{K_{triglycerides}} \tag{1.17}$$

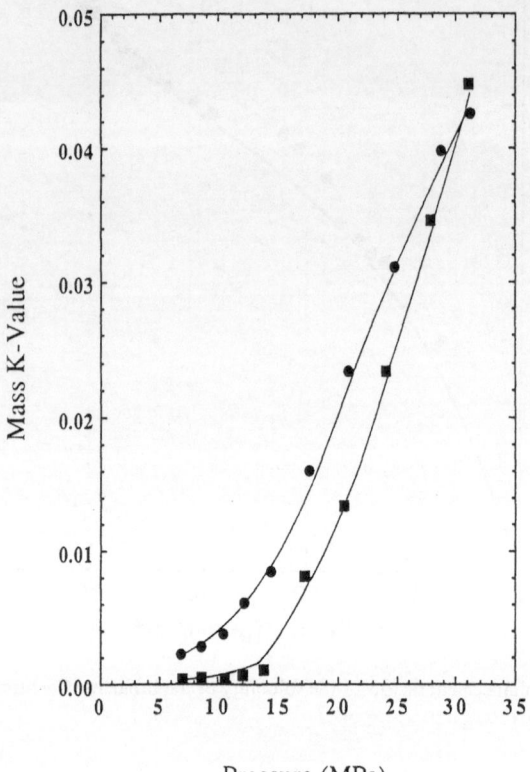

Figure 1.8   Mass K-value of cholesterol in the SC-CO$_2$ + AMF system (Yu *et al.*, 1992b). ●, 313.15 K; ■, 333.15 K.

Where $K_{cholesterol}$ and $K_{triglycerides}$ are the mass K values of cholesterol and triglycerides in SC-CO$_2$, respectively. The estimated selectivity of cholesterol over triglycerides is shown in Figure 1.9. The highest selectivity of cholesterol from anhydrous milk fat occurs at pressures between 8 and 12 MPa at 313.15 and 333.15 K.

1.3   Mass transfer operation and economics

1.3.1   Theory

The design of extraction systems requires reliable data on mass transfer and hydrodynamics. Unlike solid feeds, the mass transfer resistances associated with the morphology (e.g. external shell) of the material are absent for liquid feeds, which conceptually indicates that SCF processing of liquids must be relatively more cost effective than that of solids. However, unfortunately most of the investigations performed so far have focused on solid feed materials;

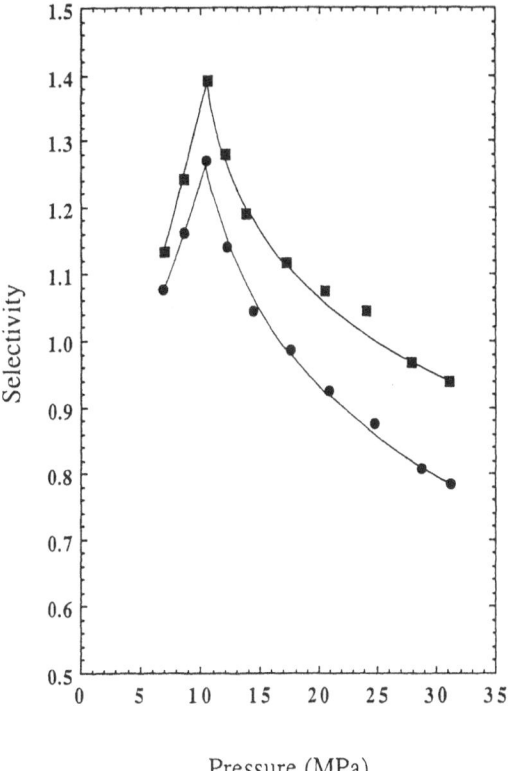

Figure 1.9 Selectivity of cholesterol compared to triglycerides in the SC-CO$_2$ + AMF system (Yu *et al.*, 1992b). ●, 313.15 K; ■, 333.15 K.

very few articles have been published on the SCF processing of liquid feeds. Examples of such studies include:

1. The mass transfer and hydraulic characteristics of SFE columns have been studied for the model systems CO$_2$/ isopopanol/ water, CO$_2$/ ethanol/ water and toluene/ acetone/ water (Rathkamp *et al.*, 1987; Seibert and Moosberg, 1988; Lahiere and Fair, 1987).
2. Cygnarowicz and Seider (1990) studied the optimization and control strategies for the extraction of β-carotene from fermentation broth.
3. de Haan *et al.* (1990) and de Haan (1991) studied the extraction of lactones from milk fat and the extraction of liquid hydrocarbon mixtures.
4. Peter and Brunner (1978) studied the separation of glycerides using compressed propane and ethylene.

Scrutiny of these articles indicates a clear need for investigations on the SCF processing of liquid feed materials; such literature contributions will be of permanent value in the areas of chemical and food engineering.

Generally two classes of columns are commonly used: packed columns, where changes in fluid composition occur continuously throughout; plate columns, where changes occur more or less in steps. A packed column is often used for continuous extraction when only a few equilibrium stages are required (Seibert and Fair, 1988). Because of higher efficiency, counter-current mode of contact is normally favored as against co-current mode for continuous processes; the mass transfer and hydrodynamic characteristics of such contacting schemes are similar to those of the traditional liquid-liquid extraction or gas absorption processes (Treybal, 1955). A major objective in these operations is to determine the number of stages required to accomplish a desired separation. While the height of the column used in any diffusional process such as absorption distillation or extraction is dependent on many factors, the calculation of the height for any given conditions is straightforward (Colburn, 1939). Thus the concepts of height equivalent to a theoretical stage (HETS) or the height of a transfer unit (HTU) can be utilized (de Haan et al., 1990; de Haan, 1991; Lahiere and Fair, 1987) to describe the efficiency of a continuous SFE process at different processing capacities. One or more of the individual mass transfer resistances in the series of diffusional processes may be important; it is rather laborious to characterize each of these resistances for a given material and that only overall mass transfer coefficients are required for the general equipment/process design. The overall gas-phase coefficient is based on the overall driving force $(Y^* - Y)$, where Y^* is the gas-phase fraction corresponding to equilibrium with the liquid-phase composition. Thus, the equation for the column height can be written as (McCabe et al., 1989)

$$Z_\mathrm{T} = \frac{V}{S\,K_\mathrm{OG}{}^\mathrm{a}} \int_\mathrm{in}^\mathrm{out} \frac{dY}{Y^* - Y} \qquad (1.18)$$

where S is the cross-section and $K_\mathrm{OG}{}^\mathrm{a}$ is the overall volumetric mass transfer coefficient. The integral in equation (1.18) is defined as the number of transfer units (N_OG). It represents the change in the vapor concentration divided by the average driving force. The subscripts show that N_OG is based on the overall driving force for the gas phase.

$$N_\mathrm{OG} = \int_\mathrm{in}^\mathrm{out} \frac{dY}{Y^* - Y} \qquad (1.19)$$

A large number of transfer units are required for a high degree of absorption. The other part of equation (1.18) is the height of transfer units (H_OG) and has units of length:

$$H_\mathrm{OG} = \frac{V}{S\,K_\mathrm{OG}{}^\mathrm{a}} \qquad (1.20)$$

Therefore Z_T can be expressed as

$$Z_\mathrm{T} = H_\mathrm{OG} N_\mathrm{OG} \qquad (1.21)$$

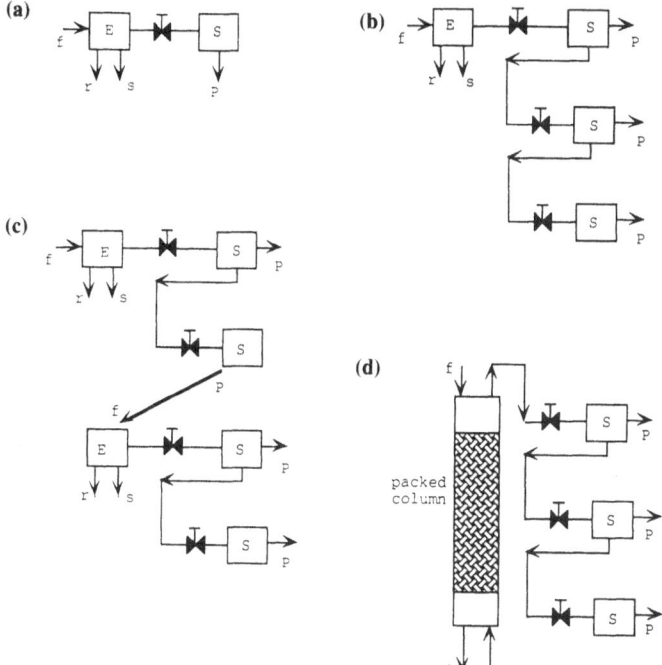

Figure 1.10 Single and multistage extraction and separation processes. (a)–(c) Batch extraction primarily for solid feed materials; (d) continuous extraction suitable for liquid feed materials. (a) Single stage extraction, single stage separation. (b) Single stage extraction, three stage separation. (c) Single stage extraction, two stage separation followed by extraction of one of the product streams. (d) Countercurrent extraction, three stage separation. s, supercritical solvent; f, feed; p, product; r, residual.

The value of H_{OG} or $K_{OG}{}^a$ is usually based on experimental data. An example of this can be seen in the work of Mizandjian and Massie (1988) who studied the performance of packed contactor to recover butyric acid from an acetic acid aqueous solution with SC-CO_2. Another example of this is the work of Bhaskar *et al.* (1993) who designed a system for continuous processing of anhydrous milk fat (AMF) with SC-CO_2. They calculated the number of transfer units for AMF and its triglyceride composition, grouped as low-, medium- and high-melting triglycerides.

1.3.2 Scale-up

Several conceptual designs and scale-up indices are possible. At the present time, the best known case in SCF extraction is the single-step batch process (Figure 1.10). For the satisfactory operation of a single-stage process, the separation factor for the desired component must be larger than the rest of the components, which will remain unextracted and rejected in the raffinate (Brunner, 1990); examples include extraction of edible oils from seeds and extraction of plant alkaloids.

Most often processes involving solid substrates are single-stage processes as well as batch processes. If the separation factors for the individual components in the SCF phase are not sufficiently large, then a multistage mode needs to be employed (Figure 1.10). Transfer of the desired components into the supercritical phase may be achieved in one or more extraction towers and the separation of the absorbed components from the SCF will involve two or more distinctive stages (or vessels) where the pressure of the SCF extract stream is reduced or the temperature is varied.

The continuous transport of solids into and out of the extraction vessels is not yet possible at the required high operating pressures (60–1000 bar). Consequently, bulky solid materials are processed either in batch, or at best, in semi-continuous operations, which often involves significant time being spent on operations other than extraction, such as packing, unloading of solids, pressurization and decompression (Eggers, 1978). As the processing time increases, the capacity of the plant decreases and the processing cost per unit volume of the product increases, which often make the economics of SFE unattractive. In the case of liquids, they can be pumped continuously at high pressures and hence the processing time can be minimized substantially by developing a suitable continuous process.

While the scale-up problem of continuous processing of liquid biomaterials can easily be attacked following the conventional approach in chemical engineering practice (Fair, 1985), the scale-up prospects of batch SCF processing of solid biomaterials are not well understood. Eggers and Sievers (1989a) have developed simple scale-up rules for the SCF processing of natural products based on the rate of external mass transfer from the solid surface to the SCF. If pilot and production plants are expected to display a similar mass transport rate, then

$$Sh = f(Re, Sc) \tag{1.22}$$

where Sh is the Sherwood number, Re is the Reynolds number, Sc is the Schmidt number, would demand equal flow velocities (of the SCF) at any scale of operation. However, equal flow velocity of solvent does not linearly increase the extraction yield between the pilot and production units because of phase equilibrium limitations. A relationship of the type

$$\frac{m_E}{m_S} = C_0 \left(1 - \exp(-kH)\right) \tag{1.23}$$

where m_E is the mass extracted, m_S is the mass flow of solvent, C_0 is the equilibrium solubility of the solute in SCF solvent, k is an empirical constant and H is the packed height of solid biomaterials, is suggested where k must be determined by a pilot-scale operation for different charge heights (H). It is further recommended that scale-up be made in accordance with the relationship between extraction efficiency (defined as $(m_E/m_0) \times 1/\tau_{ex}$) and the specific

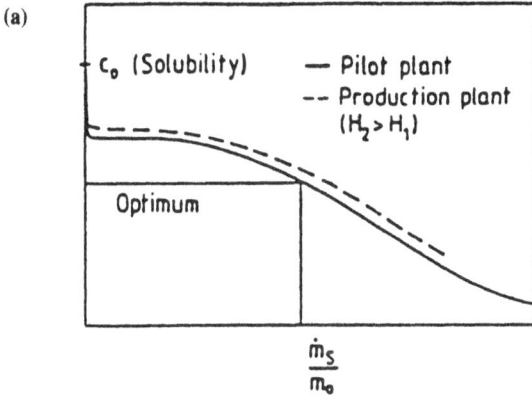

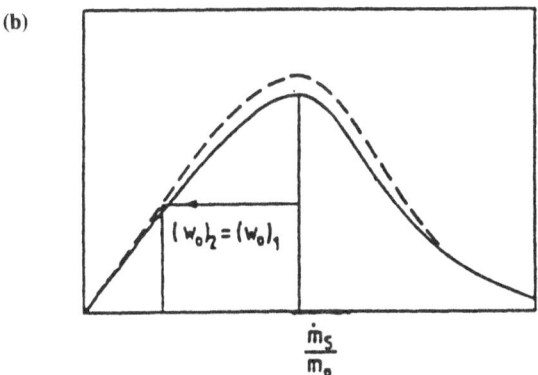

Figure 1.11 Variation in (a) extract yield and (b) solvent requirement as functions of extraction time and solid pretreatment (Eggers and Sievers, 1989b).

solvent mass flow rate (defined as m_S/m_0) gained from pilot plant operation, which may show a pronounced maximum at a given ratio of m_S/m_0 (Figure 1.11).

1.3.3 Economics

Natural products are often present in solid matrices. The internal and external mass transfer resistances from the bound state of the material of interest to release it into the SCF need to be ascertained (Jones, 1991). Little attention has been paid to the mass transfer aspects to date. These resistances may contribute to major economic demands (Eggers and Sievers, 1989b) in terms of energy and processing costs. For most natural substances, a decision must be made between (Eggers and Sievers, 1989a):

1. materials whose geometry may not be destroyed during the process (e.g. coffee, tea, tobacco);

2. materials whose geometry may be destroyed or for which even a pretreatment may be possible (e.g. oilseed, hops, plant alkaloid substrates).

There is not much choice in the hands of a process engineer concerning those materials for which the geometry must not be destroyed; the SCF has to enter the pores of the solid matrix, reach the chemical species of interest, dissolve it and then the supercritical solution has to diffuse out of the solid matrix and mix with the bulk stream. Moreover, considering the complex nature of biological substrate matrices with hard shells, multilayers with varying porosity and the bound state of the extractable material, investigators have to rely on a good deal of empiricism. The overall mass transport characteristics from solid to fluid will, in general, be similar to that of a drying process (Treybal, 1955) and an equivalent mathematical treatment is possible.

For materials whose geometry can be altered for SCF processing, the particle size and shape factor need to be selected carefully to give high mass flux rates of the biomaterial from the particle into the fluid. Lee *et al.* (1986) have reported volumetric mass transfer coefficients for the processing of canola oilseeds with SC-CO_2; but the differences in transport rate between crushed, flaked and finely chopped oilseeds have not been brought out explicitly. Data from King *et al.* (1987) show clearly that about 60% of the lipids are present in a 'free' state in rapeseeds and the balance is present in a 'bound' state. Higher mass transfer resistances are encountered to liberate the bound lipids; such high mass transfer resistances seem to have strong implications on the economic feasibility of the overall process. Higher transport resistances require high solvent circulation velocity to transfer the biomaterial into the SCF phase. Eggers and Sievers (1989b) demonstrate that the specific solvent requirement (Figure 1.12) to reduce the residual oil content of rapeseeds varies appreciably depending upon the type of pretreatment. The specific energy requirement (e_i), i.e. energy required per unit mass of starting material for each process step *i* as given by the product of specific solvent requirement (m_{CO_2}/m) and the enthalpy difference Δh_i,

$$e_i = (m_{CO_2}/m)\,\Delta h_i \qquad (1.24)$$

increases rapidly. This energy expenditure boosts up the recompression and pumping costs of CO_2 and increases the processing costs (King *et al.*, 1990). Specific examples include:

1. Economic estimations for the recovery of β-carotene from fermentation broths (Cygnarowicz and Seider, 1990) using SC-CO_2 indicate that designs requiring a high CO_2 flow rate will not be cost effective.
2. Milk fat (butter) is a low priced (about US$0.80/lb) commodity and reported to be available in surplus of about 300 million pounds in the United States (USDA, 1991). While milk fat can be fractionated with SC-CO_2 and put to specific end uses in the food industry (Fjaervoll, 1970;

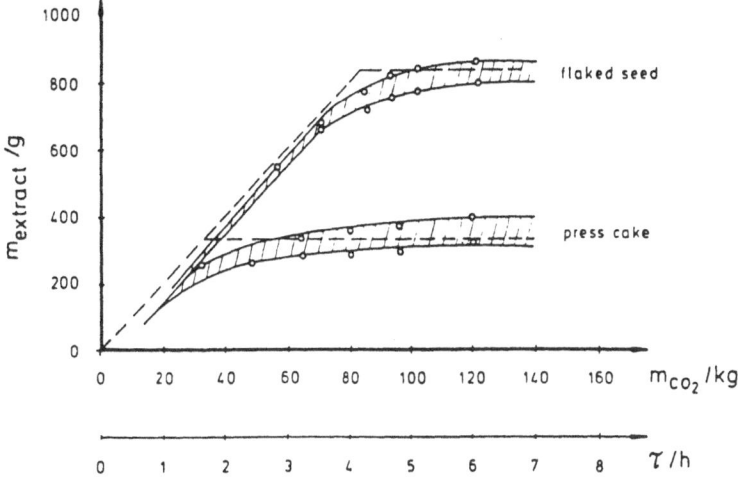

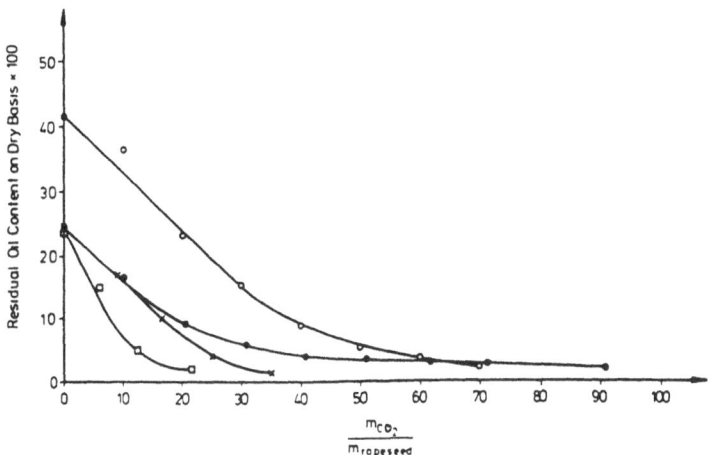

Figure 1.12 Optimization of solvent-feed ratio and scale-up of supercritical extraction of solid feed material (Eggers and Sievers, 1989a). ///, H_2O; P_{EX} = 30 MPa; t_{EX} = 40°C; – – –, saturation loading; $m_{CO2}/m \approx 10 \cdot 1/h$ (in top portion). ×, press cake, flaked; ●, press cake; ○ flaked seed (P_{EX} = 30 MPa; t_{EX} = 40°C); □, press cake, flaked (P_{EX} = 75 MPa; t_{EX} = 80°C)

Kaufmann *et al.*, 1982), the prevailing opinions (Cosgrove and Nova-kovic, 1990; Raj *et al.*, 1993) are contradictory. Boudreau and Arul (1991) indicated that the SCF process for the fractionation of milk fat may provide a higher degree of molecular separation than the melt crystallization process but is somewhat inferior to the short-path distil-lation process; moreover, the SCF process may be capital and energy

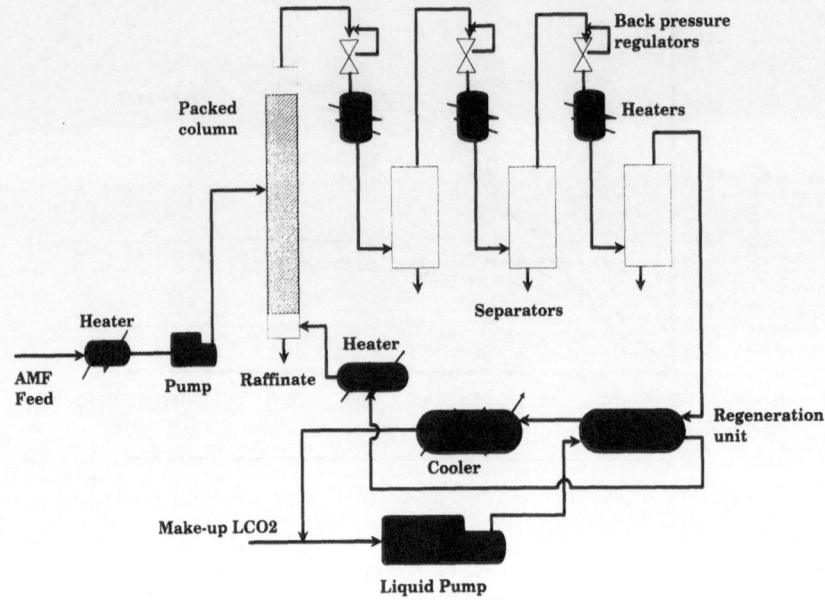

Figure 1.13   Single stage extraction and three stage fractionation of milk fat (Singh and Rizvi, 1993).

intensive. However, the process and cost analyses of Singh and Rizvi (1994) for a conceptual milk fat fractionation process (Figure 1.13) based on the available design data (Lim *et al.*, 1991; Bhaskar *et al.*, 1993) as shown in Table 1.2 indicate that the conversion cost for a plant with an annual capacity of 10 000 tons will work out at 15 cents/kg of milk fat.

3. In a specific energy assessment for the processing of rape seeds, Eggers and Sievers (1989b) showed that the recompression (for CO_2) energy expenditure would be about 450–600 kW-h per ton of rapeseed whereas solvent extraction with hexane would involve only 75 kW-h per ton.

4. In another study involving economic and energy assessment, King *et al.* (1990) conclude that near-critical extraction may not be a separation method of universal applicability but it does deserve careful consideration in some situations where non-polar or slightly polar solvents are used at the present time. This is especially important in the food industry where the currently permitted residual organic solvent levels are liable for re-examination in the future.

1.4   Biochemical reactions in supercritical fluids

In the manufacture of high-value dietary fats, those low in caloric value and high in unsaturation, esterification has been used to also modify the physico-chemical and functional properties of fats and oils. Enzyme esterification is of

Table 1.2 Summary of economic analysis for milk fat extraction and fractionation (all costs in 1000$) (Singh and Rizvi, 1994)

	Plant capacity (tons/yr)		
	800	3,200	10,000
Fixed capital	1163	2332	4373
Manufacturing expenses			
Raw materials			
AMF (@ $1.98/kg)	1584	6336	1980
CO_2 (@ $0.08/kg)	29.3	117.0	365.7
Operating expenses			
Labor	320.0	320.0	320.0
Utilities	37.8	151.3	371.8
Other			
Maintenance	17.5	35.0	65.6
Local taxes	17.5	35.0	65.6
Total manufacturing expenses	2006.3	6994.3	21088.7
Depreciation	54.2	108.8	204.1
Conversion cost (¢/kg)	55	23	15
Added price (¢/kg)	38	19	11
Revenue from sales	2304.7	7603.2	22176.0
Profit before income tax	298.5	608.9	1087.3
Profit after tax	178.2	357.6	670.5
Payback period (years)	5	5	5

great interest since it offers several advantages over chemical methods, namely, selective and positionally specific fatty acid exchange, milder reaction conditions, and tolerance of water and free fatty acids (Macrae, 1983a; Sreenivasan, 1978). The physical and functional properties of edible fats and oils can be changed by enzymatic esterification. In this process, chemical catalysts such as sodium metal, sodium-potassium alloy, sodium alkoxide or a biocatalyst such as lipase are used to accelerate the esterification process.

Lipase (glycerol ester hydrolase) comprises a group of enzymes that catalyze the hydrolysis of triacylglycerols to give free fatty acids, diacylglycerol, monoacylglycerol and glycerol. Since the reaction is reversible, lipase can conversely catalyze (i) the formation of esters from alcohols and fatty acids and acylglycerols from glycerol and fatty acids (esterification) and (ii) the exchange of fatty acids in fats and oils (transesterification and interesterification). It is possible to use enzymatic esterification to produce specific fatty acid esters similar to the flavorants used in margarine, imitation dairy products, confections and other prepared foods (Nelson, 1972). Formation of flavor esters have been catalyzed by an immobilized *Candida cylindracea* lipase in organic solvents (Marlot *et al.*, 1985; Gillies *et al.*, 1987; Langrand *et al.*, 1988). This lipase exhibits activity over a wide pH range (2–8.5) (Macrae, 1983b) and has been shown not only to esterify ethanol with the fatty acids found in butter triglycerides (Kanisawa, 1983), but also exhibit a preference towards short chain fatty acids (C_4–C_8) (Kanisawa, 1983; Gillies *et al.*, 1987). Another common and effective lipase from *Mucor miehei* has been used to synthesize

Table 1.3 Fatty acid composition (% by weight) before and after enzymatic esterification with ethanol in SC-CO_2

Fatty acid	Before	After	Change
Saturated			
C4:0	1.79	4.38	+145
C6:0	1.03	2.13	+106
C8:0	1.00	2.32	+131
C10:0	2.57	5.64	+120
C12:0	3.42	4.42	+ 29
C14:0	9.02	10.49	+ 16
C16:0	29.19	26.38	− 10
C18:0	15.58	13.27	− 15
C4:0–C8:0	3.82	8.83	+131
C10:0–C14:0	15.01	20.55	+ 37
C16:0–C18:0	44.76	39.65	− 11
Unsaturated			
C14:1	0.37	0.77	+107
C16:1	1.36	1.41	+ 3
C18:1	31.02	25.64	− 17
C18:2	2.80	2.39	− 15
C18:3	0.84	0.76	− 10
C14:1–C18:3	36.40	30.97	− 15

esters from numerous alcoholic and acidic substrates (Marlot *et al.*, 1985; Langrand *et al.*, 1988; Omar *et al.*, 1989).

Transesterification exchanges fatty acids from one fat or oil with those of another in the presence of lipase as a catalyst. For example, commercially available porcine pancreatic lipase was used to interesterify canola oil with lauric acid (Thomas *et al.*, 1988). Coleman and Macrae (1980) demonstrated interesterification of triglycerides in a mixture of coconut oil and olive oil by using *Candida cylindracea* lipase as catalyst. Wisdom *et al.* (1985, 1987) also showed that immobilized lipases *Rhizopus arrhizus* and *Aspergillus* sp. were suitable for fat interesterification.

SC-CO_2 is a unique solvent, especially for food and pharmaceutical applications and details of the process appear in many excellent reviews (Rizvi *et al.*, 1986; Bruno and Ely, 1991). The advantages of SC-CO_2 include pollution free solvents, low temperature operation, low energy cost, and high mass transfer and diffusion rates making it an excellent medium for interesterification and transesterification. SC-CO_2 has been used, for example, as a medium for reactions catalyzed by alkaline phosphatase (Randolph *et al.*, 1985), polyphenol oxidase (Hammond *et al.*, 1985) and cholesterol oxidase (Randolph *et al.*, 1988). Van Eijs *et al.* (1988) successfully used SC-CO_2 to transesterify isoamylacetate with the corresponding alcohol. Recently, Yu *et al.* (1992c) studied the synthesis of ethyl esters from milk fat at 13.6 MPa and 40°C. The esterification of fatty acids from milk fat with ethanol is shown in

Table 1.3, and compares the fatty acid composition before and after enzymatic esterification. The time for enzymatic esterification of fatty acids from milk fat was approximately 1 h in SC-CO_2. The same reaction, however, using an organic solvent, ethanol, required 120 h (Kanisawa, 1983).

This area offers unique opportunities for further work to develop novelty chemicals of premium value. In addition, by exploiting the specificity of the lipases, it is possible to produce useful glyceride mixtures which cannot be obtained by chemical interesterification processes.

References

Bhaskar, A. R., Rizvi, S. S. H. and Harriott, P. (1993) Performance of a packed column for continuous SC-CO_2 processing of anhydrous milk fat. *Biotech. Progress*, **9**, 70–4.

Boudreau, A. and Arul, J. (1991) Physical and chemical modification of milk fat. *Bull. Int. Dairy Federation* No. 260, pp. 7–10.

Brunner, G. (1990) Mass separation with supercritical gases. *Int. Chem. Eng.* **30**, 191.

Bruno, T. J. (1991) Thermophysical property data for supercritical fluid extraction design, in *Supercritical Fluid Technology*, eds. T. J. Bruno and J. F. Ely, CRC Press, Ann Arbor, MI. Ch. 7.

Bruno, T. J. and Ely, J. F. (1991) *Supercritical Fluid Technology: Reviews in Modern Theory and Applications*, CRC Press, Boca Raton, FL.

Chi, Y. M., Nakamura, K. and Yano, T. (1988) Enzymatic interesterification in supercritical carbon dioxide. *Agric. Biol. Chem.* **52**, 1541.

Chrastil, J. (1982) Solubility of solids and liquids in supercritical gases. *J. Phys. Chem.* **86**, 3016.

Colburn, A. P. (1939) Simplified calculation of diffusional process. *Ind. Eng. Chem.* **33**, 459.

Coleman, M. H. and Macrae, A. R. (1980) UK patent 1,557,933.

Cosgrove, T. and Novakovic, A. (1990) *The Milkfat Issue: Production, Processing and Marketing*, Publication no. A.E. Ext. 90–19, Department of Agricultural Economics, Cornell University, Ithaca, NY 14853, USA.

Cygnarowicz, M. L. and Seider, W. D. (1990) Design and control of a process to extract β-carotene with supercritical carbon dioxide. *Biotechnol. Prog.* **6**, 82.

de Haan, A. B. (1991) Supercritical fluid extraction of liquid hydrocarbon mixtures, Dissertation, Delft University of Technology, The Netherlands.

de Haan, A. B., de Graauw, J., Schaap, J. E. and Badings, H. T. (1990) Extraction of flavors from milk fat with supercritical carbon dioxide. *J. Supercritical Fluids* **3**, 15.

Debenedetti, P. G. (1986) Diffusion and mass transfer in supercritical fluids. *AIChE J.* **32**, 2034.

del Valle, J. M. and Aguilera, J. M. (1988) An improved equation for predicting the solubility of vegetable oils in supercritical CO_2. *Ind. Eng. Chem. Res.* **27**, 1551.

Diepen, G. A. M. and Scheffer, F. E. C. (1948) The solubility of naphthalene in supercritical ethylene. *J. Am. Chem. Soc.* **70**, 4085.

Dumont, T., Barth, D. and Perrut, M. (1991) Continuous synthesis of ethyl myristate by enzymatic reaction in supercritical carbon dioxide, in *Proc. 2nd Int. Symp. on Supercritical Fluids*, Boston, MA, pp. 150–153.

Eggers, R. (1978) Large-scale industrial plant for extraction with supercritical gases. *Angew. Chem. Int. Ed. Engl.* **17**, 751.

Eggers, R. and Sievers, V. (1989a) Present state of extraction of natural materials with supercritical fluids and developmental trends, in *Supercritical Fluid Science and Technology*, eds. K. P. Johnston and J. M. L. Penninger, ACS Symposium Series No. 406, pp. 478–498.

Eggers, R. and Sievers, V. (1989b) Processing of oilseed with supercritical carbon dioxide. *J. Chem. Eng. Jpn.* **22**, 641.

Ekart, M. P., Brennecke, J. F. and Eckert, C. A. (1991) Molecular analysis of phase equilibria in supercritical fluids, in *Supercritical Fluid Technology*, eds. T. J. Bruno and J. F. Ely, CRC Press, Ann Arbor, MI, Ch. 3.

Ely, J. F. (1986) Improved mixing rules for one-fluid conformal solution calculations. *Am. Chem. Soc. Symp. Ser.* **300**, 331.

Fair, J. R. (1985) *Scaleup of chemical processes: conversion from laboratory scale tests to successful commercial Size Design*, eds. A. Bisio and R. L. Kabel, Wiley Interscience, New York, Ch. 12, 13.

Fjaervoll, A. (1970) Anhydrous milk fat fractionation offers new applications for milk fat. *Dairy Ind.* **8**, 502.

Giddings, J. C., Myers, M. N., McLaren, L. and Keller, R. A. (1968) *Science* **162**, 67.

Giddings, J. C., Myers, M. N. and King, J. W. (1969) Dense gas chromatography at pressures to 2000 atmospheres. *J. Chromatogr. Sci.* **7**, 276.

Gillies, B., Yamazaki, H. and Armstrong, D. W. (1987) Production of flavor esters by immobilized lipase. Biotechnol. Lett. **9**, 709.

Gunawan, R. J. (1990) *Experimental Measurement of Fluid-Liquid Equilibria of Dodecane + Carbon Dioxide and Methyl Oleate + Carbon Dioxide*. M.S. Thesis, Cornell University.

Hammond, D. A., Karel, M. and Klibanv, A. M. (1985) Enzymatic reactions in supercritical gases. *Appl. Biochem. Biotechnol.* **11**, 393.

Huron, M. J. and Vidal, J. (1979) New mixing rules in simple equation of state for representing vapor-liquid equilibria of strongly non-deal mixtures. *Fluid Phase Equilibria* **3**, 255.

Inomata, H., Kondo, T., Hirohama, S., Arai, K., Suzuki, Y. and Konno, M. (1989) Vapor-liquid equilibria for binary mixtures of carbon dioxide and fatty acid methyl esters. Fluid Phase Equilibria, **46**, 41–52.

Johnston, K. P. (1984) Supercritical fluids, in *Encyclopedia of Chemical Technology*, Wiley, New York, p. 872.

Jones, M. C. (1991) Mass transfer in supercritical extraction from solid matrices, in *Supercritical Fluid Technology*, eds. T. J. Bruno and J. F. Ely, CRC Press, Ann Arbor, MI, pp. 365–381.

Kanisawa, T. (1983) Production of ethyl ester mixture from butter fat by *Candida cylindracea* lipase. *Nippon Shokuhin Kogyo Gakkaishi* **30**, 572.

Kaufmann, W., Biernoth, G., Frede, E., Merk, W., Precht, D. and Timmen, H. (1982) Fractionation of butterfat by extraction with SC-CO_2. *Milchwissenschaft* **37**, 92.

King, M. B., Bott, T. R., Barr, M. J., Mahmud, R. S. and Sanders, N. (1987) Equilibrium and rate data for the extraction of lipids using compressed carbon dioxide. *Sep. Sci. Technol.* **22**, 1103.

King, M. B., Catchpole, O. J. and Bott, T. R. (1990) Energy and economic assessment of near-critical extraction processes. I. *Chem. Eng. Symp. Ser.* **119**, 165.

Lahiere, R. J. and Fair, J. R. (1987) Mass-transfer efficiencies of column contactors in supercritical extraction service. *Ind. Eng. Chem. Res.* **26**, 2086.

Langrand, G., Triantaphylides, C. and Baratti, J. (1988) Lipase catalyzed formation of flavour esters. *Biotechnol. Lett.* **10**, 549.

Lee, A. K. K., Bulley, N. R., Faltori, M. and Meisen, A. (1986) Modelling of supercritical carbon dioxide extraction of canola oilseed in fixed beds. *J. Am. Oil Chem. Soc.* **63**, 921.

Lim, S. B., Lim, G. B. and Rizvi, S. S. H. (1991) Continuous supercritical CO_2 processing of milk fat, *Proc. 2nd Int. Conf. on Supercritical Fluids*, Boston, MA.

Macrae, A. R. (1983a) Lipase-catalyzed interesterification of oils and fats. *J. Am. Oil Chem. Soc.* **60**, 291.

Macrae, A. R. (1983b) Extracellular microbial lipases, in *Microbial Enzymes and Biotechnology* ed. W. M. Fogarty, Elsevier Applied Science, London, pp. 225–250.

Mansoori, G. A. and Ely, J. F. (1985) Density expansion (DEX) mixing rules: thermodynamic modeling of supercritical extraction. *J. Chem. Phys.* **82**, 406.

Marlot, C., Langrand, G., Triantaphylides, C. and Baratti, J. (1985) Ester synthesis in organic solvent catalyzed by lipases immobilized on hydrophilic supports. *Biotechnol. Lett.* **7**, 647.

Mart, C. J., Papadopoulos, K. D. and Donohue, M. C. (1986) Application of perturbed-hard-chain theory to solid-supercritical fluid equilibria modeling. *Ind. Eng. Chem. Proc. Design Dev.* **25**, 394.

Marty, A., Chulalaksananukul, W., Condoret, J. S., Willemot, R. M. and Durand, G. (1990) Comparison of lipase-catalyzed esterification in supercritical carbon dioxide and in n-hexane. *Biotechnol. Lett.* **12**, 11.

McCabe, W. L., Smith, J. C. and Harriott, P. (1989) *Unit Operations of Chemical Engineering*, McGraw-Hill, New York, NY.

Mizandjian, J. L. and Massie, J. F. (1988) Performance of a packed contactor in supercritical CO_2 countercurrent extraction, *Proc. Int. Symp. on Supercritical Fluids*.

Nelson, J. H. (1972) Enzymatically produced flavors for fatty systems. *J. Am. Oil Chem. Soc.* **49**, 559.

Omar, I. C., Saeki, H., Nishio, N. and Nagai, S. (1989) Synthesis of acetone glycerol acyl ester by immobilized lipase of *Mucor miehei*. *Biotechnol. Lett.* **11**, 161.

Panagiotopoulos, A. Z. and Reid, R. C. (1986) in *Equations of State: Theory and Application*, eds. K. C. Chao and R. Robinson, Jr. ACS Symposium Ser. 300, American Chemical Society, Washington, DC, p. 571.

Peng, D.-Y. and Robinson, D. B. (1976) A new two-constant equation of state. *Ind. Eng. Chem., Fundam.* **15**, 59.

Peter, S. and Brunner, G. (1978) The separation of nonvolatile substances by means of compressed gases in countercurrent processes. *Angew. Chem. Int. Ed. Engl.* **17**, 746.

Prausnitz, J. M., Lichtenthaler, R. N. and de Azevedo, E. G. (1986) *Molecular Thermodynamics of Fluid-Phase Equilibria*, 2nd edition, Prentice Hall, Englewood Cliffs, NJ.

Rainwater, J. C. (1991) Vapor-liquid equilibrium and the modified Leung–Griffiths model, in *Supercritical Fluid Science and Technology*, eds. T. J. Bruno and J. F. Ely, CRC Press, Ann Arbor, MI, Ch. 2.

Raj, C. B. C., Bhaskar, A. R. and Rizvi, S. S. H. (1993) Processing of milk fat with supercritical carbon dioxide—mass transfer and economic aspects, *Trans Inst. Chem. Eng.*, **71**, (c), 3–10.

Randolph, T. W., Blanch, H. W., Prausnitz, J. M. and Wilke, C. R. (1985) Enzymatic catalysis in a supercritical fluid. *Biotechnol. Lett.* **7**, 325.

Randolph, T. W., Clark, D. S., Blanch, H. W. and Prausnitz, J. M. (1988) Enzymatic oxidation of cholesterol aggregates in supercritical fluids. *Science* **239**, 387.

Rathkamp, P. J., Bravo, J. L. and Fair, J. R. (1987) Evaluation of packed columns in supercritical extraction processes. *Solvent Extraction Ion Exchange* **5**, 367.

Rizvi, S. S. H. (1987) Supercritical fluid extraction of dairy foods. *N. Y. Food Life Sci. Q.* **17**(3), 23.

Rizvi, S. S. H., Daniel, J. A., Benado, A. L. and Zollweg, J. A. (1986) SFE: operating principles and food applications. *Food Technol.* **40**, 57.

Schneider, G. M., Stahl, E. and Wilke, G. (1980) *Extraction with Supercritical Gases*, Verlag Chemie, Deerfield Beach, FL.

Seibert, A. F. and Fair, J. R. (1988) Hydrodynamics and mass transfer in spray and packed liquid-liquid extraction column. *Ind. Eng. Chem. Res.* **27**, 470.

Seibert, A. F. and Moosberg, D. G. (1988) Performance of spray, sieve-tray and packed contactors for high pressure extraction. *Sep. Sci. Technol.* **23**, 2049.

Serbanovic, S. P. and Djordjevic, B. D. (1987) Influence of the optimized temperature-dependent interaction parameter on vapor-liquid equilibrium binary predictions of supercritical methane with some alkanes by means of the Soave equation of state. *Ind. Eng. Chem. Res.* **26**, 618.

Singh, B. and Rizvi, S. S. H. (1994) Design and economic analysis for continuous, countercurrent processing of milk fat. *J. Dairy Sci.*, in press.

Soave, G. (1972) Equilibrium constants from a modified Relich–Kwong equation of state. *Chem. Eng. Sci.* **27**, 1197.

Sreenivasan, B. (1978) Interesterification of fats. *J. Am. Oil Chem. Soc.* **55**, 796.

Thomas, K. C., Magnuson, B., McCurdy, A. R. and GrootWassink, J. W. D. (1988) Enzymatic esterification of canola oil, *Can. Inst. Food Sci. Technol. J.* **21**, 167.

Treybal, R. E. (1955) *Mass Transfer Operations*, McGraw-Hill, New York.

USDA (1991) USDA *Agricultural Stabilization and Conservation Service*, Commodity Fact Sheet.

Van Eijs, A. M. M., de Jong, J. P. J., Doddema, H. J. and Lindeboom, D. R. (1988) Enzymatic transesterification in supercritical carbon dioxide, in *Proc. Int. Symp. on Supercritical Fluids*, Nice, France, pp. 933–942.

Wisdom, R. A., Dunnill, P. and Lilly, M. D. (1985) Enzymatic interesterification of fats: the effect of non-lipase material on immobilized enzyme activity. *Enzyme Microbiol. Technol.* **7**, 567.

Wisdom, R. A., Dunnill, P. and Lilly, M. D. (1987) Enzymatic interesterification of fats: laboratory and pilot-scale studies with immobilized lipase from *Rhizopus arrhizus*. *Appl. Biotechnol. Bioeng.* **29**, 1081.

Wong, J. M. and Johnston, K. P. (1986) Solubilization of biomolecules in carbon dioxide based supercritical fluids. *Biotechnol. Prog.* **2**, 29.

Yu, Z. R., Rizvi, S. S. H. and Zollweg, J. A. (1992a) Phase equilibria of oleic acid, methyl oleate, anhydrous milk fat in supercritical carbon dioxide. *J. Supercritical Fluids*, **5**, 112–22.

Yu, Z. R., Rizvi, S. S. H. and Zollweg, J. A. (1992b) Fluid-liquid equilibria of anhydrous milk fat with supercritical carbon dioxide. *J. Supercritical Fluids*, **5**, 123–9.

Yu, Z. R., Rizvi, S. S. H. and Zollweg, J. A. (1992c) Enzymatic esterification of fatty acid mixtures from milk fat and anhydrous milk fat with canola oil in supercritical carbon dioxide. Biotechnol. Prog., **8**, 508–13.

2 Carbon dioxide as a supercritical solvent in fatty acid refining: theory and practice

G. B. GUARISE, A. BERTUCCO and P. PALLADO

Abstract

The recovery of α-tocopherol and the removal of water from fatty acid/ triglyceride mixtures by means of carbon dioxide at supercritical conditions are discussed taking into account both phase equilibrium and mass transfer phenomena. Simulation of a semi-batch apparatus is presented; calculated and experimental results are compared for water/oleic acid/CO_2 and α-tocopherol/ oleic acid/CO_2 and the effect of entrainers is briefly discussed. A non-equilibrium model for multistage contactors is developed and simulated steady-state composition profiles are compared with ideal ones, at different solvent flow rates.

2.1 Introduction

Supercritical fluid extraction (SFE) with CO_2 has great potential in the field of vegetable and animal oil processing, as evidenced by many papers published in the last few years (e.g. Brunner and Peter, 1982; Eisenbach, 1984; Peter *et al.*, 1987; Brunetti *et al.*, 1989; Krukonis, 1989; Nilsson *et al.*, 1989), and by communications presented during recent Symposia on Supercritical Fluid Technology (e.g. Ashour and Wennersten, 1988; Berger *et al.*, 1988; Bertucco *et al.*, 1988; Wu *et al.*, 1988; Ender and Steiner, 1990; Bhrath *et al.*, 1991). In general, the role of CO_2 as a fractionating or refining agent is currently under investigation to find a technically feasible alternative to commonly adopted separation processes; the use of such a solvent would be extremely attractive and desirable owing to its favourable properties for both safety and technical reasons. Particularly, it would allow processing of materials in very mild temperature conditions, thus providing a new tool for the recovery of thermo-labile com-pounds, such as highly unsaturated glycerides mixtures (Nilsson *et al.*, 1991).

However, the results obtained so far are not as satisfactory as expected; separations often remain difficult even at high pressure and costs for SFE operation seem to be far beyond those of conventional processes.

Furthermore, the quantitative simulation of SFE in view of any feasibility analysis is difficult, due to the high and usually not well known sensitivity of all thermodynamic and transport properties to changes in pressure, temperature and composition around the critical point; therefore, experimental investigation is needed whenever a new process is being developed.

The scope of this work is first to recall and summarize some theoretical and general concepts, currently used in chemical engineering, in order to possibly predict the applicability of SFE with CO_2 to fatty acid processing and refining; second, to present specific research on a separation that turned out to be difficult even under SFE conditions.

2.2 Phase equilibria

From the chemical engineer's standpoint, SFE is a unit operation which does not really represent a new separation process; it can be considered as a different kind of solvent extraction, or as a special stripping technique. Its peculiar feature is the use of relatively high pressures to obtain fractionation. This complicates plant construction and operation, but it means that pressure can be handled as a further variable in process design. On the other hand, energy duties are likely to be reduced with respect to conventional separation systems.

If we consider the block diagrams of SFE units reported in Figures 2.1(a) and (b), they are similar to those of a common solvent extraction plant. The mixture to be treated (binary in this case) is processed with the solvent either in semi-batch or continuous extractors E, under suitable temperature and pressure conditions in order to obtain the desired separation effect. Products are recovered at the desired degree of purity in the separators S_A and S_B, then solvent make-up and recycling to E are carried out. Recovery is obtained in a series of ways: by simply flashing the supercritical solution at the selected temperature and pressure, or by means of other techniques, such as distillation, adsorption, absorption. When a continuous contactor is used (see Figure 2.1(b)), the well known advantages of continuous units are ensured, i.e. a reduced plant size for the assigned throughput and a lower solvent circulation and consumption, together with higher extraction efficiencies, due to multiple countercurrent contact; but on the other hand, investment costs are higher.

Basically, the fractionation effect is obtained as a consequence of the phase splitting caused by the solvent on the feed mixture and is therefore related to the equilibrium distribution of A and B between the two phases. This is usually described by means of the K factors, defined as

$$K_i = y_i/x_i, \quad i = A, B \tag{2.1}$$

where y_i and x_i are molar fractions in the supercritical fluid and condensed phases, respectively. The selectivity of A with respect to B is given by

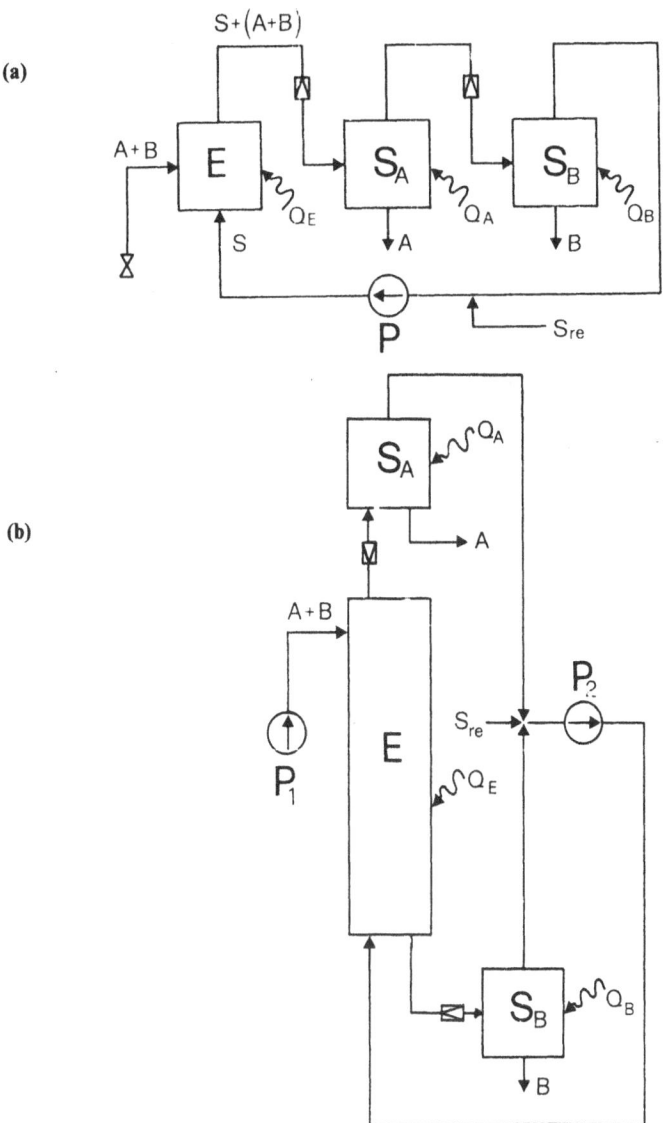

Figure 2.1 Block diagrams for SFE operations: semi-batch (a) and continuous units (b).

$$S_{A,B} = K_A/K_B \tag{2.2}$$

Knowledge of K factors is essential to evaluate the thermodynamic feasibility of the proposed separation; in all cases, $S_{A,B}$ should be greater than about 1.1 (or lower than about 0.9) to ensure a sufficient driving force for fractionation. Furthermore, in cases other than the removal of volatile impurities from a heavy substrate, the K values have to be high enough to allow good utilization of the solvent.

Calculation of K factors can be performed by expressing the chemical potential in terms of activity or fugacity; for example, according to the γ–φ approach (Van Ness and Abbott, 1982), it can be derived from

$$K_i = F_i P_i / P \tag{2.3}$$

with the enhancement factor F_i expressed by

$$F_i = \frac{\gamma_i \varphi_i^*}{\varphi_i} \exp\left[v_i (P - P_i) / RT\right] \tag{2.4}$$

where P_i is the vapor pressure of the component i, P is the total pressure, γ_i, is the activity coefficient, φ_i^* is the fugacity coefficient of pure i at T and P_i, φ_i is the fugacity coefficient of component i in the supercritical phase and v_i is the liquid molar volume.

The exponential term in equation (2.4) is known as the Poynting factor. From equations (2.3) and (2.4), one can see that the solubility of substances in supercritical media is enhanced with respect to ideal behavior (i.e. distribution controlled by vapor pressure) owing to both high Poynting correction and low fugacity coefficient in the gas phase. Indeed, when dealing with heavy compounds like fatty acids, triglycerides and tocopherols, high molar volumes and high reduced pressures are found, which mean high Poynting factors and low fugacities, so that appreciable solubilites in the gas phase are expected, in spite of their extremely low vapor pressure values. Of course, the resulting K_i values depend also on φ_i and γ_i, namely the interactions with the solvent in the gas phase and with all components in the condensed phase. The behavior of CO_2 in this respect can be improved by the use of small amounts of selective co-solvents.

An alternative way to address the problem of equilibrium distribution prediction is the equation of state approach, i.e.

$$K_i = \varphi_i^{SP} / \varphi_i^C \tag{2.5}$$

where φ_i^{SP}, φ_i^C are the fugacities of component i in the supercritical (SP) and condensed (C) phases at equilibrium.

A number of equations of state have been modified and tested for supercritical fluid applications (Brennecke and Eckert, 1989). We do not intend to pursue these issues here; a detailed analysis is necessary to be exhaustive but this is beyond the scope of the present work. However, it can be concluded that, so far, practical calculations of solubilities in supercritical CO_2 based on either equation (2.4) or (2.5) plus the related thermodynamic models do not provide quantitative results yet, because of the difficulty in accounting for molecular behavior in the critical region.

Experimental investigation on phase equilibria is always needed to evaluate the potentials of supercritical extraction in this field.

Many data have been published in the last decade, especially for binary systems (e.g. Chrastil, 1982; King *et al.*, 1983; Wong and Johnston, 1986; Ohgaki *et al.*, 1987; Peter *et al.*, 1988; Zou *et al.*, 1990a); and very recently, ternary and quaternary data are also becoming available (e.g. Bamberger *et al.*, 1988; Zou *et al.*, 1990b; Nilsson *et al.*, 1991). Unfortunately, published data are often in conflict with one another; for example, we refer to the system tristearin/CO_2 (Pearce and Jordan, 1991), where deviations between different authors are as large as two orders of magnitude. A remarkable scatter in the equilibrium solubilities is found when comparing different sources for oleic acid/CO_2 (e.g. Chrastil, 1982; King *et al.*, 1983; Peter *et al.*, 1988; Zou *et al.*, 1990a; Nilsson *et al.*, 1991) and for α-tocopherol/CO_2 (e.g. Chrastil, 1982; Ohgaki *et al.*, 1988; Zehnder and Trepp, 1991). An interesting analysis of these discrepancies is provided by Nilsson *et al.* (1991). Moreover, extensive data as a function of pressure and temperature would be necessary to perform analyses for process development.

To better explain this fact, let us consider as an example the separation of α-tocopherol from tripalmitin and oleic acid; both cases have already been discussed in the literature (e.g. Ohgaki *et al.*, 1987; Bertucco *et al.*, 1988), but here we only want to show how solubility data can be used to address these problems qualitatively. Since no ternary data are available, we refer to binary information from the work of Chrastil, although the reliability of these data has been strongly questioned (Nilsson *et al.*, 1991); solubilities in CO_2 for many compounds and correlated data points to the solvent density were measured by means of a logarithmic expression,

$$c = dA_1 \exp(A_2/T + A_3) \tag{2.6}$$

where c is the solubility (g/l), d is the solvent density (g/l), T is the temperature (K), and A_1, A_2 and A_3 are component constants. According to equation (2.6) and using a suitable equation of state for the evaluation of volumetric properties of CO_2 (the one proposed by Siewers (1984) has been adopted), the curves reported in Figures 2.2(a,b) can be calculated. Here the solubilities as a function of pressure at constant temperatures are presented for comparison between α-tocopherol/tripalmitin (Figure 2.2(a)) and α-tocopherol/oleic acid (Figure 2.2(b)).

It is evident from Figure 2.2(a) that differences in solubilities for the first system can be made large enough to ensure a satisfactory separability, by a suitable selection of the extraction temperature and pressure. On the contrary, for the second system (Figure 2.2(b)), the difference in solubility is less favourable. To be precise, K factors should be compared rather than solubilities, but the results would be similar.

A second feature is also remarkable, concerning the crossover pressures, that is the pressures ranges where the inversion of solubility dependence on temperature takes place. Pennisi and Chimowitz (1986) showed how this effect could be exploited in order to obtain an ideally infinite actual selectivity; no matter what is extracted in the extraction vessel, deposition of only one

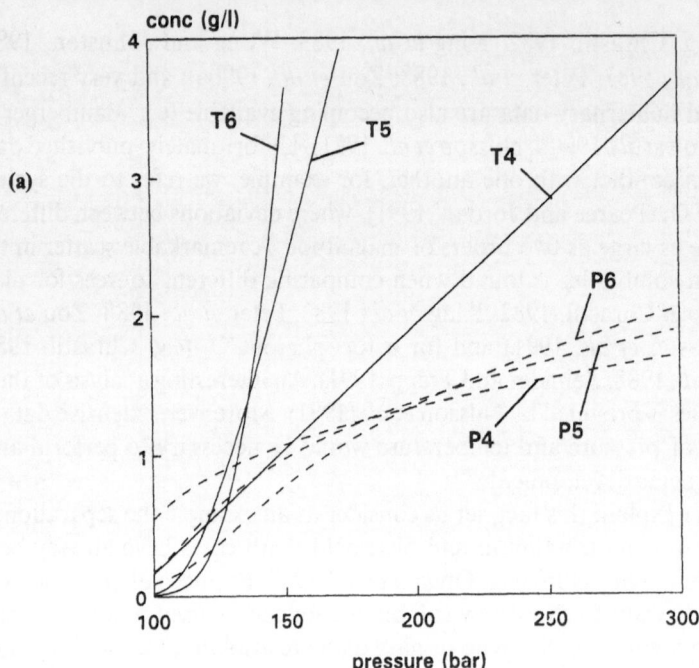

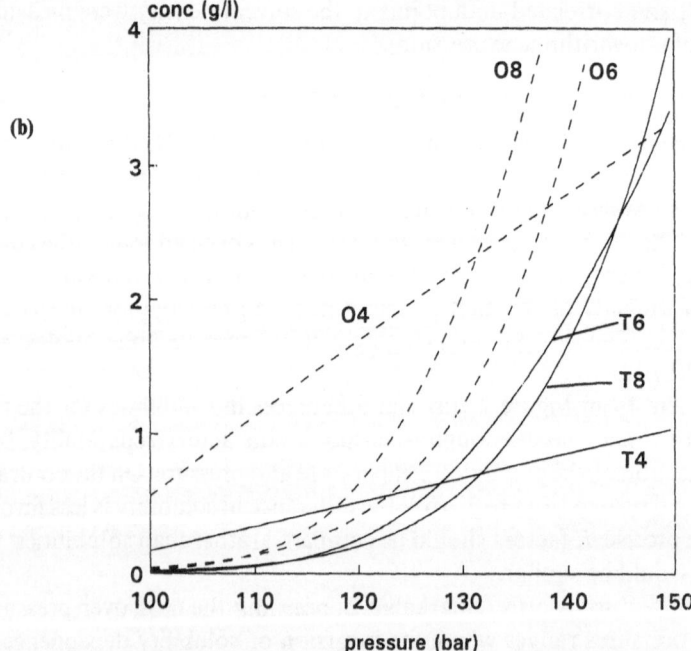

Figure 2.2 Concentration dependence on pressure and temperature for: (a) α-tocopherol/CO_2 and tripalmitin/CO_2; (b) α-tocopherol/CO_2 and oleic acid/CO_2. T4, P4, O4 = 313 K; T5, P5 = 323 K; T6, P6, O6 = 333 K; T8, O8 = 353 K.

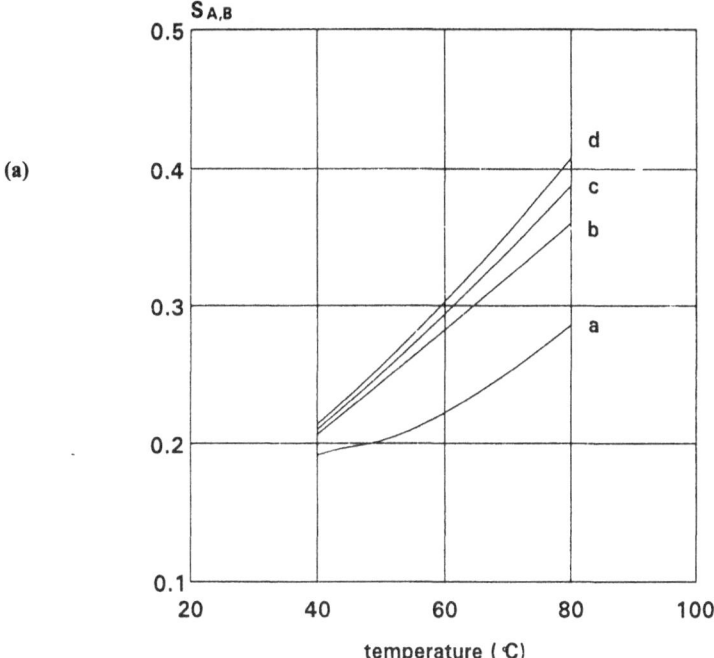

(a)

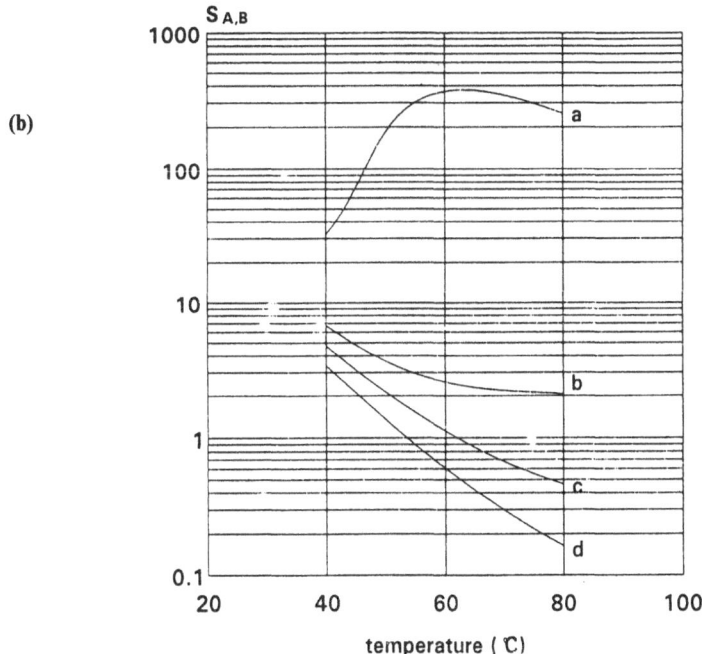

(b)

.**Figure 2.3**   Selectivity of α-tocopherol versus oleic acid (a) and water versus oleic acid (b) as a function of temperature and pressure. Isobars: a, 100 bar; b, 160 bar; c, 200 bar; d, 260 bar.

component can be obtained in the separator, provided it can be operated at a pressure that is intermediate between the two crossover pressures. It is clear that, in principle, one can take advantage of this fact for the first system, while α-tocopherol and oleic acid present overlapping crossover pressure ranges.

Finally, curves of selectivity as a function of temperature and pressure can be calculated starting from extensive solubility data; an example is reported in Figures 2.3(a,b), where the selectivities of α-tocopherol and water (component A) with respect to oleic acid (component B) in supercritical CO_2 are calculated according to Chrastil by

$$S_{A,B} = C_A \, MW_B \, / \, (C_B \, MW_A) \tag{2.7}$$

In equation (2.7), C_A and C_B are evaluated from Chrastil's data (equation (2.6)) and MW_A, MW_B are the molecular weights. It can be seen that α-tocopherol is heavier than oleic acid, while stripping of water is in principle easy to obtain at moderate conditions.

It must be noted that the above remarks and conclusions are only qualitative, because interactions are neglected when moving from binary to ternary systems, i.e. a limiting assumption is made. Figures 2.3(a) and (b) can be used only as a starting point for further analysis; from the methodological standpoint, however, they are a powerful tool for exploiting the potentials related to the use of supercritical CO_2 as an extraction solvent.

2.3 A case study

The possibility of purifying fatty acids from water and unsaponifiable compounds has been partially examined in a previous work (Bertucco et al., 1988), to which we refer concerning motivations for addressing the problem and an introductory overview.

The system selected for investigation is a model quaternary system, water/α-tocopherol/oleic acid/CO_2, which is representative of a class of separations which are both interesting for possible applications and clearly difficult to obtain.

We reiterate the need for pure fatty acid mixtures for industrial catalytic hydrogenation (Carturan et al., 1983) and, above all, the attractive possibility of recovering high purity α-tocopherol from vegetable oil deodorization scums in the food processing industry (Shishikura et al., 1988).

On the other hand, a qualitative insight into the separability of this system (see Figures 2.3(a) and (b)) shows that water can be easily extracted under proper operating conditions, while α-tocopherol purification appears troublesome. Of course, multicomponent phase behavior should be examined rather than information coming from binary data only; unfortunately, they are not available, either for the quaternary system or for the two ternary systems, water/oleic acid/CO_2 and α-tocopherol/oleic acid/CO_2. In these cases, some alternatives are possible, such as determination of extensive and precise equilibrium data, or simulation of equilibrium distribution with better models than the one which neglects multicomponent interactions, or extraction experiments.

Since the first choice was beyond the scope of our research project, a combined approach of the second and third tools was adopted: component equilibrium distributions were evaluated using equations (2.3), (2.4) together with binary data, and extraction runs were carried out in a conventional semi-batch apparatus for supercritical extraction (Bartolomeo *et al.*, 1986); the effect of different liquid solvents as entrainers has been also considered.

As far as equilibrium calculations are concerned, the general behavior reported in Figures 2.3(a,b) is confirmed, but the non-ideality related to the low concentration of both water and α-tocopherol enhance their K factors, with positive effect only on water selectivity; indeed, for α-tocopherol, the predicted value of selectivity is closer to 1.

Detailed information about materials, analytical techniques and operating procedures for experimental runs are reported elsewhere (Bertucco *et al.*, 1988; Dahir, 1989). Pressure in the range 100–230 bar was investigated, with temperatures spanning between 313 and 333 K. Results are summarized in Figures 2.4(a,b) for selected conditions.

From Figure 2.4(a), it is confirmed that the selectivity of α-tocopherol is less than but close to 1, so that its recovery from oleic acid solutions is practically impossible (see square symbols); the use of small amounts of entrainers such as ethanol and acetone increases α-tocopherol volatility; also water behaves as an entrainer, and the best effect is obtained by water ethanol mixtures (for example, points corresponding to upward and downward triangles), in accordance with results from Ohgaki *et al.* (1988). As a consequence, α-tocopherol becomes more volatile than oleic acid, but its selectivity still remains close to 1.

On the other hand, it can be seen from Figure 2.4(b) that an acceptable extent of purification can be obtained for water in a short extraction time at 115 bar and 333 K, so that supercritical extraction with CO_2 can be suitably applied to fatty acid dehydration; it is noteworthy, however, that operating pressure must not be increased, in order to keep high water selectivity.

It must be pointed out that experimental results can be affected by mass transfer limitations, as discussed in the next section; therefore it seems more correct to refer to 'actual' rather than 'thermodynamic' selectivities. On the basis of experimental extraction curves, actual selectivities were calculated between 1.1 and 1.25 (values around 0.8 were found without entrainers at the same temperature and pressure), and could be surely improved by means of higher amounts of co-solvents. It is well known that such values make α-tocopherol recovery feasible only with high efficiency multistage contactors; again, capability of estimation of effects due to mass transfer limitations has to be achieved before considering possible process applications in continuous supercritical extraction systems.

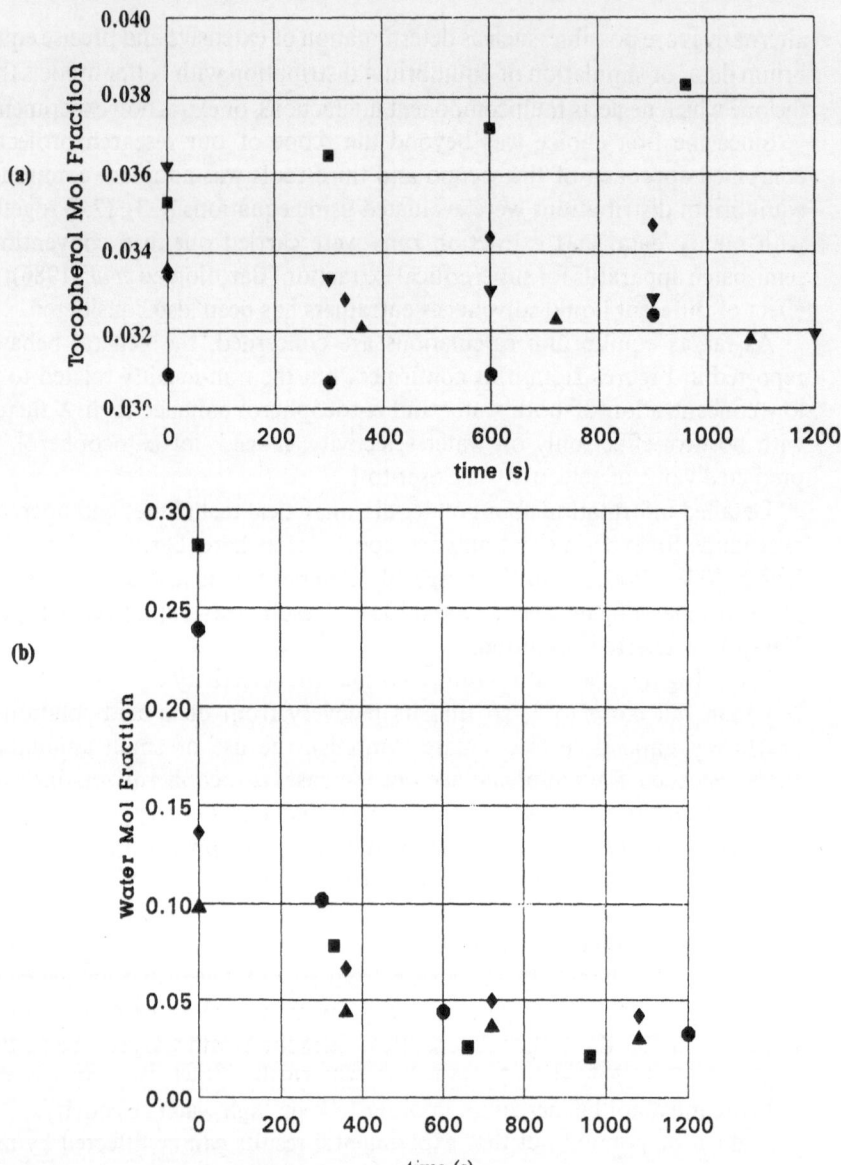

Figure 2.4 Extraction profiles for: (a) α-tocopherol (solvent free basis): $T = 333$ K, $P = 160$ bar, no entrainers (■); $T = 333$ K, $P = 160$ bar, 8wt% ethanol (♦); $T = 343$, $P = 100$ bar, 9wt% acetone (●); $T = 333$ K, P = 110 bar, 2wt% water + 4wt% ethanol (▼); $T = 333$ K, $P = 110$ bar, 0.7% water (▲). (b) Water (solvent free basis): $T = 333$ K, P = 115 bar (■); $T = 333$ K, P = 110 bar (●,▲); $T = 333$, P = 160 bar (♦).

2.4 Mass transfer

The development of a model for simulating the real performance of a super-critical extraction contactor involves hydrodynamics and mass transfer

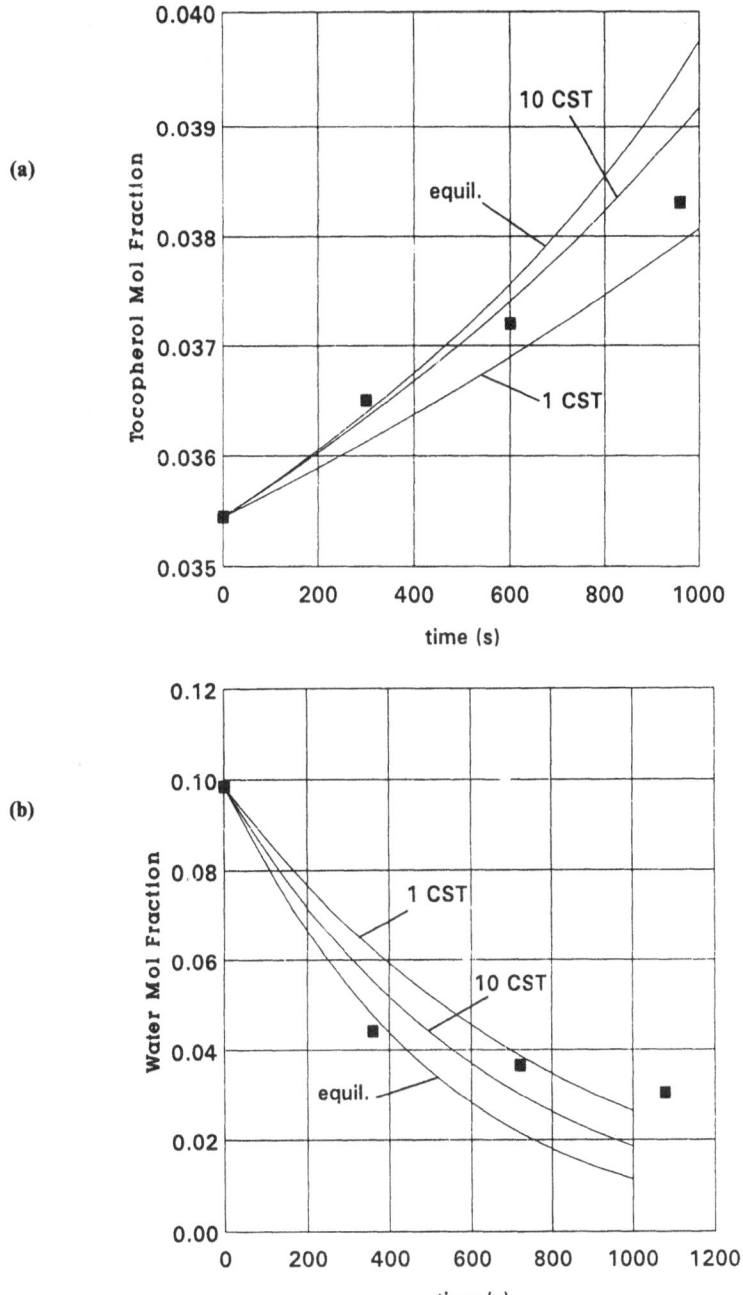

Figure 2.5 Comparison between experimental and calculated extraction profiles (solvent free bases): (a) α-tocopherol/oleic acid; (b) water/oleic acid.

phenomena. It can be addressed following both a gas-liquid and a liquid-liquid approach, as discussed in a previous paper (Bartolomeo *et al.*, 1990). In accordance with the results outlined there for aqueous mixtures, the gas-liquid model was shown to be more suitable and has also been applied to the semi-batch extraction of the solutions under current investigation. At the same time, it has been extended to continuous countercurrent contactors.

We refer to Bartolomeo *et al.* (1990) for details concerning the development of correlations needed for the calculations reported here.

Figures 2.5(a,b) show two examples of comparison between simulated and experimental results from the semi-batch extraction. In Figure 2.5(a), three curves are reported to represent the α-tocopherol enrichment in the liquid phase: the ideal profile (thermodynamic equilibrium at any time) and single-film simulations at two different degrees of mixing. The agreement with data points (■) is acceptable, so that the single-film approach can be usefully adopted for this kind of system.

Figure 2.5(b) shows water composition profiles in the liquid phase; simulation of experimental results is not so good as in the previous case. This fact could be ascribed to the inadequacy of the equilibrium model and to the difficulty of analytical determinations at very low water content.

Due to the impossibility of predicting phase equilibria for systems with more than three components, which is related to the unpredictable effects of entrainers over α-tocopherol, no simulation could be performed in those cases.

2.5   Continuous multistage simulation

On the basis of previous considerations, two multistage models have been developed for simulation of continuous supercritical extraction at steady-state conditions: one is for ideal trays, the second accounts for non-equilibrium effects. As already shown for different separation unit operations, we note that non-equilibrium models are the correct way to represent departures from ideality, since definition of Murphree efficiency appears troublesome for multicomponent systems.

Under the hypothesis of constant pressure and temperature throughout the column, energy balances can be neglected, so that the resulting sets of equations for each tray are as reported below. Although both models have been tailored for simulation of ternary systems, they can be easily extended for use with multicomponent systems, provided suitable equilibrium relationships are available.

1. *Ideal case*: the overall number of equations and variables per trays is $NC + 1$:

$$L_{n-1}x_{n-1,i} - L_n x_{n,i} + V_{n+1}K_{n+1,i}x_{n+1,i} - V_n K_{n,i}x_{n,i} = 0 \qquad (2.8)$$
$$(NC - 1 \text{ equations})$$

$$L_{n-1} - L_n + V_{n+1} - V_n = 0 \qquad (2.9)$$

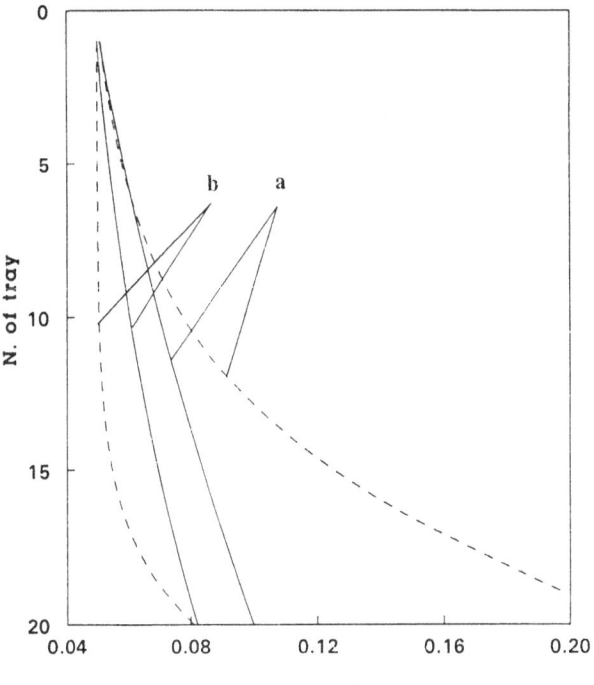

Figure 2.6 Simulated solvent free liquid composition profiles in a continuous supercritical extraction column for the system α-tocopherol/oleic acid/CO_2: ideal model (dashed lines) and mass transfer model (full lines); $V = 540$ mol/h (a), $V = 468$ mol/h (b).

$$1 - \Sigma\, K_{n,i}x_{n,i} - K_{n,NC}(1 - \Sigma\, x_{n,i}) = 0 \qquad (2.10)$$

Summations are extended to the first $NC - 1$ components.

2. *Non-equilibrium*: only equations for single-film are reported, the extension to the two-film case being straightforward. The overall number of equations and variables per tray is $2 \times NC$.

$$L_{n-1}x_{n-1,i} - L_n x_{n,i} + V_{n+1}K_{n+1,i}x^*_{n+1,i} - V_n K_{n,i}x^*_{n,i} = 0 \qquad (2.11)$$
$$(NC - 1 \text{ equations})$$

$$L_{n-1}x_{n-1,i} - L_n x_{n,i} + N_{n,i}S = 0 \qquad (NC - 1 \text{ equations}) \qquad (2.12)$$

$$L_{n-1} - L_n + V_{n+1} - V_n = 0 \qquad (2.13)$$

$$1 - \Sigma\, K_{n,i}x^*_{n,i} - K_{n,NC}(1 - \Sigma\, x^*_{n,i}) = 0 \qquad (2.14)$$

Summations are extended to the first $NC - 1$ components. Fluxes across the interface $N_{n,i}$ are calculated as reported previously (Bartolomeo *et al.*, 1990).

Once feed stream flow rates and compositions and number of trays are assigned, the problem is completely defined and the overall system of equations can be solved numerically to provide composition profiles in both liquid

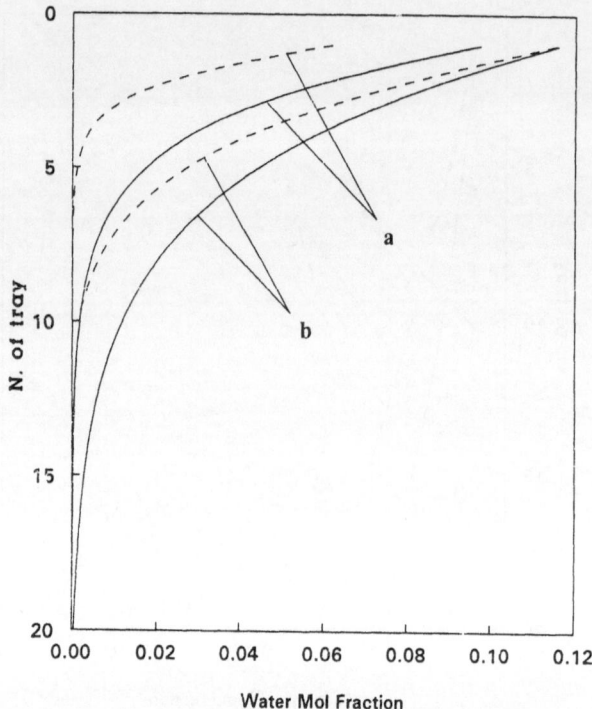

Figure 2.7 Simulated solvent free liquid composition profiles in a continuous supercritical extraction column for the system water/oleic acid/CO_2: ideal model (dashed lines) and mass transfer model (full lines); $V = 540$ mol/h (a), $V = 288$ mol/h (b).

and fluid phases. A structured algorithm has been used for this stage (Buzzi and Tronconi, 1986).

Both models refer to a multistage countercurrent configuration as represented in Figure 2.1(b), without a top reflux stream; in practice, a top reflux stream is often employed, for example running the top of the column at a different temperature (as with a 'hot finger' device). In this case, equations for the top tray have to be replaced accordingly, involving different equilibrium factors. A further degree of freedom is introduced when operating in this way; it can be saturated by assigning, for example, the split ratio of the top reflux.

Some simulation profiles are reported in Figure 2.6 for an α-tocopherol/oleic acid feed and in Figure 2.7 for a water/oleic acid feed; operating conditions are summarized in Table 2.1 and are consistent with the dimensions of the pilot plant under construction in our laboratory. Ideal (dashed lines) and single-film non-equilibrium (continuous curves) models are compared at two different solvent flow rates.

It can be noted that:

1. Mass transfer effects cannot be neglected, the performance being markedly worse than those obtained under equilibrium assumptions.

Table 2.1 Operating conditions for simulations of countercurrent supercritical CO$_2$ extraction of α-tocopherol and water

	T (°C)	P (bar)	No. of trays	V_{CO_2} (mol/h)	F (mol/h)	X_{feed}	X_{oleic}
Tocopherol	60	160	20	468–540	1.656	0.05	0.95
Water	60	115	20	288–540	2.448	0.15	0.85

Moreover, a CO$_2$ solubility consistently lower than ideal values is calculated in liquid phases
2. Oleic acid dehydration can be achieved within a reasonable number of stages.
3. α-Tocopherol concentration in the raffinate requires a higher number of trays.
4. In any case, the product purity is extremely sensitive to the solvent-to-feed flow rate ratio.

2.6 Conclusions

The possibility of profitable applications of supercritical CO$_2$ for refining and fractionation of lipid and glyceride mixtures has been considered both as a general issue and with respect to extraction from oleic acid of α-tocopherol and water.

It has been underlined that equilibrium information is the most crucial parameter to be addressed in the field. On the other hand, mass transfer phenomena may also play an important role. Simulation of a real extraction unit, either semi-batch or continuous, has been developed taking into account mass transfer effects and calculation results have been presented.

With respect to the specific separation problem considered, experiments showed that, while dehydration of oleic acid can be easily achieved, CO$_2$ does not provide acceptable selectivity for the purification of α-tocopherol; in this case, a possible process application requires the use of entrainers and a continuous multistage contactor.

In conclusion, experimental data are still lacking and better models have to be developed in order to allow predictions of supercritical extraction devices.

Acknowledgements

Financial support from the Italian Ministero della Ricerca Scientifica is gratefully acknowledged. Thanks are due to Dr M. Dahir for helpful discussions.

References

Ashour, I. and Wennersten, R. (1988) Supercritical fluid extraction, a potential method for the separation of fatty acids, in *Proc. Int. Symp. on Supercritical Fluids*, ed. M. Perrut, INPL, Nice, pp. 115–124.

Bamberger, T., Erickson, J. C., Cooney, C. L. and Kumar, S. K. (1988) Measurements and model prediction of solubilities of fatty acids, pure triglycerides, and mixture of triglycerides in supercritical carbon dioxide. *J. Chem. Eng. Data* **33**, 327.

Bartolomeo, G., Bertucco, A., Guarise, G. B. and Rienzi, S. A. (1986) Estrazione con fluidi in condizioni supercritiche: impianto sperimentale. *Riv. Combustibili* **40**, 185.

Bartolomeo, G., Bertucco, A. and Guarise, G. B. (1990) Mass transfer in a semibatch supercritical fluid extraction contactor for liquid mixtures. *Chem. Eng. Commun.* **95**, 57.

Berger, C. Jusforgues, P. and Perrut, M. (1988) Purification of unsaturated fatty acid esters by preparative supercritical fluid chromatography, in *Proc. Int. Symp. on Supercritical Fluids*, ed. M. Perrut, INPL, Nice, pp. 397–404.

Bertucco, A., Guarise, G. B., Dahir, M. and Navazio, G. (1988) Refining fatty acids with supercritical carbon dioxide, in *Proc. Int. Symp. on Supercritical Fluids*, ed. M. Perrut, INPL, Nice, pp. 791–798.

Bhrath, R., Adschiri, T., Inomata, H., Arai, K. and Saito S. (1991) Separation of fatty acids with supercritical CO_2, in *Proc. 2nd Int. Symp. on Supercritical Fluids*, ed. M. McHugh, Boston, MA, pp. 288–291.

Brennecke, J. F. and Eckert, C. A. (1989) Phase equilibria for supercritical fluid process design. *AIChE J.* **35**, 1409.

Brunetti, L., Daghetta, A., Fedeli, E., Kikic, I. and Zanderighi, L. (1989) Deacidification of olive oils by supercritical carbon dioxide. *J. Am. Chem. Oil. Soc.* **66**, 209.

Brunner, G. and Peter, S. (1982) State of art of extraction with compressed gases. *Ger. Chem. Eng.* **5**, 181.

Buzzi Ferraris, G. and Tronconi, E. (1986) BUNLSI—A Fortran program for solution of systems of nonlinear algebraic equations. *Comp. Chem. Eng.* **10**, 129.

Carturan, G., Facchin, G. and Navazio, G. (1983) C18 fat hydrogenation with Pd catalysts: selectivity in function of metal dispersion, metal support interactions and features of support porosity. *Chim. Ind.* **65**, 688.

Chrastil, J. (1982) Solubility of solids and liquids in supercritical gases. *J. Phys. Chem.*, **86**, 3016.

Dahir, M. M. (1989) Purificazione degli Acidi Grassi con CO_2 Supercritica, PhD Dissertation, University of Padova.

Eisenbach, W. (1984) Supercritical fluid extraction: a film demonstration. *Ber. Bunsenges. Phys. Chem.* **88**, 882.

Ender, U. and Steiner, R. (1990) Separation of triglyceride mixtures by high pressure extraction, in *Proc. 2nd Int. Symp. on High Pressure Chemical Engineering*, ed. G. Vetter, GVC, Erlangen, pp. 259–264.

King, M. B., Alderson, D. A., Fallah, F. H., Kassim, D. M., Kassim, K. M., Sheldon, J. R. and Mahmud, R. S. (1983) in *Chemical Engineering at Supercritical Fluid Conditions*, eds. M. Paulaitis, J. M. L. Penninger, R. D. Gray and P. Davidson, Ann Arbor Science, Ann Arbor, MI, pp. 31–80.

Krukonis, V. J. (1989) Supercritical fluid processing of fish oils: extraction of polychlorinated biphenyls. *J. Am. Chem. Oil. Soc.*, **66**, 818.

Nilsson, W. B., Gauglitz, E. J. and Hudson, J. K. (1989) Supercritical fluid fractionation of fish oil esters using incremental pressure programming and a temperature gradient. *J. Am. Chem. Oil. Soc.*, **66**, 1596.

Nilsson, W. B., Gauglitz, E. J. and Hudson, J. K. (1991) Solubilities of methyl oleate, oleic acid, oleyl glycerols, and oleyl glycerol mixtures in supercritical carbon dioxide. *J. Am. Chem. Oil. Soc.* **68**, 87.

Ohgaki, K., Tsukahara, I., Semba, K. and Katayama, T. (1987) A fundamental study of super-critical fluid extraction. *Kagaku Kogaku Ronbunshu*, **13**, 298.

Ohgaki, K., Nishikawa, M., Furuichi, T. and Katayama, T. (1988) Entrainer effect of water and ethanol on α-tocopherol extraction by compressed CO_2. *Kagaku Kogaku Ronbunshu* **14**, 342.

Pearce, D. L. and Jordan, P. J. (1991) Solubility of triglycerides in carbon dioxide, in *Proc. 2nd Int. Symp. on Supercritical Fluids*, ed. M. McHugh, Boston, MA, pp. 478–481.

Pennisi, K. J. and Chimowitz, E. H. (1986) Solubilities of solid 1, 10-decanediol and a solid mixture of 1,10-decanediol and benzoic acid in supercritical carbon dioxide. *J. Chem. Eng. Data*, **31**, 285.

Peter, S., Schneider, M., Weidnwer, E. and Ziegelitz, R. (1987) The separation of lecithin and soya oil in a countercurrent column by near critical fluid extraction. *Chem. Eng. Technol.*, **10**, 37.

Peter, S., Seekamp, M. and Bayer, A. (1988) Dissolution of oleic acid in dense gases, in *Proc. Int. Symp. on Supercritical Fluids*, ed. M. Perrut, INPL, Nice, pp. 99–106.

Shishikura, A., Fujimoto, K., Kaneda, T., Arai, K. and Saito, S. (1988) Concentration of tocopherols from soybean sludge by supercritical fluid extraction. *J. Jpn. Oil. Chem. Soc.* **37**, 8.

Siewers, U. (1984) *Die Thermodynamischen Eigenschaften von Kholendioxide*, VDI-Verlag, Dusseldorf.

Van Ness, H. C. and Abbott, M. M. (1982) *Classical Thermodynamics of Non Electrolytyle Solutions*, McGraw-Hill, New York, pp. 289–299.

Wong, J. M. and Johnston, K. P. (1986) Solubilization of biomolecules in carbon dioxide based supercritical fluids. *Biotechnol. Prog.* **2**, 29.

Wu, A. H., Stammer, A. and Prausnitz, J. M. (1988) Extraction of fatty-acid methyl esters with supercritical carbon dioxide, in *Proc. Int. Symp. on Supercritical Fluids*, ed. M. Perrut, INPL, Nice, pp. 107–114.

Zehnder, B. and Trepp, C. (1991) Mass transfer coefficients and equilibrium solubilities for fluid-supercritical solvent systems by on-line NIR-spectroscopy, in *Proc. 2nd Int. Symp. on Supercritical Fluids*, ed. M. McHugh, Boston, MA, pp. 329–332.

Zou, M., Yu, Z. R., Rizvi, S. S. H. and Zollweg, J. A. (1990a) Fluid-liquid phase equilibria of fatty acids and fatty acid methyl esters in supercritical carbon dioxide, *J. Supercritical Fluids* **3**, 23.

Zou, M., Yu, Z. R., Rizvi, S. S. H. and Zollweg, J. A. (1990b) Fluid-liquid equilibria of ternary systems of fatty acids and fatty acid esters in supercritical carbon dioxide, *J. Supercritical Fluids* **3**, 85.

3 Mass transfer phenomena in supercritical carbon dioxide extraction for production of spice essential oils

K. UDAYA SANKAR and B. MANOHAR

Abstract

Supercritical carbon dioxide extraction at pressures of 8 MPa, 10 MPa, 12 MPa and temperatures of 40°C, 50°C, 60°C was used to extract ginger. The physical and chemical characteristics of the oil were determined and revealed minimal amounts of non-volatile matter. Thus, an extraction with supercritical carbon dioxide at selective pressures and temperatures can be used to produce essential oils from spice materials. The mass transfer phenomena were studied using an unsteady state model. For ginger, the mass transfer coefficients were found to vary from 1.581×10^{-8} to 43.24×10^{-8} kg/m^2 s. The mass transfer coefficient was found to decrease with increase in particle size.

3.1 Introduction

The advent of food processing as a modern industry has ushered in the use of food extractives rather than raw materials. The food extractives of spices, called oleoresins, are used in the food processing industry for their appealing flavour and to improve product quality. Solid extraction is a heterogenous, non-stationary mass transfer operation. To concentrate small quantities of substances present in a solid matrix or to remove unwanted principles, extraction as a separation technique is used. Solvents are generally used for extraction of spices, sometimes with prior steam distillation, to yield essential oils that can be mixed with the solvent extracted resin to make the oleoresin.

With the development of supercritical fluid extraction, the liquid solvents are replaced by supercritical gases. Using carbon dioxide, industrial plants are in operation for extraction of hops, coffee and tea (Parkinson and Johnson, 1989). Selectivity of the extraction can be achieved using supercritical carbon dioxide and the essential oils of pepper and ginger can be extracted without much contamination of non-volatile matter by suitable selection of the extraction pressure and temperature of carbon dioxide (Rao *et al.*, 1988; Sankar and Manohar, 1988). The quality of the extracts obtained are true to natural in flavour and odour.

Mathematical modelling of mass transfer during solvent extraction of spices is scarcely available. Recently, a study of mass transfer using a steady-state model on extraction of peppers by ethanol (Aguilera *et al.*, 1987), ginger by acetone (Spiro and Kandiah, 1989), ginger by dichloromethane, ethanol, 2-propanol and acetone/water mixtures (Spiro *et al.*, 1990), ginger by super-critical carbon dioxide (Kandiah and Spiro, 1990), has been carried out. Attempts have been made on mathematical modelling of mass transfer of foods (Schwartzberg, 1975; Schwartzberg and Chao, 1982; Sankar *et al.*, 1983; Aguerri, 1985) based on Fick's second law of diffusion. The earlier studies are mainly on equilibrium mass transfer of foods based on the kinetic model of extraction. Mass transfer phenomena in fixed bed extraction with multi stages (Schwartzberg, 1980) or just a single stage for paprika and turmeric (Houser *et al.*, 1975) have been reported. The rate equations are derived based on concentration–time data using the kinetic approach in the latter, while in the former, the equilibrium approach is used.

3.2 Theory

The extraction of essential oil of ginger was characterized by the fact that:

1. At the selected conditions of extraction using carbon dioxide, only essential oil was preferentially extracted and the solubility of the non-volatile components was small.
2. The essential oil concentration in the solid phase was very low and so also the concentration of the essential oil content in the gas phase.
3. The fixed bed consisted of particles of ginger of about the same size.
4. Although the cell walls were ruptured, the oil sacs were very minute in size and fairly equally distributed in the average particle of ginger.

In the present study, for the extraction process, a fixed bed of spheres was considered. The mass transfer in a single sphere was investigated and the results were transferred to a fixed bed.

The extraction process for this case can be described by a coupled set of differential equations which take into account

1. time-dependent concentration and consequently a time-dependent driving force and time-dependent mean transport coefficient and
2. the influence of the flow of the gas.

The mass transport process was split into two steps. For the mass transport in the solid phase, an inside transport coefficient k_i was defined and the concentration difference was the mean concentration of ginger oil in the ginger particles minus the concentration of the oil at the surface. The outer transport coefficient k_o was defined as usual.

The oil is embedded in the sphere. The mean concentration in the solid sphere decreases until the mass transfer stops as the essential oil concentration in the spherical particles reaches zero or as it approaches equilibrium concentration in the interface. Extraction in most food materials is controlled by the unsteady state mass transfer rate in the solid phase.

Mass transfer is considered to be an unsteady state. Approximate solutions can be obtained for the differential equations representing the quasi steady state of the problem. In the present study, the second approach is followed (Brunner, 1984):

$$m = k_i\, A\, (C_m - C_i) \tag{3.1}$$

$$m = k_o\, A\, (C_i - C_g) \tag{3.2}$$

where m is the amount of oil extracted per time, A is the mass transfer area, C_m is the concentration of oil in the solid phase, C_g is the concentration of oil in the gas phase, C_i is the concentration of oil in the interface, k_i is the solid phase mass transfer coefficient and k_o is the gas phase mass transfer coefficient.

Mass balance relates the time-dependent mean concentration to the quantity extracted as

$$m = -m_s\, C_o\, (dC/dt) \tag{3.3}$$

where m_s is the amount of solids, C_o is the initial oil concentration in the solid and t is the time. By defining an overall mass transfer coefficient K and combining the equations, a differential expression for the time-dependent concentration is obtained. If K is assumed to be independent of time, the equation is integrated readily.

$$1/K = 1/k_i + 1/k_a$$

From equations (3.2) and (3.3)

$$K\, A\, (C_t - C_g) = - (m_s\, C_o)\, dC/dt$$

On integrating,

$$dC/(C_t - C_g) = - K\, A/(m_s\, C_o)\, dt$$

we obtain

$$(C_t - C_g) / (C_o - C_g) = \exp(- K\, A\, t/ m_s\, C_o) \tag{3.4}$$

The assumption of a constant transport coefficient implies that an average mean transport coefficient must be used. For transport of the essential oil, which is much less in the solid phase and embedded in the solid phase, the resistance to mass transfer lies in the solid (inner) phase. Therefore

$$1/K \ll 1/k_i \text{ and } K = k_i$$

The results obtained for a single sphere were transferred to a fixed bed of spheres by the enhancement function, where the void volume (U) was taken into account.

Here the solvent phase mass transfer resistance is less and the resistance lies entirely in the solid phase.

Plotting $-\ln(C_t - C_g)/(C_o - C_g)$ versus time t, the slope gives $KA/m_s C_o$. The regression value for the best fit was obtained using a standard computer program.

The specific surface area and the mean diameter of the particle D was obtained by differential sieve analysis (McCabe and Smith, 1976).

3.3 Materials and methods

A 4-l high-pressure extraction facility fabricated by NOVA SWISS, Switzerland, was used. The extraction vessel has a diameter of 0.092 m and height of 0.33 m. The carbon dioxide can be recirculated after separation at subcritical conditions. A constant flow of carbon dioxide is ensured by controlling the suction pressure and temperature of the gas. The carbon dioxide in circulation is monitored by an on-line flow measurement system using a calibrated flowmeter, temperature and pressure recorder (Figure 3.1).

The carbon dioxide was drawn from the cylinder G, which was kept on a platform balance and compressed to the extraction pressure by the diaphragm compressors C_1 and C_2 and heated to the extraction temperature by using the heat exchangers TC_1. The extraction pressure and temperature were measured by pressure gauge PI_2 and Pt resistance thermometer TR_1. The supercritical carbon dioxide was brought in contact with the material to become enriched with the solute. The enriched carbon dioxide was brought to the subcritical separating conditions of temperature (16–20°C) and pressure (4–5.5 MPa) by the pressure controlled regulating valve (PCR). The pneumatically controlled solenoid valve opens up to allow the carbon dioxide to pass through, once the transducer indicates the set pressure is reached with the help of a Eurotherm Controller (not shown in figure). The carbon dioxide in the separator was further heated to vaporisation and passed through a pre-calibrated flow meter and filter before being compressed back to the extraction pressure. As the separator was being maintained in vapour/liquid equilibrium, the complete separation of the material was ensured. The solubility of the solute in the vapour of carbon dioxide was negligible. The entrained material if any, was filtered using a microporous filter (MF) of pore size 10 Å. The extracts in the separator were drawn at regular time intervals and the carbon dioxide used for the extraction was manipulated using the flow meter (FM). The carbon dioxide used initially for loading into the

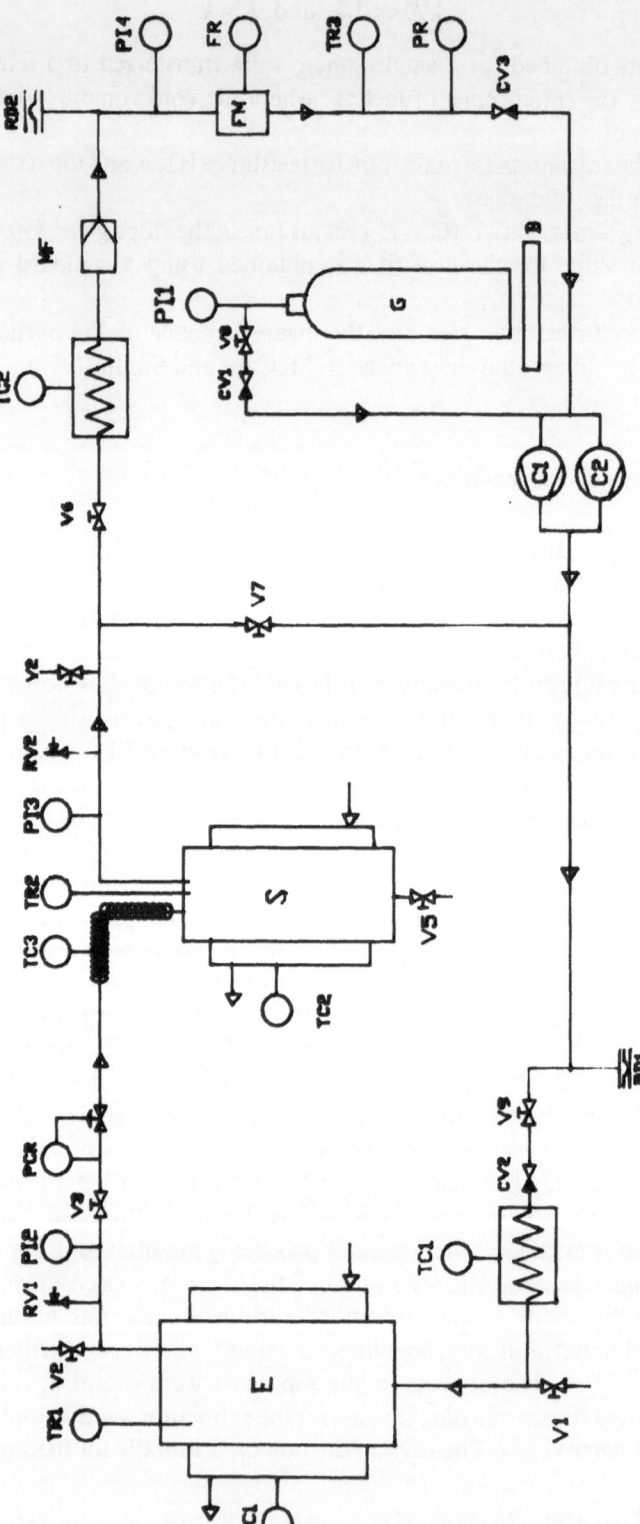

Figure 3.1 Supercritical CO_2 extraction apparatus. G, CO_2 cylinder; B, balance; RD1–2, rupture disc assemblies; MF, micro filte; PCR, micro metering valve; FM, flow meter; TC1–3, heat exchangers; CV1–3, check valve assemblies; V1–9, valves; PI1–4, pressure indicators; TR1–3, temperature recorder; PR, pressure recorder; FR, flow recorder; RV1–2, relief valve assemblies.

Table 3.1 Regression values of the concentration of ginger oil for the extraction of ginger at different pressures and temperatures (specific surface area = 45.23 m^2/kg)

Sample no.	Time (s) × 10^4	Concentration of the oil in the bed (× 10^{-2})	
		Experimental	Calculated
Experiment 1[a]			
1	1.032	1.43	1.186
2	1.44	1.115	0.989
3	1.65	0.977	0.900
4	1.919	0.796	0.798
5	2.399	0.524	0.642
Experiment 2[b]			
1	4.5	2.364	2.311
2	7.59	2.047	2.020
3	10.41	1.836	1.787
4	13.229	1.625	1.581
5	16.03	1.414	1.400
6	19.41	1.203	1.208
7	23.06	0.992	1.031
Experiment 3[c]			
1	8.399	1.43	1.335
2	13.2	1.158	1.109
3	16.8	0.977	0.965
4	20.09	0.716	0.852
Experiment 4[d]			
1	0.921	2.341	2.199
2	1.507	2.01	1.881
3	1.926	1.79	1.683
4	2.429	1.57	1.472
5	2.931	1.35	1.288
6	3.35	1.13	1.152
7	3.978	0.91	0.975
8	4.857	0.69	0.771
9	6.114	0.58	0.552
Experiment 5[e]			
1	3.190	2.278	2.218
2	4.776	1.911	1.972
3	6.90	1.666	1.685
4	8.972	1.421	1.445
5	12.08	1.176	1.147
Experiment 6[f]			
1	1.44	1.430	1.45
2	2.751	1.158	1.163
3	3.604	0.977	1.008
4	4.796	0.724	0.825
5	7.322	0.453	0.539
6	14.64	0.181	0.158
Experiment 7[g]			
1	4.71	2.021	2.02
2	6.28	1.802	1.815
3	7.06	1.584	1.719

Table 3.1 *continued*

Sample no.	Time (s) $\times 10^4$	Concentration of the oil in the bed ($\times 10^{-2}$) Experimental	Calculated
4	9.42	1.149	1.459
5	10.59	0.931	1.345
6	12.59	0.713	1.141
7	20.8	0.495	0.661
8	40.2	0.278	0.173
Experiment 8[h]			
1	7.406	2.078	2.192
2	11.52	1.526	1.843
3	14.81	1.159	1.603
4	22.22	0.791	1.172
5	53.48	0.423	0.313

[a] $P = 8$ MPa; $T = 40°C$; $N_{Re} = 2$; $C_0 = 1.883 \times 10^{-2}$ kg/kg; $m_s = 1$ kg; $K = 1.862 \times 10^{-8}$ ms.
[b] $P = 8$ MPa; $T = 50°C$; $N_{Re} = 4.31$; $C_0 = 2.81 \times 10^{-2}$ kg/kg; $m_s = 1$ kg; $K = 2.7 \times 10^{-8}$ ms.
[c] $P = 8$ MPa; $T = 60°C$; $N_{Re} = 2.61$; $C_0 = 1.848 \times 10^{-2}$ kg/kg; $m_s = 1$ kg; $K = 1.581 \times 10^{-8}$ ms.
[d] $P = 10$ MPa; $T = 40°C$; $N_{Re} = 2.76$; $C_0 = 2.81 \times 10^{-2}$ kg/kg; $m_s = 1$ kg; $K = 1.654 \times 10^{-7}$ ms.
[e] $P = 10$ MPa; $T = 50°C$; $N_{Re} = 3.22$; $C_0 = 2.81 \times 10^{-2}$ kg/kg; $m_s = 1$ kg; $K = 4.605 \times 10^{-8}$ ms.
[f] $P = 10$ MPa; $T = 60°C$; $N_{Re} = 1.66$; $C_0 = 1.848 \times 10^{-2}$ kg/kg; $m_s = 1$ kg; $K = 1.868 \times 10^{-8}$ ms.
[g] $P = 12$ MPa; $T = 50°C$; $N_{Re} = 2.71$; $C_0 = 2.81 \times 10^{-2}$ kg/kg; $m_s = 1$ kg; $K = 4.324 \times 10^{-7}$ ms.
[h] $P = 12$ MPa; $T = 60°C$; $N_{Re} = 2.79$; $C_0 = 2.998 \times 10^{-2}$ kg/kg; $m_s = 1$ kg; $K = 2.8 \times 10^{-9}$ ms.

system was found by the difference in weight of the cylinder using the platform balance (B).

The ginger was procured from a local market, cleaned thoroughly in water, and cut into 3–6 mm slices using a Stephen mill. The slices were dried in a cabinet drier with optimum tray load at 50–60°C, then stored in a cold room at 4°C before powdering to prevent loss of volatiles during storage and grinding. The ginger was powdered before extraction in a hammer mill. The mean particle size of ginger was found to be 1.10×10^{-4} m.

The extractions were carried out on 0.9–1.1 kg of material and the extract was drawn at different intervals of time to monitor the rate of extraction. The extraction was carried out at 8–12 MPa and 40–60°C. It was found that under these conditions, essential oil alone was collected in the separator, which was maintained at subcritical conditions of 5–6 MPa at 20°C. The extract drawn at various stages of the extraction was weighed using a Mettler balance. Care was taken to withdraw the extract from the separation vessel without spilling. The carbon dioxide was circulated to the compressor after separation for recompression and recirculation.

3.4   Results and discussion

The experimental data of concentration versus time for 8 MPa and 50°C are presented in Table 3.1. The application of equation (3.4) to the experimental

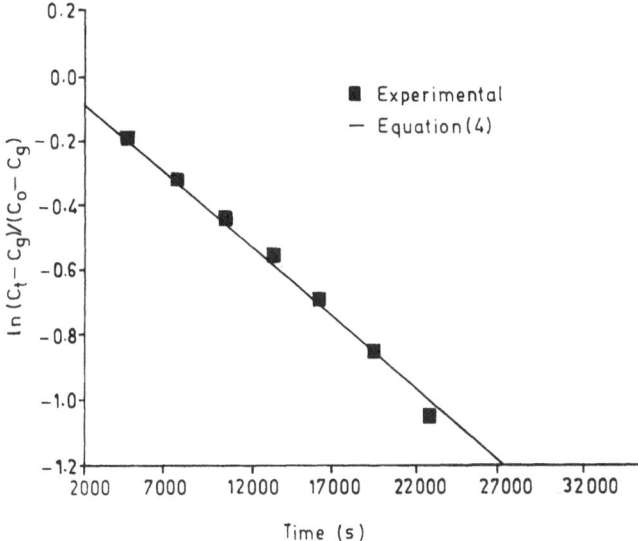

Figure 3.2 Application of equation (3.4) to the CO_2 extraction of ginger for production of volatile ginger oil at 8 MPa and 50°C. ■, Experimental; ——, equation.

extraction data (■) for the same condition of extraction is illustrated in Figure 3.2. The slope of the straight line in Figure 3.2 gives the mass transfer coefficient. Using regression, with the help of a computer program based on least squares fit, the slopes were obtained for different experimental conditions. The experimental data were regressed to obtain the theoretical data using equation (3.4), and the correlation coefficients obtained using a least squares fit for each set of experiments (Table 3.2). The correlation coefficient obtained (0.95–0.99) was of high degree of significance. In Table 3.3, mass transfer coefficients for the entire range of experimental conditions are presented. In the studies of extraction of 6-gingerol from rhizomes of ginger at 12.8–19.7 MPa and 50–65°C of carbon dioxide, a fast diffusion stage followed by a slower

Table 3.2 Parameters of equation (3.4) with correlation coefficient for different extraction conditions

P (MPa)	T (°C)	Slope ($\times 10^{-4}$)	Correlation coefficient
8	40	0.447	0.958
8	50	0.435	0.998
8	60	0.387	0.989
10	40	2.662	0.992
10	50	0.741	0.997
10	60	1.681	0.995
12	50	6.960	0.951
12	60	4.225	0.950

Table 3.3 Mass transfer characteristics of ginger extraction for volatile oil

P (MPa)	T (°C)	Density (kg m^{-3})	Diffusion coefficient[a] (kg/m^3 s $\times 10^{-8}$)	Linear velocity (cm/min)	Reynolds number	K (kg/m^2 s $\times 10^{-8}$)
8	40	299.7	8.840	8.23	2.00	1.862
8	50	222.6	12.050	23.16	4.31	2.700
8	60	192.6	14.950	15.10	2.61	1.581
10	40	611.6	3.487	3.21	2.76	16.540
10	50	405.5	5.728	10.07	3.22	4.605
10	60	294.1	9.789	6.95	1.66	6.868
12	50	484.4	4.737	9.47	2.71	43.240
12	60	418.6	6.939	9.47	2.79	28.000

[a]Diffusion coefficient of carbon dioxide.

diffusion stage was observed, resulting in two rate constants (Kandiah and Spiro, 1990).

It has been observed in our laboratory that the relatively non-volatile organic components such as 6-gingerol are not being extracted in experiments conducted under the conditions mentioned in the present work. The essential oil constituted most of the steam-distillable volatile fraction. In equilibrium studies conducted by Kandiah and Spiro (1990), a faster rate of extraction of 6-gingerol was found initially, because of the entrainer effect of the essential oil. The essential oil, 6-gingerol and carbon dioxide form a ternary equilibrium. Hence, the 6-gingerol is solubilised in the essential oil, resulting in a faster rate of extraction of 6-gingerol in the initial stage of extraction.

Experiments conducted at various Reynolds numbers at the same pressure and temperature showed that mass transfer coefficients did not have any particular order with Reynolds number.

The mass transfer coefficients were found to decrease with increase in particle size. When the particle size of the bed increased from 0.110 mm to 0.184 mm, the overall mass transfer coefficient decreased from 28.0×10^{-8} to 26.2×10^{-8} kg/m^2 s.

References

Aguerri, R. J., Gabitto, J. F. and Chieife, J. (1985) Utilisation of Fick's second law for the evaluation of diffusion coefficients in food processes controlled by internal diffusion. *Int. J. Food Technol.*, **20**, 623.

Aguilera, J. M., Escobar, G. A., Del Valle, J. M. and San Martin, R. (1987) Ethanol extraction of red peppers minetic studies and microstructure. *Int. J. Food Sci. Technol.* **22**, 225.

Brunner, G. (1984) Mass transfer from solid material in gas extraction. *Ber. Bunsenges. Phys. Chem.* **88**, 887.

Houser, T. J., Biftu, T. and Hsieh, P. F. (1975) Extraction rate equations for paprika and turmeric with certain organic solvents. *J. Agric. Food Chem.* **23**, 353.

Kandiah, M. and Spiro, M. (1990) Extraction of ginger rhizome: kinetic studies with supercritical carbon dioxide. *Int. J. Food Technol.* **25**, 328.

McCabe, W. and Smith, W. (1976) *Unit Operations of Chemical Engineering*, 4th Edition, McGraw-Hill, New York, pp. 821.

Parkinson, G. and Johnson, E. (1989) Supercritical processes win CPE acceptance. *Chem. Eng.* **96**(7), 36.

Rao, V. S. G., Mukhopadya, M. and Sankar, K. U. (1988) Selective extraction of spice oil constituents by supercritical carbon dioxide, presented at Indian Institute of Chemical Engineers, Baroda.

Sankar, K. U., Raghavan, C. V., Rao, P. N. S., Rao, K. L., Kuppuswamy, S. and Ramanathan, P. K. (1983) Studies on extraction of caffeine from coffee beans. *J. Food Sci. Technol. (India)* **20**(3&4), 64.

Sankar, K. U. and Manohar, B. (1988) Extraction of essential oils using supercritical carbon dioxide, in *Proc. Int. Symp. on Supercritical Fluids*, ed. P. M. Vandeouvre, Vol. 2, pp. 807–814.

Schwartzberg, H. G. (1975) Mathematical analysis of solubilisation kinetics and diffusion in foods. *J. Food Sci.* **40**, 211.

Schwartzberg, H. G. (1980) Continuous counter current extraction in the food industry. *Chem. Eng. Prog.* **76**(4), 67.

Schwartzberg, H. G. and Chao, R. Y. (1982) Solute diffusivities in leaching processes. *Food Technol.* **36**(2), 73.

Spiro, M. and Hunter, J. F. (1985) The kinetics and mechanism of caffeine infusion from coffee: the effect of roasting. *J. Sci. Food Agric.* **36**, 871.

Spiro, M. and Kandiah, M. (1989) Extraction of ginger rhizome: kinetic studies with acetone. *Int. J. Food Sci. Technol.* **24**, 589.

Spiro, M., Kandiah, M. and Price, W. (1990) Extraction of ginger rhizome: kinetic studies with dichloromethane ethanol, 2-propanol and acetone-water mixture. *Int. J. Food Sci. Technol.* **25**, 157.

4 Biochemical reactions in supercritical fluids

K. NAKAMURA

Abstract

Acidolysis reactions in supercritical carbon dioxide (SC-CO$_2$) were conducted in a continuous reactor, and the effect of temperature, pressure, residence time and feed concentration on product constituents and productivity was examined. The equilibrium product constituents could be well estimated by the equations derived. There was optimum water concentration which shifted to a larger value as the substrate concentration increased or the residence time decreased. The effect of temperature on the product constituents was not remarkable when the pressure was as high as 30 MPa, while the triglycerides were less hydrolyzed at lower pressure. The reaction seemed to be influenced by internal mass transfer, so that a combination of low substrate concentration and low pressure was favorable for better productivity when the residence time was too short to bring the reaction to the equilibrium state.

4.1 Introduction

Supercritical fluids (SCF), which have been gaining recognition as suitable solvents for extraction of biomolecules, also show promise for use as novel media for enzymatic reactions. Their advantages, compared with organic solvents, include easily controllable solubility of components, high diffusivity, low toxicity and improved reaction rates. A review on biochemical reactions in supercritical fluids describes the solubility of substrate and water in SCF, the stability of enzymes in SCF and examples of enzymatic reactions in SCF (Nakamura, 1990a).

Among the enzymatic reactions in SCF investigated so far, the use of lipase shows most commercial promise. Many opportunities exist for applications in biomolecular synthesis and modification. SC-CO$_2$ has a high capacity for dissolving water, and its solubility depends on the choice of temperature and pressure. A supercritical CO$_2$/H$_2$O mixture may be used as a reaction medium for either hydrolytic or synthetic reactions catalyzed by lipase and other appropriate hydrolases. The author and colleagues have studied the acidolysis of triolein with stearic acid, a model reaction for production of cocoa butter substitute, and have reported the results of batch reaction in SC-CO_2 (Nakamura *et al.*, 1986; Chi *et al.*, 1988). The study has been extended to the continuous reaction, and the effects of several factors such as water concentration, temperature, pressure and flow velocity are investigated. These factors can influence the reaction in SCF directly or indirectly because the constants of the reaction and mass transfer such as rate constant, solubility, effective diffusivity, mixing diffusivity and mass transfer coefficient depend on temperature and pressure as well as flow velocity.

4.2 Acidolysis reaction

The reaction considered is the acidolysis of triolein with stearic acid which is catalyzed by 1,3-regiospecific lipase. The lipase is a hydrolytic enzyme, and the mono- and di-glycerides are also produced even though the reaction is conducted under micro-aqua conditions.

$$\text{Substrates} \left(\mathsf{E}^{\mathrm{o}}_{\substack{\mathrm{o}\\\mathrm{o}}}, \text{ stearic acid, } H_2O \right)$$

$$\text{Products} \left(\mathsf{E}^{\mathrm{o}}_{\substack{\mathrm{o}\\\mathrm{s}}}, \mathsf{E}^{\mathrm{s}}_{\substack{\mathrm{o}\\\mathrm{s}}}, \mathsf{E}^{\mathrm{o}}_{\mathrm{o}}, \mathsf{E}_{\mathrm{o}}, \text{ oleic acid} \right) \quad (4.1)$$

where E is the glycerol moiety, and the letters o and s denote oleate and stearate, respectively.

The total reaction consists of interconnected reaction pathways, and the time course of the reaction can be simulated if the rate equation is assumed for each reaction step (Kyotani *et al.*, 1988; Nakamura, 1988). The optimum value of many rate constants, however, is difficult to achieve with the simulation. On the other hand, the equilibrium constituent of the product can be simply calculated using the following assumptions:

1. The substrate-specificity of lipase is not different between the two fatty acids, stearic acid and oleic acid.
2. Water participates in the reaction and is bound to the glycerol moiety K times as efficiently as either of the two fatty acids.

Then the residual triglyceride and the extent of interesterification can be calculated at the equilibrium state of the reaction (Nakamura, 1990b).

$$T_\infty = \frac{\left(E_O^O + E_S^O + E_S^S\right)_{equil.}}{\left(E_O^O\right)_{ini.}} = \frac{(2C_t + C_s)(C_t + C_s)}{(2C_t + C_s + KC_w)(C_t + C_s + KC_w)} \tag{4.2}$$

$$E_\infty = \frac{\left(E_O^O + 2E_S^O\right)_{equil.}}{3 \times \left(E_O^O + E_S^O + E_S^S\right)_{equil.}} = \frac{2}{3} \times \frac{C_s}{2C_t + C_s} \tag{4.3}$$

where C_t, C_s and C_w are the initial concentration of triolein, stearic acid and water, respectively. The productivity is defined by equation (4.4), and its equilibrium value can also be expressed with the initial concentration of substrates as described by Nakamura (1990b):

$$P = \frac{\left(E_S^O + E_S^S\right)}{\left(E_O^O\right)_{ini.}} \times \frac{C_t^F}{M} \tag{4.4}$$

where F is the flow rate and M is the amount of immobilized enzyme packed in the column reactor.

4.3 Materials and methods

4.3.1 Reactor

The continuous reactor developed is schematically shown in Figure 4.1. It was set up with pumps, back-pressure regulator valves, column oven, etc., most of which were the same as those used in commercial supercritical fluid extraction and supercritical chromatography, JASCO SUPER-200. One of the substrates, stearic acid (6), was contacted with flowing carbon dioxide and dissolved or sublimed at a chosen temperature. The other substrate, triolein, was fed separately with the syringe pump (2) to the mixer (9) and mixed homogeneously together with the SC-CO_2/stearic acid phase. The concentra-

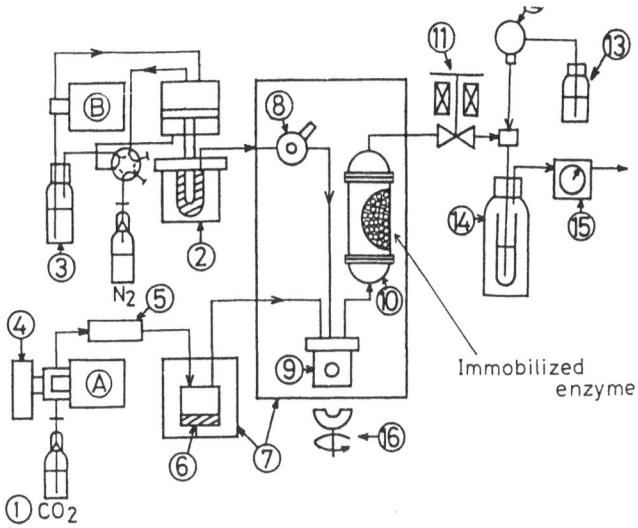

Figure 4.1 Scheme of experimental apparatus. A, B, HPLC pump; 2, syringe pump; 3, IPA tank; 4, cooling unit; 5, molecular sieves 13 ×; 6, stearic acid; 7, constant temperature bath; 8, injector (water); 9, mixing cell; 10, column reactor; 11, pressure regulator; 12, pump; 13, ethanol; 14, trap; 15, flow meter; 16, stirrer.

tion of water was controlled by feeding it manually at regular intervals. The reactor (10) was a small column, 4.6 mm in inner diameter, and its length was 50, 100, 150 or 200 mm. The packed amount of immobilized lipase (particle size 440 μm) was 0.3 g in the case of the shortest column, and it increased in the other columns in proportion to their length. The product was recovered in the trap (14) connected to the exit of the pressure regulator (11), and the flow rate of carbon dioxide was measured at atmospheric pressure with a gas meter (15).

4.3.2 Reaction conditions

The temperature was changed from 40°C to 70°C, and the pressure was changed from 20 MPa to 30 MPa. The range of flow velocity corresponded to the particle Reynolds number from 2 to 8 or to the residence time from 15 to 100 s. The feed concentration of triolein and stearic acid was less than 10 mM, and the concentration of water was changed from 2 to 20 mM.

4.3.3 Materials

The substrate, triolein and stearic acid, were purchased from Sigma (purity 99%). The immobilized *Mucor miehei* lipase was a commercial product of Novo, Lipozyme IM20 which contained 0.14 g of protein/g of carrier.

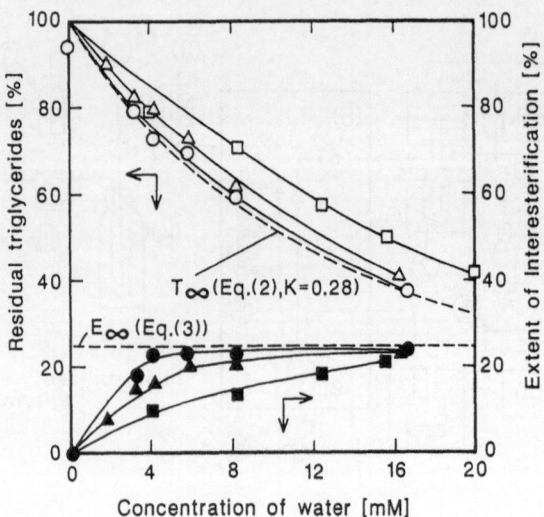

Figure 4.2 Effect of water on residual triglyceride and extent of interesterification. Residence time θ (s): □ ■, 22; △ ▲, 45; ○ ●, 91.

4.3.4   Analysis

The product was recovered in ethanol which was fed by a pump into the exit tube of a pressure regulator (11). Ethanol was vacuum-evaporated at 35°C, and the residue was analyzed by HPLC as reported previously (Chi *et al.*, 1988).

4.4   Results and discussion

4.4.1   Steady state

The elution pattern of the products resembled the transient response curve of the feed when the conditions of temperature and pressure were favorable to dissolve the products and simultaneously the flow rate was large enough. The steady state of the product concentration, however, came in some cases much later than that of the feed concentration probably due to the adsorption of products on the immobilized enzyme particles and the saturation of the biocatalyst with water. The steady state of the reactions was analyzed and the influence of temperature, pressure, residence time and substrate concentration is discussed in the following sections.

4.4.2   Effect of water

Water is a reactant in hydrolytic reactions as well as an important factor in

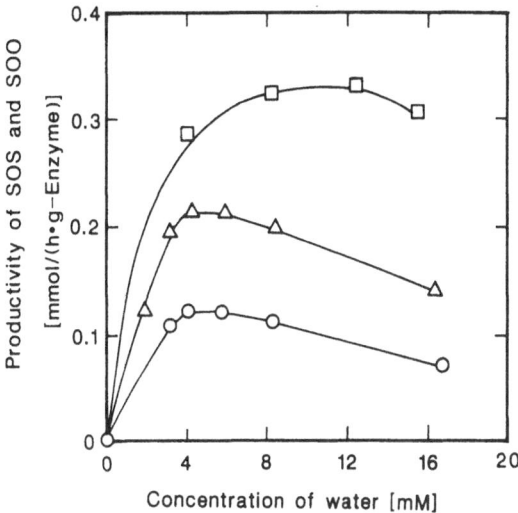

Figure 4.3   Effect of water on productivity of product triglycerides SOS and SOO. θ (s): □, 22; △,45; O, 91.

controlling the catalytic activity of enzyme. The concentration of water should be optimized for maximum productivity in an acidolysis reaction.

An example of the experimental results is shown in Figure 4.2. These results were obtained at 50°C and 29.4 MPa, and the feed concentration of triolein and stearic acid was 2.6–2.7 mM and 3.3–3.5 mM, respectively. It can be seen from the results of Figure 4.2 that the residence time of 45 s was almost enough to bring the reaction into equilibrium when the water concentration was larger than 16 mM.

Figure 4.3 shows the results of productivity which correspond to the results of residual triglyceride and the extent of interesterification shown in Figure 4.2. The optimum water concentration shifted from 4 mM to 12 mM when the residence time changed from 22 s to 91 s. The higher water concentration is preferable to promote the reaction at the short residence time, while the hydrolysis should be suppressed at the longer residence time by use of low water concentration.

4.4.3   Effect of temperature and pressure

The higher concentration of substrates could contribute to the better results of residual triglycerides and the extent of interesterification when the reactions are conducted under the same conditions of water concentration and residence time. The residence time, however, should be long enough to make the reaction almost complete. This expectation proved to be true experimentally.

The effect of temperature was examined under the same conditions of pressure and substrate concentration. The pressure was as high as 29.4 MPa,

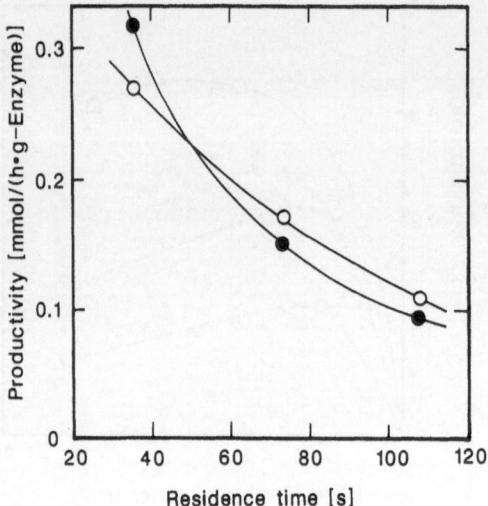

Figure 4.4 Effect of residence time on productivity of product triglycerides SOS and SOO (triolein 1.7 mM, water 9.7 mM). Pressure (MPa), Cs (mM): ●, 19.6, 4.5; ○, 29.4, 10.2.

and the product constituent did not differ at either 50°C or 70°C. The other reactions were conducted with different combinations of pressure and stearic acid concentration while the concentration of triolein as well as water was kept constant. The results of productivity are compared in Figure 4.4, and it is shown that a combination of low substrate concentration and low pressure resulted in larger productivity than the opposite combination at the short residence time of 37 s. This rather unexpected result could be due to the effect of mass transfer on the reaction since the increase in pressure leads to an increase in SC-CO$_2$ density and to a decrease in substrate diffusivity.

4.5 Conclusions

Acidolysis reactions in SC-CO$_2$ were conducted in a continuous reactor, and the effects of temperature, pressure, residence time and feed concentration were examined. There was an optimum water concentration which changed under different conditions of residence time and substrate concentration. Temperature and pressure should be high enough to dissolve substrates and products well in SC-CO$_2$. The effect of temperature on the product constituent, however, was not remarkable when the pressure was as high as 30 MPa. The reaction seemed to be influenced by the mass transfer, so that a combination of low substrate concentration and low pressure was favorable for better productivity when the residence time was too short to bring the reaction to the equilibrium state.

Acknowledgements

The author expresses his thanks to Mr H. Fujii and Mr T. Noda for their contribution to the experimental work and also to The Skylark Food Science Institute for financial support.

References

Chi, Y. M., Nakamura, K. and Yano, T. (1988) Enzymatic interesterification in supercritical carbon dioxide. *Agric. Biol. Chem.* **52,** 1541.

Kyotani, S., Fukuda, H., Nojima, Y. and Yamane, I. (1988) Interesterification of fats and oils by immobilized fungus at constant water concentration. *J. Ferment. Technol.* **66,** 567.

Nakamura, K. (1988) Supercritical fluid bioreactor for triglyceride interesterification. *ISF-JOCS World Congress 1988*, Tokyo, Japan, Abstracts p. 287.

Nakamura, K. (1990a) Biochemical reactions in supercritical fluids. *Trends Biotechnol.* **8,** 288.

Nakamura, K. (1990b) Supercritical fluid bioreactor, *Proc. APBioChEC'90*, Kynju, Korea, pp. 480–483.

Nakamura, K., Chi, Y. M., Yamada, Y. and Yano, T. (1986) Lipase activity and stability in supercritical carbon dioxide. *Chem. Eng. Commun.* **45,** 207.

5 Use of semi-preparative supercritical fluid chromatography for the separation and isolation of flavor and food constituents

I. FLAMENT, U. KELLER and L. WÜNSCHE

Abstract

Supercritical fluid chromatography (SFC) and its application to separations of food and flavor has been scarcely used in this field although it offers appealing advantages over other chromatographic methods. The value of semi-preparative SFC is stressed and exemplified by structure elucidation of non-volatile constituents of coffee. Coupling of SFC with thin layer chromatography results in a powerful method for identification and isolation of trace substances such as phenolic antioxidants.

5.1 Introduction

Although supercritical fluids have already been extensively applied to the supercritical fluid extraction (SFE) of food constituents, the use of supercritical fluid chromatography (SFC) for separation of less volatile food products is still uncommon. One reason is that SFC may not offer an advantage over existing chromatographic methods. As an example of this in the field of fats and oils, Perrin and Prevot (1988) found no particular advantage of SFC over GLC or HPLC. Nevertheless, many successful applications of this technique to the analysis of triglycerides in butter fat and fish oil have been described (Huopalahti *et al.*, 1988; Kallio *et al.*, 1989). Since the birth of SFC, analytical chemists have hoped that supercritical chromatography would fill a gap between GLC and HPLC by improving and accelerating the separation of compounds of medium or low volatility. For instance, such constituents are relatively abundant in spices: SFC of non-derivatized thyme oil gave about the same percentage composition of the main compounds as capillary GC does after silylation, thus eliminating one step in the sample preparation (Manninen *et al.*, 1990). For a complex mixture of peppermint oil or basil oil, SFC also seems to give more reliable quantification than capillary GC, especially for oxygenated compounds, although the separation of monoterpene hydrocarbons is, as expected, much better done by GC than by SFC. The application of various chromatography-spectroscopy couplings, including SFC-FTIR, has allowed identification of 42 terpenoid constituents of pepper essential oil (Pichard *et al.*, 1990). Another example of successful application of SFC has been given by Raynor *et al.* (1989), who analyzed distilled beverage flavors and separated aromatic acids, such as vanillic acid. As in the previous cases, the corresponding GC separation would necessitate a derivatization into more volatile and less polar esters, an operation which is time consuming and sometimes impractical. HPLC methods, on the other hand, often require the use of complex eluent gradients.

Applications of coupled SFE-SFC to natural products have also been frequently described. A direct on-line coupling of small subcritical and super-critical fluid extractors with packed column SFC has been developed by Jahn and Wenclawiak (1988). Skelton *et al.* (1986) have described a stopcock system that allows collection of supercritical extracts and injection of them directly onto a chromatography column without returning them to atmospheric pressure. Several devices coupling SFE with various chromatographies have been described (Thiebaut *et al.*, 1989) and even commercialized (Kumar *et al.*, 1988). A combined technique, SFE-CC-SFC including cryogenic collection (CC), has been evaluated by Ashraf-Khorassani *et al.* (1990). In previous papers (Flament and Keller, 1987; Flament *et al.*, 1987; Flament and Keller, 1988), we have also described the use of a multifunctional chromatographic system coupling SFE-loop condensation and separation on packed columns. On-line coupling of SFE and SFC with chromatographic analyses has been reviewed by Veuthey *et al.* (1990). The combination of SFE-SFC with fraction collection has been described by Xie *et al.* (1989): fractions of the effluent were collected from a frit restrictor at the column outlet in vials that contained a preselected solvent. Direct coupling of CO_2 fluid extraction with capillary supercritical fluid chromatography has been used by Anton *et al.* (1988) to obtain rapid qualitative sample information of complex matrices of a plastic material, a coffee powder and a hydrocarbon test mixture.

SFC separation technology has been mainly directed towards open-tubular columns. However, Greibrokk *et al.* (1989) have emphasized the advantages of packed columns: higher loadability, wider commercial choice of column selectivities and higher efficiency per unit time. Schoenmakers and Uunk (1989) have also compared the reciprocal advantages of open tubular versus packed SFC columns. The main benefit of the former is a much larger number of plates for the same pressure drop, allowing higher total plate numbers to be reached. In return, packed columns are characterized by better efficiency per unit of time, larger sample capacity and easier flow control.

Schwartz *et al.* (1988) have also shown the advantages of packed over capillary columns when the separation of relatively polar compounds requires the use of modifiers. Packed column SFC/MS offers obvious advantages over HPLC/MS in terms of speed of analysis and solvent elimination. Major findings in the evolution of the efficiency of packed columns have been discussed by Taylor and Chang (1990) and Berger and Deye (1990). Applicability of different stationary phases for packed column SFC has been described by Schoenmakers *et al.* (1990). Sample introduction and elution methods for preparative SFC have been studied by Yamauchi *et al.* (1990). Some practical aspects of column design for packed column SFC have been studied by Dean and Poole (1989a). The adverse effects of the strength of the injection solvent on packed column SFC band broadening have also been demonstrated by Dean and Poole (1989b) who have introduced a solventless injection system that eliminates these effects, the analyte being delivered to the column in a solvent-free plug of supercritical fluid mobile phase.

Thanks to the low flow rates of mobile phase that they require, small-bore columns can easily be coupled to mass (MS) or infrared (IR) spectrometers for compound identification. For instance, Shah and Taylor (1989) have described the separation of herbicide precursors on a microbore packed column connected with a FTIR flow cell. A similar technique has also been used by Ikushima *et al.* (1989).

In the field of flavor and food analysis, packed column SFC has scarcely been used. Separations with packed columns of triglycerides (Perrin and Prevot, 1988; Schwartz *et al.*, 1988), fatty acids (Geiser *et al.*, 1988; Schwartz *et al.*, 1988; Nizery *et al.*, 1989) and their derivatives (Gorner and Perrut, 1989; Nomura *et al.*, 1989) as well as carbohydrates (Gere, 1983; Herbreteau *et al.*, 1990) have been reported, although these papers often do not deal with food analysis, but use the compounds as test analytes. Vitamins and tocopherols were also analyzed by this technique. Early on, Gere (1983) used packed column SFC to separate carotenoids (vitamin precursors) from paprika oleoresin. Packed capillary separations (with mass spectrometric detection) of water and fat soluble vitamins were carried out by Matsumoto *et al.* (1986). Saito and co-workers reported semi-preparative separations of tocopherols using either SFE-SFC coupling (Saito *et al.*, 1989) or recycle SFC (Saito and Yamauchi, 1990). On the flavor side, separation of a grapefruit oil on a packed capillary column shows the potential of this system to separate complex samples, but no identification was given (Taylor and Chang, 1990). Yamauchi and Saito (1990) successfully fractionated lemon peel oil into compound types by semi-preparative packed column and performed the identifications by subsequent GC-MS analysis. We have also reported the use of semi-preparative columns (coupled with TLC) to analyse black pepper, clove (Flament and Keller, 1988; Keller and Flament, 1989) and coffee extracts (Wünsche *et al.*, 1991). To our knowledge, Saito's work and ours are the only studies where small scale preparative SFC has been used for the analysis of food and flavor constituents. On the other hand, large scale SFC has been used in industrial production. Both large and small scale preparative SFC have been reviewed by Berger and Perrut (1990).

This short review shows that SFC and particularly packed column SFC emerges as a very valuable technique, whose potential has been neglected for food and flavor analysis. The present work reports examples of analysis of food constituents which were carried out successfully by the use of semi-preparative SFC.

5.2   Materials and methods

The samples used in this work come from natural sources. The detailed procedures for their extraction and fractionation are beyond the scope of this paper.

The capillary SFC separations were run on a SFC 3000 system (Carlo Erba Instruments, Rodano, Italy) equipped with a fused silica capillary column (DB-5, 10 m × 0.05 mm; J & W Scientific Inc., Rancho Cordova, CA, USA). SFC grade CO_2 was obtained from Scott Specialty Gases (Plumsteadville, PA, USA).

The packed column system has already been described in detail (Keller and Flament, 1989; Wünsche *et al.*, 1991). The separations were performed on C_{18} columns (25 cm × 10 mm i.d., 5 µm, Supelco, Bellefonte, PA, USA) and silica gel TLC plates (20 cm × 20 cm, Merck, Darmstadt, Germany) using CO_2 with *c.* 2–3% of modifier.

Other chromatographic conditions are given in the legends of the figures.

5.3   Results and discussion

Among the various problems which need the use of semi-preparative separations in food and flavor analysis, obviously the most common is structure elucidation. When dealing with flavor mixtures of unknown composition, identification of their constituents relies primarily on mass spectrometry. Indeed, this technique presents the major advantages of being highly sensitive, and easily coupled with chromatography. Although mass spectra give important information on the structures of the molecules, they are often poorly informative about the fine details of the molecular shapes. As a result, mass spectral identifications are most often carried out by comparison with spectra libraries, a method which obviously fails when no corresponding reference spectrum exists in the library or when it has an equivocal fragmentation pattern. In these instances, ^{1}H or ^{13}C nuclear magnetic resonance spectrometry (NMR) is the method of choice. However, the large amount of structural information it provides is gained at the expense of a 10^4 to 10^6 decrease in sensitivity compared with mass spectrometry, leaving the analyst with the problem of isolating milligram amounts of pure substances; a task which obviously cannot be achieved with analytical (capillary) scale separation techniques.

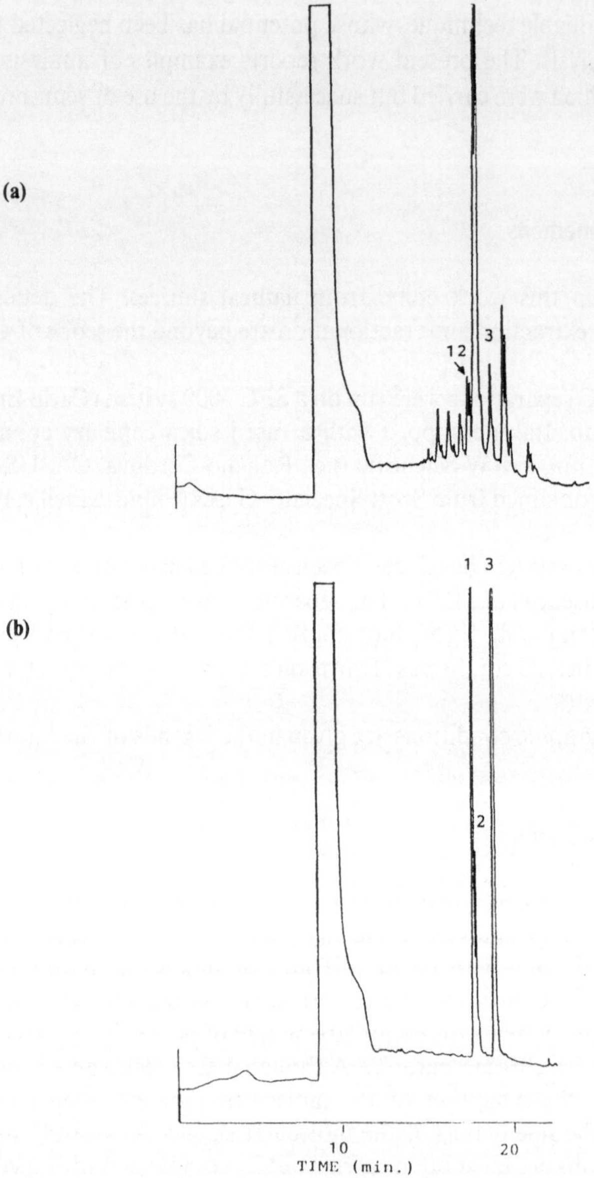

(a)

(b)

TIME (min.)

Figure 5.1 Capillary SFC of the non-volatile part of a coffee extract: (a) neutral fraction; (b) medium polar subfraction of (a). SFC conditions: column DB-5, 10 m × 0.05 mm, pressure program from 9 MPa for 5 min to 30 MPa at a rate of 1.5 MPa/min.

Such a case is illustrated with the analysis of the non-volatile part of the neutral fraction of a coffee extract. Figure 5.1(a) shows the capillary SFC of the starting fraction. Initial separation by liquid chromatography led to a subfraction which still contains three components as shown on the capillary

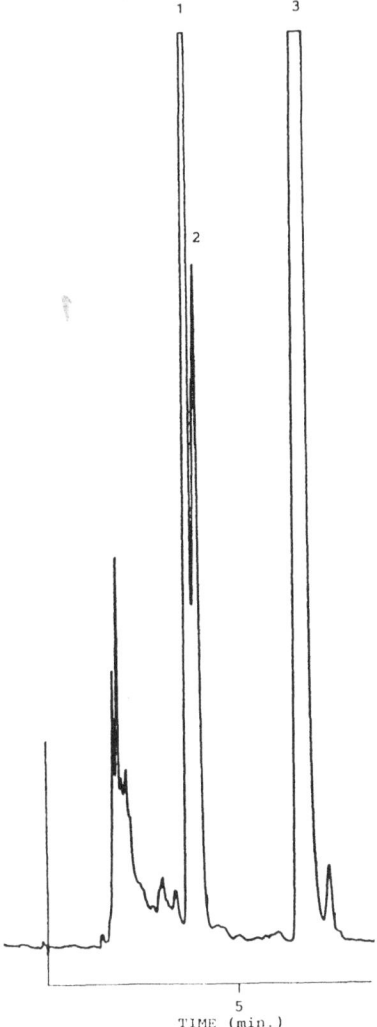

Figure 5.2   Semi-preparative SFC of fraction (b) of Figure 5.1. SFC conditions: column SPLC-18, CO_2 + MeOH at 20 MPa, 40°C, detector UV at 210 nm, 2.0 aufs.

SFC of Figure 5.1(b). Analysis of this fraction by mass spectrometry showed for peak 3 the typical fragmentation pattern of a C_{28} analogue of tocopherol (m/z = 151, 169, 416), but mass spectrometry failed to give the exact position of the methyl groups. Using semi-preparative supercritical fluid chromatography (Figure 5.2), we were able to isolate a few milligrams of this compound in a pure form. Subsequent NMR analysis led to its unequivocal structure elucidation as the beta isomer of tocopherol. These chromatograms also show that the semi-preparative separation compares favorably with capillary separation, especially in terms of efficiency per unit of time. However, capillary

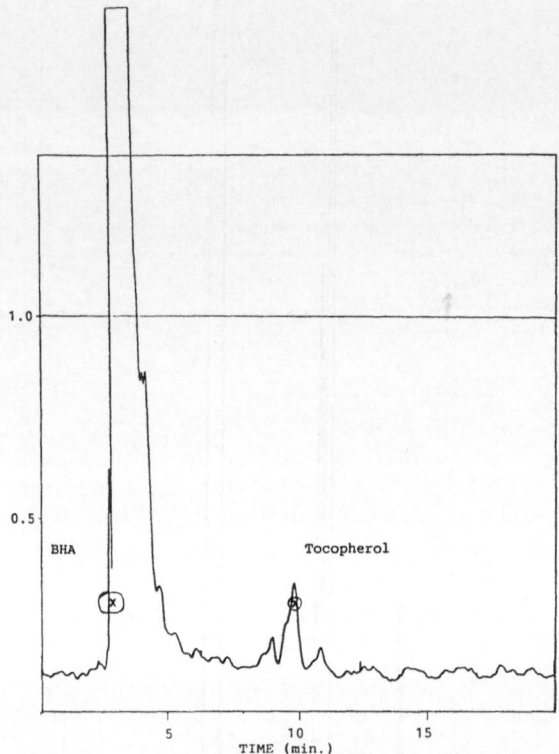

Figure 5.3 SFC-TLC of natural limonene stabilized with BHA and spiked with tocopherol. SFC conditions: column SPLC-18, CO_2 + MeOH at 20 MPa, 40°C, detector UV at 215 nm, 1.0 aufs. TLC conditions: 20 cm × 20 cm × 0.25 mm silica gel plate eluted with hexane/ethyl acetate (9:1) and sprayed with a 1% solution of 2,6-dichloroquinone chlorimide in ethanol.

analysis remains necessary to obtain accurate quantification using FID detection.

A major advantage of semi-preparative SFC over its HPLC counterpart lies in the volatility of the mobile phase, facilitating on-line connection with spectroscopic techniques (MS, FTIR, etc.), as well as with hyphenated separation methods. This is particularly true for coupling with thin layer chromatography. Several HPLC-TLC systems have been described. However, in order to prevent the flooding of the TLC plate, they are restricted to microbore packed columns (Fijimoto *et al.*, 1988; Hofstraat *et al.*, 1988) or require splitting of the eluent before deposition onto the plate (Jänchen, 1988).

We have already described an SFC-TLC system, which can accommodate packed columns up to 22 mm i.d. (Keller and Flament, 1989; Wünsche *et al.*, 1991). In this system, the eluent from the SFC column is directed towards a TLC plate, which is slowly moved by a conveyor under the jet of CO_2. After SFC separation, the eluent is thus deposited on the plate as a sharp line. The TLC plate is then developed and the spots are visualized.

One can restrict the semi-preparative scale to the SFC part, and use the TLC plate basically as a detector. TLC plates are indeed convenient for detection; they allow either general or specific detection depending on the type of the visualization reagent used. Furthermore, the retention factors of the spots give additional information on the nature of the analytes.

An example of selective detection is given by the identification of antioxidants in food products. In the past, 2-tert-butyl-4-methoxyphenol (BHA) has been widely used as antioxidant in a broad range of food products. Nowadays, there is concern about possible adverse effects associated with the use of this compound. Natural tocopherols derived from vegetable oil seeds are also effective antioxidants, and tend to replace BHA in many products.

We have taken advantage of the specificity offered by the TLC visualization reagents to detect the presence of these phenolic antioxidants in citrus oils. In a preliminary experiment, limonene stabilized with BHA (1000 ppm) was spiked with tocopherols (350 ppm). Forty microliters of the crude sample were injected into the system. After SFC separation and subsequent deposition of the eluent on the TLC plate, the plate was developed and eventually treated with 2,6-dichloroquinone chlorimide; a reagent specific for phenols. The result (Figure 5.3) shows that the two antioxidants are selectively detected and can easily be distinguished by their different retention times on SFC. The specificity of the TLC reagent allows their detection in trace amounts even when they co-elute with other constituents of the sample.

In Figure 5.4, two samples of cold pressed orange oil were concentrated 10 times by distillation and analyzed the same way. Figure 5.4(a) clearly shows a spot with SFC and TLC retentions corresponding to those of tocopherol, while no spot characteristic for tocopherol or BHA appears in the second sample (Figure 5.4(b)), indicating stabilization with tocopherol for the former oil, and the absence of phenolic antioxidant in the latter.

Alternatively TLC can be used as a general detection method. A striking advantage in this case lies in the ability of TLC to check the purity of each SFC fraction. As an example of this technique, Figure 5.5 shows a SFC separation of a paprika extract. Here the plate is treated with anisic aldehyde, a non-specific reagent. The result indicates that the main peaks are still mixtures, thus requiring further separations before NMR identifications can be carried out. Similar analyses of black pepper and clove extracts were reported elsewhere (Flament and Keller, 1988; Keller and Flament, 1989).

Many TLC plates resulting from such separations of natural products show a complicated pattern of spots. This shows first that complex mixtures are rarely totally resolved by packed column SFC alone, and secondly that SFC-TLC coupling strongly enhances the overall separation. Provided both chromatographic methods can handle milligram amounts of samples and a non-destructive method of spot visualization is applied, deposition of the totality of the SFC eluent onto TLC will allow two-dimensional semi-preparative separations to be carried out. The first problem is easily resolved by

(a)

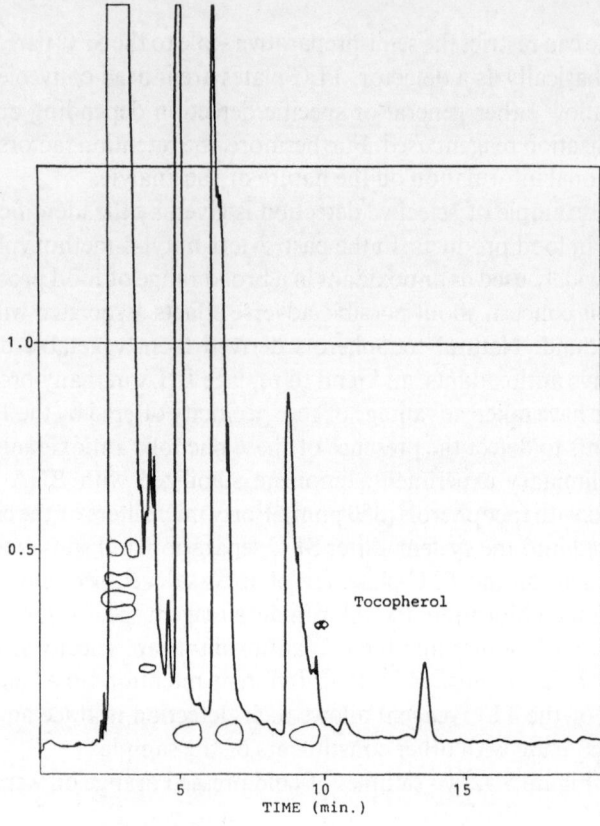

(b)

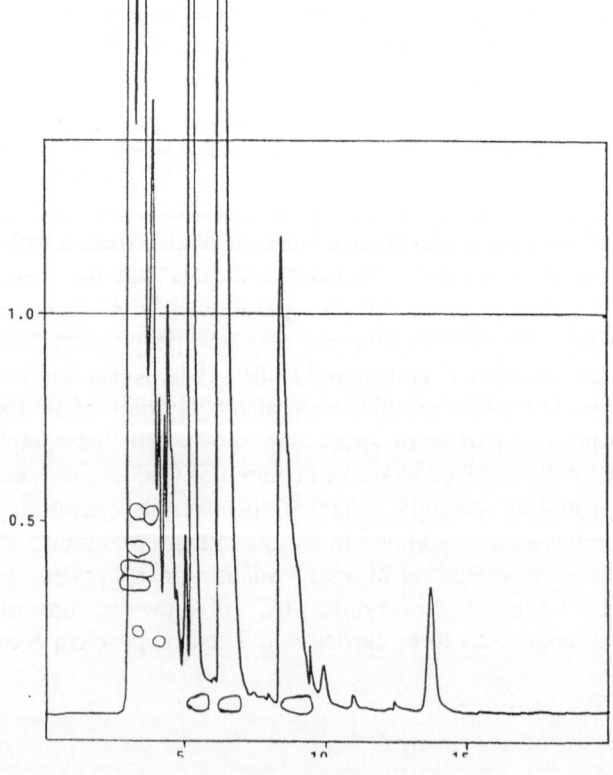

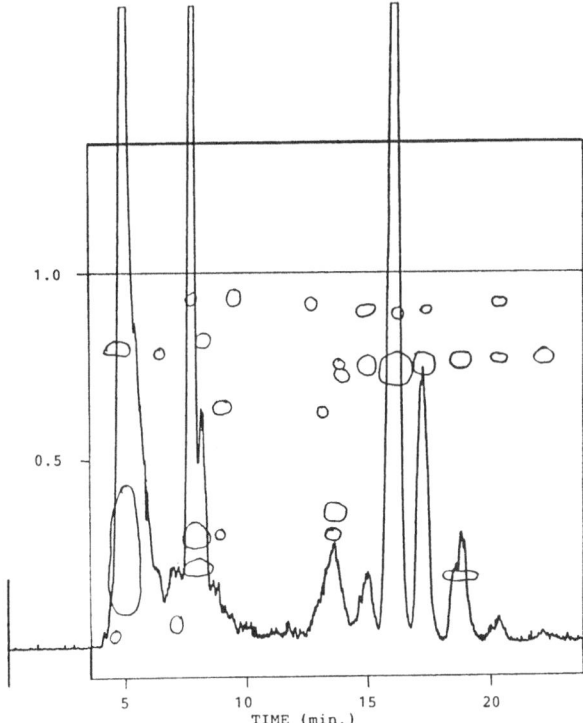

Figure 5.5 Bidimensional recording of a paprika extract. SFC conditions: column SPLC-18, CO_2 + EtOH at 30 MPa, 40°C, laser light-scattering detector at 150°C. TLC conditions: 20 cm × 20 cm × 0.25 mm silica gel plate eluted with hexane/ethyl acetate (4:1) and sprayed with 0.5% anisaldehyde in acidic methanol.

the use of 10–20 mm i.d. packed columns associated with TLC plates of 2 mm adsorbant thickness. The sample capacity of both systems is adequate. On the other hand, two different non-destructive visualization methods have been tested with success. The first method uses berberine chloride. This reagent makes organic compounds UV detectable, even for molecules without any chromophore. As berberine chloride is insoluble in diethyl ether, it stays on the adsorbant when the analyte is recovered by washing the silica with this solvent. In the second method, a soft (plastic) TLC sheet is pressed against the still wet thick one. Thus, tiny fractions of the compounds are transferred. The plastic sheet is then treated with any reagent, to give the mirror image of the main plate. This technique leads to a unique two-dimensional semi-preparative separation. It has been described in a recent publication (Wünsche *et al.*, 1991), which also shows applications on the identification of jasmine and coffee constituents.

Figure 5.4 (*opposite*) SFC-TLC of two different batches of concentrated cold pressed orange oil, one stabilized with tocopherol (a), the other without antioxidant (b). SFC conditions: column SPLC-18 + MeOH, CO_2 at 20 MPa, 40°C, detector UV at 215 nm, 1.0 aufs. TLC conditions: 20 cm × 20 cm × 0.25 mm silica gel plate eluted with hexane/ethyl acetate (9:1) and sprayed with a 1% solution of 2,6-dichloroquinone chlorimide in ethanol.

5.4 Conclusion

Although SFC has been used only occasionally for food and flavor analysis, we consider the technique of great value in this field. For analytical separations, open tubular capillary columns offer large plate numbers and allow accurate quantifications through FID detection. However, analytical SFC should not be misused; for stable volatile compounds, GC is definitively easier and more efficient, while SFC competes badly with HPLC for separations of very polar molecules. Between these extremes, there is a large range of compounds falling in the application domain of SFC, for which this technique offers significant benefits. For separations of molecules of medium or low volatility, capillary packed and open tubular columns do not require a derivatization step as in GC and provide often shorter retention times or better efficiency than HPLC. For large scale separations, the advantages of SFC are even emphasized. The low cost of CO_2 and its non-toxicity make it a very attractive solvent for preparative systems which consume a large quantity of mobile phase. Furthermore, CO_2 is gaseous at atmospheric pressure and room temperature, and is thus easily eliminated after decompression. This renders straightforward analyte recovery or on-line coupling with subsequent identification systems or separation techniques. The latter advantage is particularly relevant for coupling SFC with thin layer chromatography. Indeed, packed column SFC can be coupled with both analytical or preparative TLC plates and the detection of spots carried out either in a general or specific manner. This leads to a very versatile two-dimensional separation system, which has proved its value for structure identification of flavor constituents as well as detection and characterization of trace of antioxidants in food samples.

References

Anton, K., Menes, R. and Widmer, H. M. (1988) Direct coupling of CO_2 fluid extraction with capillary SFC. *Chromatographia* **26**, 221.

Ashraf-Khorassani, M., Kumar, M. L., Koebler, D. J. and Williams, G. P. (1990) Evaluation of coupled supercritical fluid extraction–cryogenic collection–supercritical fluid chromatography (SFE-CC-SFC) for quantitative and qualitative analysis. *J. Chromatogr. Sci.* **28**, 599.

Berger, C. and Perrut, M. (1990) Preparative supercritical fluid chromatography. *J. Chromatogr.* **505**, 37.

Berger, T. A. and Deye, J. F. (1990) Effect of mobile phase density gradients on efficiency in packed column supercritical fluid chromatography. *Chromatographia* **30**, 57.

Dean, T. A. and Poole, C. F. (1989a) Some practical aspects of column design for packed-column supercritical-fluid chromatography. *J. Chromatogr.* **468**, 127.

Dean, T. A. and Poole, C. F. (1989b) Solventless injection for packed column supercritical fluid chromatography. *J. High Resolut. Chromatogr. Chromatogr. Commun.* **12**, 773.

Fijimoto, C., Morita, T., Jinno, K. and Shafer, K. H. (1988) Micro-HPLC/TLC/FTIR. *J. High Resolut. Chromatogr. Chromatogr. Commun.* **11**, 810.

Flament, I. and Keller, U. (1987) Application of SFC-TLC coupling to the analysis of flavour and essential oil constituents, in *Proc. on Supercritical Fluids Colloq.*, Pont-à-Mousson, France, ed. M. Perrut, Institut National Polytechnique de Lorraine, pp. 389–396.

Flament, I. and Keller, U. (1988) Application of SFC chromatography to the analysis of flavour and essential oil components. Characterization of spices by two-dimensional coupling SFC-TLC, in *Proc. Int. Symp. on Supercritical Fluids*, Nice, France, ed. M. Perrut, Institut National Polytechnique de Lorraine, pp. 465–472.

Flament, I., Chevallier, C. and Keller, U. (1987) Extraction and chromatography of food constituents with supercritical CO_2, in *5th Weurman Symp.*, Oslo, Norway, *Flavour Science and Technology*, ed. M. Martens, G. A. Dalen and H. Russwurm, Wiley, New York, pp. 151–163.

Geiser, F. O., Yocklovich, S. G., Lurcott, S. M., Guthries, J. W. and Levy, E. J. (1988) Water as a stationary phase modifier in packed-column supercritical fluid chromatography. I. Separation of free fatty acids. *J. Chromatogr.* **459**, 173.

Gere, D. R. (1983) Separation of paprika oleoresins and associated carotenoids by supercritical fluid chromatography. Application Note AN 800-5, Hewlett Packard Co, Avondale, PA, USA.

Gorner, T. and Perrut, M. (1989) Separation of unsaturated fatty acid methyl esters by supercritical fluid chromatography on a silica column. *LC-GC* **7**, 36.

Greibrokk, T., Doehl, J. and Lundanes, E. (1989) Current use of packed columns in SFC. *Prog. HPLC* **4**, 53.

Herbreteau, B., Lafosse, M., Morin, A. L. and Dreux, M. (1990) Analysis of sugars by supercritical fluid chromatography using polar packed columns and light-scattering detection. *J. Chromatogr.* **505**, 299.

Hofstraat, J. W., Griffionen, S., van der Nesse, R. J., Brinkman, U. A. Th., Gooijer, C. and Velthorst, N. H. (1988) Coupling of narrow-bone column liquid chromatography and thin layer chromatography. *J. Planar Chromatogr.* **1**, 220.

Huopalahti, R., Laakso, P., Saaristo, J., Linko, R. and Kallio, H. (1988) Preliminary studies on triacylglycerols of fats and oils by capillary SFC. *J. High Resolut. Chromatogr. Chromatogr. Commun.* **11**, 899.

Ikushima, Y., Saito, N., Hatakeda, K., Ito, S. and Goto, T. (1989) Development of extraction and separation monitoring in situ with on-line SFE-SFC/FTIR system. *Chem. Lett.* 1707.

Jahn, K. R. and Wenclawiak, B. (1988) Direct on-line coupling of small subcritical and supercritical fluid extractors with packed column supercritical fluid chromatography. *Chromatographia* **26**, 345.

Jänchen, D. (1988) Direct coupling of column liquid chromatography and thin-layer-chromatography. *GIT Suppl.* **3**, 78.

Kallio, H., Laakso, P., Huopalahti, R. and Linko, R. R. (1989) Analysis of butter fat triacylglycerols by SFC/electron impact mass spectrometry. *Anal. Chem.* **61**, 698.

Keller, U. and Flament, I. (1989) A direct and practical method of coupling packed column supercritical fluid chromatography with thin-layer chromatography. *Chromatographia* **28**, 445.

Kumar, M. L., Houck, R. K. and Winwood, H. R. (1988) Coupling of extraction and supercritical fluid chromatography. *Spectra 2000* **16**, 47.

Manninen, P., Riekkola, M. L., Holm, Y. and Hiltunen, R. (1990) SFC in analysis of aromatic plants. *J. High Resolut. Chromatogr. Chromatogr. Commun.* **13**, 167.

Matsumoto, K., Tsuge, S. and Hirita, Y. (1986) Fundamental conditions in pressure-programmed supercritical fluid chromatography-mass spectrometry and some applications to vitamin analysis. *Chromatographia* **21**, 617.

Nizery, D., Thiebaut, D., Caude, M., Rosset, R., Lafosse, M. and Dreux, M. (1989) Improved evaporative light-scattering detection for supercritical fluid chromatography with carbon dioxide-methanol mobile phases. *J. Chromatogr.* **467**, 49.

Nomura, A., Yamada, J., Tsunoda, K., Sakaki, K. and Yokochi, T. (1989) Supercritical fluid chromatographic determination of fatty acids and their esters on an ODS-silica gel column. *Anal. Chem.* **61**, 2076.

Perrin, J. L. and Prevot, A. (1988) Chromatography in supercritical phase: impacts in the field of fats and oils. *Rev. Fr. Corps Gras* **35**, 485.

Pichard, H., Caude, M., Morin, P., Richard, H. and Rosset, R. (1990) Identification of pepper essential oil constituents by various chromatography – spectroscopy couplings. *Analysis* **18**, 167.

Raynor, M. W., Kithinji, J. P., Davies, I. L. and Bartle, K. D. (1989) Chromatography with supercritical fluid, in *Distill. Beverage Flavour, Proc. Int. Symp., 1988*, ed. J. R. Piggott and A. Paterson, Ellis Horwood, Chichester, UK, pp. 75–83.

Saito, M. and Yamauchi, Y. (1990) Isolation of tocopherols from wheat germ oil by recycle semi-preparative supercritical fluid chromatography. *J. Chromatogr.* **505**, 257.

Saito, M., Yamauchi, Y., Inomata, K. and Kottkamp, W. (1989) Enrichment of tocopherols in wheat germ by directly coupled supercritical fluid extraction with semipreparative supercritical fluid chromatography. *J. Chromatogr. Sci.* **27**, 79.

Schoenmakers, P. J. and Uunk, L. G. M. (1989) Mobile and stationary phases for supercritical fluid chromatography, in *Advances in Chromatography*, Vol. 30, ed. J. C. Giddings, E. Grushka and P. R. Brown, Marcel Drekker, New York, pp. 1–80.

Schoenmakers, P. J., Uunk, L. G. M. and Janssen, H. G. (1990) Comparison of stationary phases for packed-column supercritical fluid chromatography. *J. Chromatogr.* **506**, 563.

Schwartz, H. E., Barthel, P. J., Moring, S. E., Yates, T. L. and Lauer, H. H. (1988) Comparison of packed and capillary columns for practical SFC separations. *Fresenius Z. Anal. Chem.* **330**, 204.

Shah, S. and Taylor, L. T. (1989) On-line SFC/FTIR analysis of agriculturally-related compounds. *J. High Resolut. Chromatogr. Chromatogr. Commun.* **12**, 599.

Skelton, R. J., Johnson, C. C. and Taylor, L. T. (1986) Sampling considerations in supercritical fluid chromatography. *Chromatographia* **21**, 3.

Taylor, L. T. and Chang, H. C. K. (1990) Packed column development in supercritical fluid chromatography. *J. Chromatogr. Sci.* **28**, 357.

Thiebaut, D., Chervet, J. P., Vannoort, R. W., DeJong, G. J., Brinkman, U. A. T. and Frei, R. W. (1989) SFE of aqueous samples and on-line coupling to SFC. *J. Chromatogr.* **477**, 151.

Veuthey, J. L., Caude, M. and Rosset, R. (1990) On-line coupling of SFE and SFC with chromatographic analyses. *Analusis* **18**, 103.

Wünsche, L., Keller, U. and Flament, I. (1991) Combination of supercritical fluid chromatography with thin-layer chromatography on a semi-preparative scale. *J. Chromatogr.* **552**, 539.

Xie, Q. L., Markides, K. E. and Lee, M. L. (1989) SFE-SFC with fraction collection for sensitive analytes. *J. Chromatogr. Sci.* **27**, 365.

Yamauchi, Y. and Saito, M. (1990) Fractionation of lemon-peel oil by semi-preparative supercritical fluid chromatography. *J. Chromatogr.* **505**, 237.

Yamauchi, Y., Kuwajima, M. and Saito, M. (1990) Sample introduction and elution method for preparative supercritical fluid chromatography. *J. Chromatogr.* **515**, 285.

6 Separation of oil from fried chips by a supercritical extraction process: an overview of bench-scale test experience and process economics

S. VIJAYAN, D. P. BYSKAL and L. P. BUCKLEY

Abstract

The extraction of high-value biomaterials from solid and liquid matrices by supercritical fluid processing is a rapidly expanding technology. The lack of progress in techniques and hardware to process feedstock consisting of solid matrices has hindered technology penetration in large-volume processing of low-value biomaterials. This paper reviews the current state of development, including our experience with supercritical processing to remove oils from different fried-chip feedstocks. Analysis of a semi-continuous process plant system, designed from laboratory-scale experiments, shows the process has economic viability for commercial applications involving intermediate-value products.

6.1 Introduction

Many applications are presently known for supercritical fluid extraction (SFE) to separate biomaterials in the food, pharmaceutical and chemical sectors (Stahl *et al.*, 1988; Novak and Robey, 1989; Vijayan *et al.*, 1989, 1990). While many applications have been demonstrated in laboratory-scale experiments, and over 100 process patents filed, very few applications have been tested on engineering-scale equipment. Only some half a dozen commercial plants have gone into production since the late 1970s. Examples of SFE application areas in biomaterial processing with special reference to food components separations have been reviewed by Rizvi *et al.* (1986) and Vijayan *et al.* (1989), and show the ever-increasing interest in the technology. In the majority of applications, carbon dioxide has been used over other gases because it is relatively inert, has a low critical temperature, is relatively inexpensive and is abundant. Unlike organic solvents, it is low in toxicity and is compatible in most instances with biomaterials.

In a number of applications, the supercritical state is preferred to the liquid state because there is an increase in solubility of the substances to be separated, the extraction rate is much higher because of the higher diffusivity, and there is a wide range of possible operating temperatures (Williams, 1981).

In this paper, a review of SFE application to different product segments involving biomaterial separations is provided. Our experience in developing a SFE process, called the LOWCA (LOW CAlorie) process, for removing excess oil from potato chips, is discussed, and an overview of the process economics is presented.

6.2   Background

Several biomaterial processing applications are currently known for supercritical fluid extractions, and they are gradually entering into commercial-scale applications. Vijayan *et al.* (1989) have broadly classified the process applications into three product segments: (1) high-value, low-volume products; (2) intermediate-value, intermediate-volume products; and (3) low-value, high-volume products. Some examples of these product segments, together with their economics, are summarized below.

6.2.1   High-value, low-volume product

Because of the high value of the product extracted and the low volume of operation, the economics of this product segment have been found to be very favorable for the exploitation of the SFE technology. A variety of extracts are being produced commercially by small to large industries (Basta, 1985). The types of food processed include:

1. flavor, fragrance, spice extracts and essential oils from plants, animal and other materials;
2. hop extraction to produce alpha- and beta-acids and essential oil;
3. purification and fractionation of aroma constituents.

The demand for a wide range of these products requires most extraction plants to not be dedicated exclusively to one product.

6.2.2   Intermediate-value, intermediate-volume product

The decaffeination of coffee has been by far the most popular commercial application (Caragay, 1981). Other examples include:

1. decaffeination of tea;
2. xanthines – removal from cocoa;
3. deodorization of vegetable and animal fats and oils;
4. extraction of excess oil and fat from fried foods and vegetable matter.

The extraction of excess oil from fried foods is an emerging process, but its viability for commercial application is not evident at present. The experience gained in the development of a SFE process and the process economics are discussed in a later section.

6.2.3   Low-value, high-volume product

Practical applications for this product category have not been developed. Examples include:

1. oil extraction from vegetable seeds;
2. processing of grain flours to improve quality.

Although the SFE process has been identified for oil seed extraction since the 1950s (Mangold, 1983), there are some doubts as to whether the process will be competitive with traditional methods of processing oil seeds based on liquid hexane extraction.

6.3   Laboratory-scale study of LOWCA process

Many deep fried foods, including potato chips, are prepared by frying raw vegetable materials in hot oil/fat. Deep frying causes the food products to pick up and retain relatively large amounts of oil/fat. Typically, regular potato chips contain about 30–40% (w/w) oil/fat, 5% fiber, 55–58% protein and carbohydrate, and about 2% moisture. It appears that the demand for snacks lower in fat and hence lower in calories is growing steadily. In addition, the storage of high oil content snacks presents serious problems because of rancidity due to oxidative degradation of the oil/fat. By removing at least a portion of the oil/fat from potato chips, it is possible to obtain chips with increased nutrient value, lower in calories (Schneider and Haussener, 1984) and it may be possible to reduce rancidity and increase the shelf-life.

The objective of this study was to assess the technical feasibility of the LOWCA process using supercritical CO_2 extraction to remove controlled amounts of oil/fat present in potato chips as a result of deep frying. The specific objectives were:

1. to test and develop an SFE process to reduce the oil/fat content from regular potato chips to a target of about 33% (w/w) of the original level;
2. to evaluate optimum processing conditions by laboratory-scale experiments using different types of potato chips as the basis for a conceptual design and cost study of a commercial-scale plant.

6.3.1   Experimental

The carbon dioxide used in the tests was greater than 99% purity from Union Carbide Linde. Four types of potato chip were used:

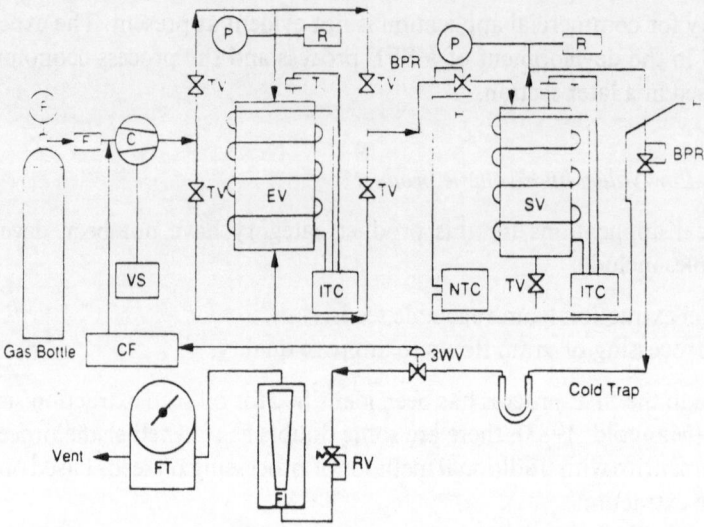

Figure 6.1 Flow diagram of experimental apparatus. BPR, back pressure regulator; C, diaphragm compressor; EV, extraction vessel; F, filter; FI, flow indicator; FT, flow totalizer; H, heater; ITC, indicating temperature controller; CF, activated carbon column; R, rupture disc; RV, relief valve; SV, separation vessel; T, thermocouple; NTC, non-indicating temperature controller; TV, two-way through valve; P, pressure gauge; VS, variable speed controller; 3WV, three-way valve.

Type I: unsalted and unspiced, having an average of 32% oil (directly from the manufacturer);

Type II: salted and unspiced, with 34% (ave.) oil (directly from the manufacturer);

Type III: salted and unspiced, with 29.6% oil (from supermarket); and

Type IV: salted and unspiced, with 38% oil (from supermarket).

6.3.2 Apparatus

The SFE apparatus employed was a laboratory-scale system from Newport Scientific Inc., as illustrated in Figure 6.1. The extraction vessel had a 1-l capacity and the separation vessel had a volume of 0.5 l. The pressure vessels were electrically heated to achieve the desired operating temperature. The CO_2 delivery system comprised a gas cylinder, a particulate filter and a variable-speed control diaphragm compressor. The fluid recycle system consisted of a three-way valve and a high-pressure activated-carbon column.

The CO_2 flow rate was measured at ambient conditions by a rotameter and a dry test meter. The flow rate was altered either by varying the speed of the compressor motor or by changing the inlet pressure to the compressor. All temperatures were measured by thermocouples and pressures by pressure gauges. The control of temperature and pressure was achieved by electronic relays. The feed containing untreated chips was manually loaded into the extraction vessel by means of a cylindrical basket constructed from stainless-steel mesh. The extracted oil was sampled through the sampling valve located in the separator.

6.3.3   Procedure

Fried potato chips ('regular' untreated chips) of known weight, typically 50–75 g, were batch fed to the extraction vessel. A stream of CO_2 at a predetermined temperature, pressure and flow rate was passed through the extractor. The extractant stream containing oil was passed through the separator at a lower pressure to separate the dissolved oil as a liquid. The CO_2 leaving the separator was either recycled via the compressor or vented via the flow meters.

The extraction was carried out for a specified period of time, after which the treated chips were removed and weighed. For a given operating condition, the difference between the untreated and treated chips represented the amount of oil that was removed. As the water content of the chips was only about 1.8% (w/w), the weight-loss method of measuring the amount of oil removed was adequate. For transient oil-removal experiments, the extracted oil samples collected in the separator at different time intervals were withdrawn and weighed. The cumulative weight of the oil samples withdrawn over a period was within 5% of the loss-in-weight of untreated and treated potato chips. This accuracy was determined to be sufficient for process optimization tests.

The initial content of oil in the untreated feedstock was determined by Soxhlet extraction using methylene chloride. The apparatus was periodically cleaned with propanol and dried thoroughly by circulating with CO_2 under pressure.

The majority of experiments were performed in a once-through mode. A few experiments were also carried out in which the pressure-expanded CO_2 in the separator at about 6.9 MPa, greater than the cylinder pressure, was recycled back to the compressor inlet.

6.4   Results and discussion

A total of 65 extraction tests were performed using four types of feedstock for a range of temperature, pressure, CO_2 volume and extraction time. The range of values for these operating variables were chosen based on the results reported for oil seed extraction using CO_2 (e.g. Fattori *et al.*, 1988). The weight percent of oil removed for different operating conditions is shown in Figures 6.2–6.5.

6.4.1   Effect of extraction time

In Figure 6.2 the percent oil removed was plotted against the amount of CO_2

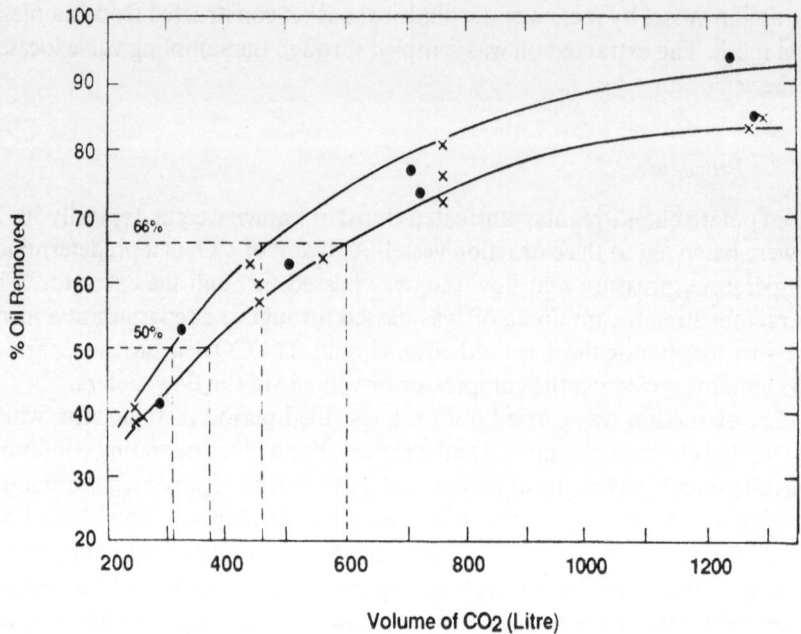

Figure 6.2   Percent oil removed as a function of volume of CO_2 used. *Feed:* Type I, ●, unsalted; II, ×, salted. *Extraction:* P = 6000 ± 100 psig, T = 55 ± 1°C. *Separation:* P = 14.7 psi; T = 30± 3°C.

used. The oil removed increased steeply up to about 60% level and then followed a gradual increase up to 70% level, eventually reaching an asymptotic value of about 90%. The trend is similar for salted and unsalted chips of Type I and Type II. The extraction rate of oil is slower in the case of salted chips than unsalted chips. Within the experimental error of about 5% in the percent oil removed, the difference in extraction rate between salted and unsalted chips is not significant. The results suggest that the most favorable region of operation is the portion of the extraction curve showing a steep rise in the percent oil removed.

Figure 6.3 depicts percent oil removal for Type III and Type IV feedstocks as compared with the oil extraction behavior of Type II chips.

6.4.2   Effect of pressure

A series of extraction tests, using Type II feedstock, was performed at pressures of 27.6 MPa, 41.4 MPa and 55.2 MPa while keeping the temperature at 55°C and CO_2 flow at about 7.5 l/min. The extraction behavior shown in Figure 6.4 suggests that for a given extraction time, the percent oil removed is the lowest at 27.6 MPa (4000 psig) and the highest at 41.4 MPa (6000 psig), whereas the data at 55.2 MPa (8000 psig) lie between the two extraction pressures. The presence of an optimum pressure in SFE processing is not generally common.

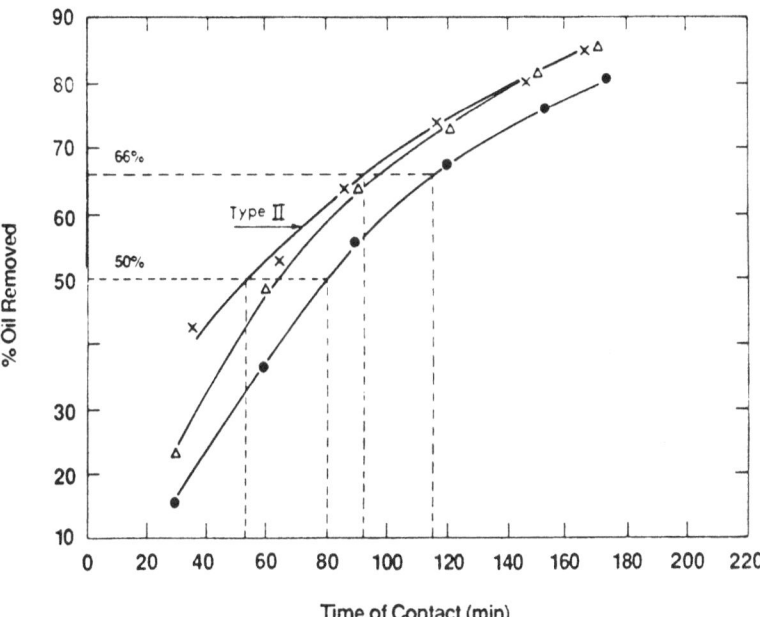

Figure 6.3 Percent oil removed as a function of average time of contact in the extractor. *Feed:* Type II, ×; III, △; IV, ●. *Extraction:* P = 6000 ± 100 psig; T = 55 ± 1°C. *Separation:* P = 500 ± 50 psig; T = 30 ± 2°C.

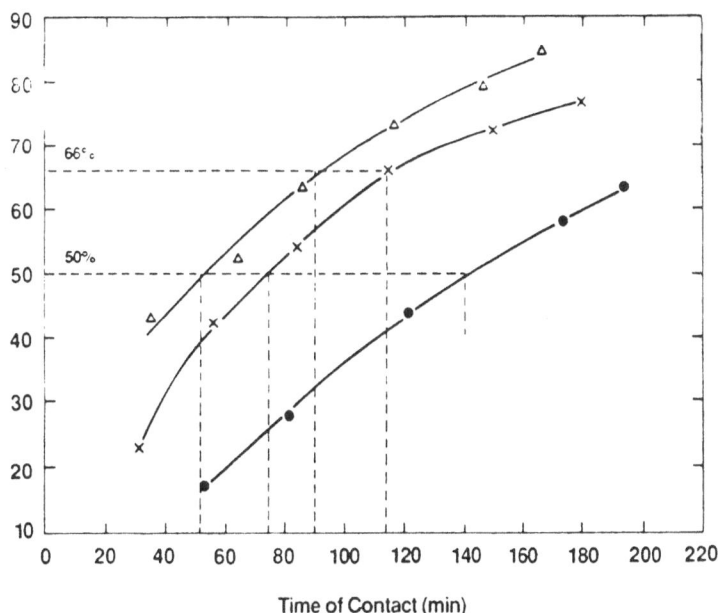

Figure 6.4 Effect of extraction pressure on percent oil removed. *Feed:* Type II. *Extraction:* P = △, 6000 ± 100 psig (CO_2 flow 8.3 l/min); ×, 8000 ± 100 psig (7.5 l/min); ●, 4000 ± 100 psig (7.8 l/min). T = 500 ± 1°C. *Separation:* P = 500 ± 50 psig; T = 30 ± 2°C.

It has been reasonably established (Paulaitis *et al.*, 1983) that the solubility of organic components increases in proportion to an increase in the CO_2 density as pressure is increased. The current results are obviously not consistent with the above solubility-density mechanism.

6.4.3   Effect of temperature

The effect of temperature on the percent oil removal was evaluated at the optimum pressure of 6000 psig at temperatures of 40°C, 55°C and 70°C. The results illustrated in Figure 6.5 reveal that at extraction times greater than about 30 min, the amount of oil removed is appreciably higher at 70°C than at 40°C, although the difference in the oil removal is insignificant at short extraction times.

At constant pressure, the density of CO_2 decreases as the system temperature is increased. Extending the solvent power of CO_2 as being almost monotonically related to its density, one would expect a decrease in the solubility of oil in CO_2 as the extraction temperature is increased. The results are contrary to this predicted behavior. Similar observations have also been reported in oil seed extraction studies (e.g. Fattori *et al.*, 1988). The explanation is that a rise in temperature provides an exponential increase in the vapor pressure of the oil. Near the critical point of CO_2 (~7.6 MPa or ~1100 psig), the density changes rapidly with temperature. A small change in temperature in this region leads to a large change in CO_2 density and hence a corresponding change in the oil solubility as shown in Table 6.1. As a result, the net effect is an overall increase in the removal of oil from the chips.

The temperature effect on the amount of oil removed from the chips at an extraction pressure of 41.4 MPa (6000 psig) suggests that an extraction temperature between 55°C and 70°C is the optimum condition. For the development of a process flowsheet, an extraction temperature around 50–60°C may be used.

6.4.4   Effect of type of feedstock

The oil extraction experiments were performed by manually filling the chips in a meshed basket and then batch-loading the basket into the extraction vessel. The basket was filled with the selected feedstock to obtain a maximum packing density with a random packing configuration. The Type I and Type II feedstock were generally flat and rounded, and thus provided the highest packing density, whereas the Type III and Type IV feedstock were curled and elliptical in shape. As a result, in most instances, the Type III and IV chips were stacked in the basket. The extraction results given in Table 6.2 show that the extraction times for 50% and 66% level oil removals are the highest for Type IV feedstock and the lowest for Type II. The extraction time for Type III

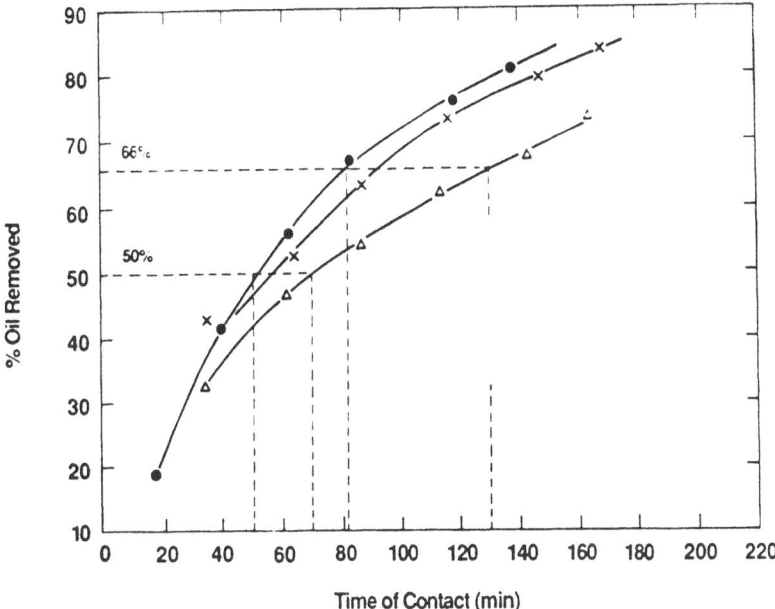

Figure 6.5 Effect of extraction temperature on percent oil removal as a function of time of contact in the extractor. *Feed*: Type II. *Extraction*: P = 6000 ± 100 psig; T = ●, 70°C ± 2°C; ×, 55°C ± 1°C; △, 40°C ± 2°C. *Separation*: P = 500 ± 50 psig; T = 30 ± 2°C.

Table 6.1 Effect of pressure and temperature on oil extraction time

Extraction		CO$_2$ density (g/ml)	Average extraction time (min) for percent oil removal at	
Pressure MPa (psig)	Temperature (°C)		50%	66%
27.6 (4000)	55	~0.8	140	>220
41.4 (6000)	40	~0.96	70	130
41.4 (6000)	55	~0.9	55	90
41.4 (6000)	70	~0.85	50	80
55.2 (8000)	55	~0.95	75	115

Table 6.2 Effect of type of feedstock on extraction time

Feedstock type	Shape and size	Ave. initial content (wt%)	Average extraction time (min) for percent oil removal at	
			50%	66%
Type I + II	Mostly flat and round (5 × 4.5 cm)	34.0	55	90
Type IV	Curled both ends and elliptical (6 × 5 cm)	38.0	80	115
Type III	Same as Type IV	29.6	65	100

feedstock lies between Type II and Type III. An explanation for the observed results may be given based on the feedstock packing characteristics, geometrical configuration and initial oil content of the feedstock.

6.4.5   Effect of CO_2 flow rate

The effect of supercritical solvent flow rate on extraction efficiency of organic components has not been established in past investigations. Results reported from oil seed extraction studies (e.g. Mangold, 1983) do not adequately address this issue for reasons of: (i) limitations in the laboratory-scale apparatus; (ii) preoccupation by a majority of the past studies directed mainly to equilibrium solubility measurements.

In the current experiments, the apparatus employed had similar limitations. However, the CO_2 flow could be varied from about 6 l/min to about 10 l/min by changing a combination of the CO_2 delivery pressure to the compressor and the speed of the compressor motor.

A dramatic reduction was observed in the extraction time of about 50% by an increase in CO_2 flow rate by about 50%. For a specified oil removal target of 66%, it is possible to lower the extraction time requirement by a careful selection of the solvent flow rate. This observation illustrates a transport mechanism, in addition to oil solubility and vapor pressure effects, as a factor for enhancing oil extraction at short extraction times. Evidently, more optimization studies are needed to correlate the flow rate effect.

By combining the results of the CO_2 flow rate effect with that of pressure and temperature, the extraction time needed to remove 66% oil from untreated feedstock at an optimum pressure of about 41 MPa (6000 psig) and a temperature of 50–60°C was reduced by about 50%. This means that the highest value for the extraction time of 115 min obtained for the removal of 66% of the oil is reduced to less than 60 min by optimizing the oil extraction with respect to CO_2 flow rate.

6.4.6   Physical characteristics of oil-reduced chips and extracted oil

The oil-reduced chips generally showed a lighter color and had a relatively drier appearance than untreated chips. The salted chips were found to retain essentially all the original salt, provided the amount of oil extracted was less than about 75%. It is believed that for the target oil removal of 66%, additional grinding (to that practiced currently) of salt and other ingredients may not be necessary before they are applied to the oil-removed chips.

The recovered oil in the separation vessel was clear and golden yellow in color, similar to fresh vegetable oils. The recovered oil did not have an apparent new odor when compared with fresh sunflower and canola oils, indicating that the recovered oil can be recycled back to the fryer for reuse.

Table 6.3 Process parameters and conditions

Process parameter	Process condition	
Extraction	Pressure:	41 MPa
	Temperature:	55°C
Separation	Pressure:	6.9 MPa
	Temperature:	40°C
CO_2 requirement	Mass rate:	8000–10000 kg/h
	Grade:	Commercial grade with 99.5% purity
Contact/cycle time	Loading/unloading of chips to	
	and from extraction vessel:	1 h
	Compression (or pump up) time:	1 h
	Oil extraction time:	1 h
	Vessel decompression time:	1 h
Mode of operation	Batch loading; two extraction vessel configuration to achieve semi-continuous operation	
Target product quality	1/2 and 2/3 oil-reduced chips	
Annual feed rate	Regular chips at 900×10^6 g	
Annual production rate	1/2 oil-reduced product at 738×10^6 g	
	2/3 oil-reduced product at 684×10^6 g	

6.5   Conceptual design and cost study of LOWCA process

6.5.1   LOWCA process flowsheet

From the results of the laboratory-scale evaluation, a basic flowsheet for the LOWCA process was designed to have the process conditions as given in Table 6.3. The guidelines reported by Marentis (1988) were factored in the conceptual design of the process plant.

A process flow diagram for the commercial-scale plant is shown in Figure 6.6. A possible scheme for integrating the LOWCA process into a typical potato-chips manufacturing plant is shown in Figure 6.7. The process arrangement in Figure 6.6 is as follows. Carbon dioxide gas leaves the storage vessel and enters a reservoir. The gas is then compressed by a compressor and enters either of the two extraction vessels. The extraction vessel, prior to charging with CO_2, is filled with the feedstock. The gas containing the extracted oil flows through the pressure reducer and enters the separator. The reduction of pressure separates the oil from the gas. The oil, after leaving the separator, is recycled to the fryer (Figure 6.7) for reuse, or stored. The gas exits from the top of the separator and is filtered before it enters the reservoir for recycling. A back pressure control valve is installed between the separator and the filters to maintain a minimum pressure of 1.1 MPa (160 psig) at 60°C in the separator.

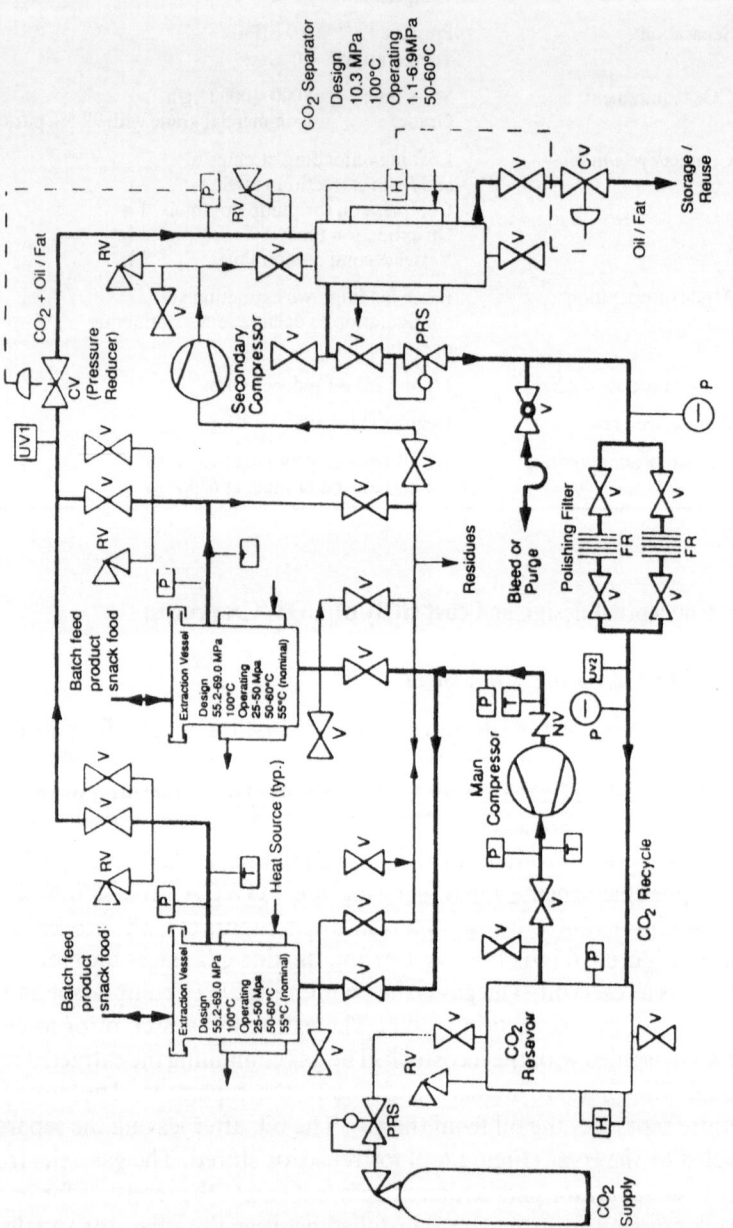

Figure 6.6 LOWCA process flow diagram.

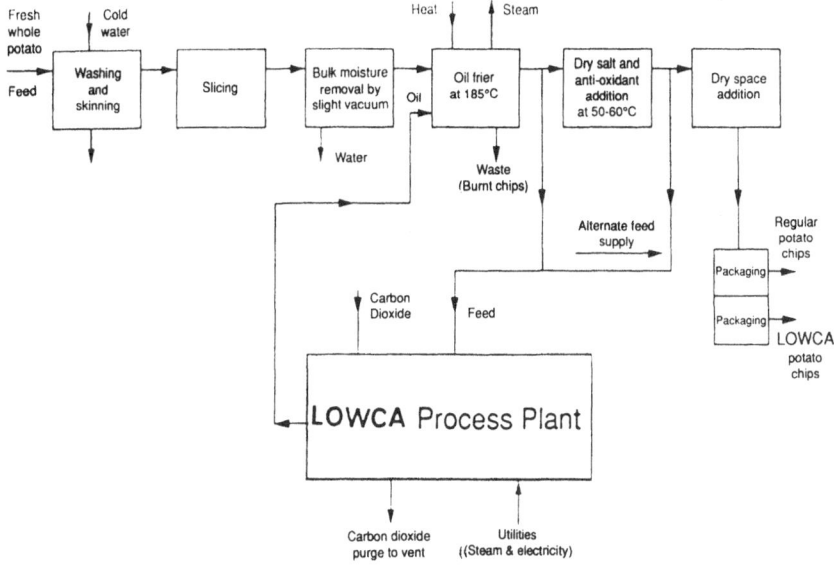

Figure 6.7 An integrated process schematic including LOWCA process within existing regular chips-manufacturing plant.

An overall mass balance of the LOWCA plant on an annual basis is given in Figure 6.8.

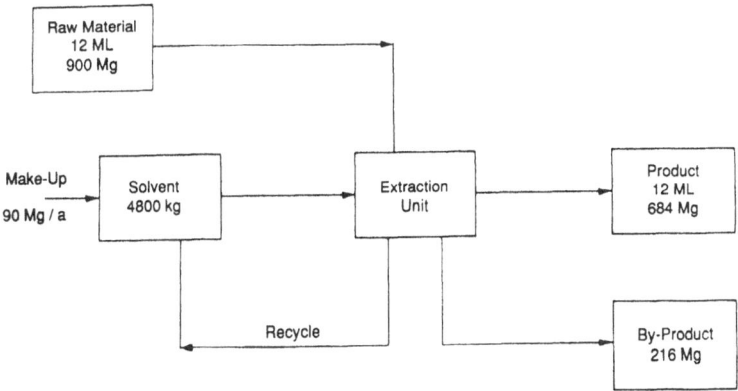

Figure 6.8 Overall material balance for LOWCA process. Basis: 1 l = 75 g raw material. 100 g raw material has 36% by-product. 100 g solvent is required to extract 1.5 g by-product. 2/3 by–product is removed. (One year—300 days).

Table 6.4   Capital cost of LOWCA process plant (in 1988 Canadian dollars)

Component	Quantity	Volume (l)	Estimated cost (in million dollars)	
			Low	High
Extraction vessel	2	4 000	2.10	4.00
Separation vessel	1	1 000	0.03	0.05
Reservoir	1	16 000	0.21	0.22
Compressor	2		0.08	0.20
Filter Bank	2		0.12	0.13
Electrical and instrumentation			0.15	0.20
Subtotal			2.69	4.80
Total plant cost[a]			~ 4.00	~ 6.00

[a] A factor of 25% for the high estimate and about 50% for the low estimate was applied to the subtotal to obtain the total plant cost. The factors represent an average of about 2.5-times the major equipment cost.

Table 6.5   Annual operating cost (in 1988 Canadian Dollars)

Item	Unit	Unit cost ($/unit)	Usage	Annual cost ($)	Total cost ($)
Raw materials					50 000
Carbon dioxide	kg	0.5	100 000	50 000	
Utilities					220 000
Steam	GJ	7.13	23 800	170 000	
Electricity	kW-h	0.03	1 600	50 000	
Labor					320 000
Operating	person	40 000	8	320 000	
Maintenance and repair					250 000
Labor: 2% of $5M capital investment				100 000	
Material: 2% of $5M				100 000	
Overhead: 50% of labor for repair				50 000	
Supplies and Miscellaneous					
1% of $5M capital				50 000	50 000
Production cost (in addition to regular chips manufacturing)					890 000

Table 6.6   Estimated price of Type II and Type III chips

Type of chip	Price per 200 g (in 1988 Canadian dollars)		
	1[a]	2[b]	3[c]
II (regular)	1.35	1.00	0.75
III (premium)	3.30	2.40	1.80

[a] The price a consumer pays at the convenience store (about 37.5% higher than price 2).
[b] The price a consumer pays at the supermarket.
[c] The price a supermarket purchases from the manufacturer (about 25% less than price 2).

6.5.2 Cost estimate

A summary of the capital and operating costs is presented in Tables 6.4 and 6.5, respectively.

The following major assumptions were used to arrive at the capital and operating costs of the plant system:

1. The extraction vessels are specialty items and are presumed to be supplied as turnkey units.
2. The operating cost presented in Table 6.5 covers only the additional costs introduced by the LOWCA process. Other costs, such as raw potatoes, oil, salt and spices, packaging, advertising, etc., are part of the existing process for producing regular (untreated) potato chips, reflected in the price of regular chips.
3. The operating cost estimate includes equipment cost, leasing and bulk CO_2 cost at \$0.30/kg (Canadian dollars).
4. Costs associated with supervision, building, transportation, overhead, depreciation, taxes, etc., are not included as most of these items are covered by the conventional plant producing regular chips.

6.5.3 Estimation of price for LOWCA-processed chips

According to the snack product properties, as stipulated by the Canadian Food and Drugs Act (FDAR, 1986; Cheney, 1988), the oil-reduced chips produced by the LOWCA process can bear the labels, 'Fat Reduced' (a 50% reduction of oil compared to regular chips) and 'Lower-in-Calorie' (a 25% reduction of calories compared to regular chips). For marketing considerations, it is assumed that the oil-reduced chips can be sold as 'premium chips' (refer to FIND/SVP, 1987 for definition).

As an example, a comparison of the supermarket price for consumers of Type II and Type III chips is given in Table 6.6. The Type III chip contains about 13% less oil than the Type II. The price estimate is based on the actual price of potato chips sold at the supermarket and convenience stores, using average profit values as reported by FIND/SVP (1987).

In the integrated chips manufacturing plant, shown in Figure 6.7, the regular chips are first produced by the conventional frying process at a production rate of 900×10^6 g/annum. The chips contain an average oil content of 36% (w/w). The selling price of the chips by the manufacturer to the supermarket is \$0.75 per 200 g.

Consider that the same plant integrated with the LOWCA process now produces 684×10^6 g/annum of chips containing 2/3 less oil and a by-product consisting of reusable frying oil at a rate of 216×10^6 g/annum. Assuming that the oil-reduced chips are sold as 'premium chips' by the manufacturer to the supermarket at a price of \$1.80/200 g (see Table 6.6), the added value of the oil-reduced chips compared with the regular chips would be US\$(1.80 − 0.75), or \$1.05/200 g.

It is this gain, as a result of producing the value-added product, which is to

Table 6.7 Total costs and sales revenue for LOWCA-processed chips

Description of cost and revenue components	Amount
1. Production rate of oil-reduced chips ($\times 10^6$ g/annum)	684
2. Production rate reusable frying oil (by-product) ($\times 10^6$ g/annum)	216
3. Additional sales revenue (using the net gain in sales at $1.05/200 g of oil-reduced chips over the price of regular chips) ($\times 10^6$ \$/annum)	2.78
4. Total operating cost ($\times 10^6$ \$/annum)	0.89
5. Cost savings as a result of recovering 216×10^6 g of reusable oil priced at $1.25/kg ($\times 10^6$ \$/annum)	0.27
6. Net operating cost ($0.89 – $0.27) ($\times 10^6$ \$/annum)	0.62
7. Capital cost of LOWCA plant ($\times 10^6$ \$/annum)	
Low estimate	4.00
High estimate	6.00

be absorbed against the additional capital and operating costs of the LOWCA plant.

In estimating the selling price of oil-reduced chips on the weight basis, no allowance has been made for the larger number of oil-reduced chips per 200 g bag as compared with a 200 g bag containing regular chips.

6.5.4   Economic analysis

The economic analysis considers only the additional cost associated with the LOWCA plant against the additional revenue resulting from the sale of oil-reduced chips. A breakdown of the total costs and sales revenue is presented in Table 6.7. The analysis does not include such factors as inflation and new capital expenditures beyond a certain period of operation.

For the low capital cost estimate of $4 million and a plant operating period of 6 years, both equity and debt financing give internal-rate-of-return on net profit of 18% and 21%, respectively, with a payout period of about 3 years. For the case of the high capital cost estimate of $6 million, both financing methods give internal-rate-of-return on net profit of about 10%, a payout period of about 4 years and a plant operating period of about 8.5 years. The analysis suggests that, provided the capital cost is in the range of 4–6 million dollars and the estimated selling price of the oil-reduced chips can be supported by the market, the LOWCA process can be a viable processing route for producing 'fat-reduced' and 'lower-in-calorie' potato chips.

6.6   Technology penetration issues

Despite the many positive attributes of SFE, the penetration of this process technology in biomaterial separations involving intermediate-value, intermediate-volume and low-value, high-volume products has been slow. The nature of the high-pressure vessels and the relatively high capital investments required to deploy the technology is expected to cause some initial resistance. The following is a list of issues generally encountered by the decision-makers:

1. technology that is more risky and complex;
2. relatively high capital, space and labor requirements, especially for batchwise processing;
3. a scaled-up version of the process could present quality control problems;
4. patent licenses may have to be bought from a variety of institutions and individuals;
5. there is a lack of design and cost data; this poses a problem if a full-scale plant is to be designed without pilot-plant tests;
6. continuous processing of solids is not possible because a high-pressure solids feed/removal system technology does not yet exist.

6.7 Conclusions

1. Laboratory-scale studies for the removal of excess oil from potato chips by supercritical CO_2 have demonstrated that up to about 66% (w/w) of oil present in regular chips is readily removed at a pressure of about 41 MPa and temperature of 55–60°C.
2. The extraction efficiency has been found to be very sensitive to small changes in the CO_2 density above 0.8 g/ml, and appears to be strongly influenced by an increase in the vapor pressure of oil at higher temperatures, coupled with the favorable diffusional transport properties of oil into the flowing stream of CO_2 in the extraction vessel.
3. The separation of oil from CO_2 has been best achieved below the critical pressure of CO_2 of about 6.9 MPa.
4. The effect of CO_2 flow rate on oil extraction efficiency has indicated that higher flow rates enhance the extraction of oil.
5. The 50% level oil-reduced chips readily meet the requirements of the Canadian Food and Drug Act and Regulations to bear the labels, 'Fat-Reduced' and 'Lower-in-Calorie'.
6. The conceptual design study has generated a commercial-scale process flowsheet for the LOWCA process with a production capacity of 684×10^6 g/year.
7. A preliminary economic analysis has shown that the internal-rate-of-return is in the order of 10–20%, with a payout period of 2–4 years, provided the capital cost of the plant is within 4–6 million dollars, and the product can be sold as 'premium chips'.
8. Given the assumptions, the preliminary design and cost study has revealed the feasibility of the LOWCA process for commercial applications involving intermediate-value, intermediate-volume products.

Acknowledgements

The work reported in this paper was performed during 1988 at AECL Research, Whiteshell Laboratories, Pinawa, Manitoba, Canada R0E 1L0. We wish to thank D. R. McLean, F. Bilsky, A. Jindal and C. P. Chung for their contribution and support, which made this work possible.

References

Basta, N. (1985) Supercritical fluids: still seeking acceptance. *Chem. Eng.* 14.

Caragay, A. B. (1981) Supercritical fluids for extraction of flavors and fragrances from natural products, *Perfume Flavorist* **6**, 43.

Cheney, M. (1988) Private communications.

Fattori, M., Bulley, N. R. and Meisen, A. (1988) Carbon dioxide extraction of canola seed: oil solubility and effect of seed treatment, *J. Am. Oil Chem. Soc.* **65**, 968.

FDAR (1986) The Food and Drugs Act and Regulations, Supply and Services Canada, Ottawa, Cat. No. H41-1/1985E.

FIND/SVP (1987) The Market for Salted Snacks, A Competitive Intelligence Report, New York.

Mangold, H. K. (1983) Liquefied gases and supercritical fluids in oil seed extraction, *J. Am. Oil Chem. Soc.* **60**, 226.

Marentis, R. T. (1988) Steps to developing a commercial supercritical carbon dioxide processing plant, *Supercritical Fluid Extraction Chromatography, ACS Symp. Ser. 366*, ACS, pp. 127–143.

Novak, A. R. and Robey, R. J. (1989) Supercritical extraction of flavoring material: design and economics, in *AC Soc. Symp. Ser. 406*, eds. K. P. Johnston and J. M. L. Pennenger, ACS, Ch. 32.

Paulaitis, M. E., Krukonis, V. J., Kurnik, R. T. and Reid, R. C. (1983) Supercritical fluid extraction, *Rev. Chem. Eng.* **1**, 179.

Rizvi, S. S. H., Daniels, J. A., Benado, A. L. and Zollweg, J. A. (1986) Supercritical fluid extraction: operating principles and food applications. *Food Technol.* **7**(6), 57.

Schneider, F. and Haussener, E. (1984) De-fatting roasted convenience food by extraction with non-toxic gaseous solvent preferably CO_2 in supercritical condition and recovering oil and/or fat, Swiss Patent 643713, assigned to E. Haussener, Gumligen.

Stahl, E., Qiurin, K. W. and Gerard, D. (1988) *Dense Gases for Extraction and Refining*, Springer-Verlag, New York, 1988.

Vijayan, S. and Byskal, D. P. (1990) Development of a supercritical extraction process for removing excess oil from fried chips, *Int. Solvent Extraction Conf. (ISEC-90)*, Kyoto, Japan.

Vijayan, S., Singh, D. and Hickson, B. E. (1989) Separation of food components by fluid extraction at subcritical and supercritical conditions, *Proc. 2nd Int. Conf. on Separations Sci. Tech.*, Toronto, Canada, Vol. 1, ISBN 0-921763-04-2.

Williams, D. F. (1981) Extraction with supercritical gases, *Chem. Eng. Sci.* **36**, 1769.

7 Selecting a pump for supercritical fluid service

S. W. VANCE

Abstract

Criteria and design considerations for selection of pumping systems for supercritical fluid (SCF) processes are described. Centrifugal and reciprocating pumps each have advantages and disadvantages for a given system. Each type has specific design details which must be considered to provide satisfactory and economical pumping for SCF systems.

7.1 Introduction

This chapter deals with criteria for selection of pumps for carbon dioxide, frequently chosen as the solvent for SCF extraction applications. Several of these criteria may be unique to SCF extraction process requirements. Both isobaric and pressure let-down operations are considered.

Supercritical fluid extraction (SFE) processes require transport equipment to return the fluid from the low pressure (separation) section to the high pressure (extraction) section (Figure 7.1). At the extraction section, solute concentration in the supercritical solvent is increased by mass transfer from the feed material (either a solid substrate or a liquid) to the solvent. In the separation section, solute concentration is reduced by changing the operating pressure and/or operating temperature, thus precipitating solute from the solvent stream. Alternatively, sorption of the solute by, for example, activated carbon adsorption or water or oil absorption can be used. Separation can occur at either supercritical or subcritical conditions. Repressuring of the solvent can be accomplished by compressors or by pumps.

7.2 Fluid criteria

Three sets of properties of the extraction fluid have an impact on pump selection. These are: lubricity; thermodynamic and thermophysical properties; and materials compatibility.

SCF SCHEMATIC

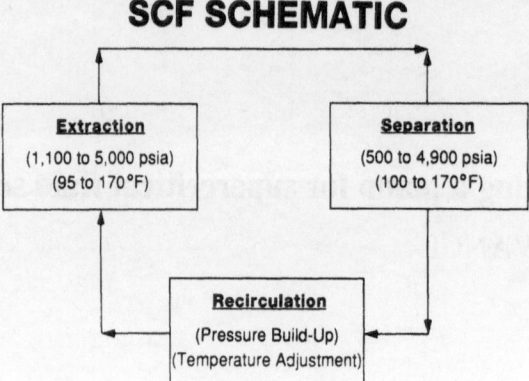

Figure 7.1 Supercritical fluid extraction schematic.

7.2.1 Lubricity

Lubricating properties of the fluid determine whether external lubrication or low-friction seals and wear parts must be provided. Carbon dioxide is a non-lubricating fluid. Selection of packing lubricants must be carefully made, however, to avoid the possibility of the lubricant dissolving in the carbon dioxide and precipitating at undesirable locations or contaminating the product being extracted.

7.2.2 Thermodynamic and thermophysical properties

7.2.2.1 Compressibility. Although liquid carbon dioxide is, like most liquids, relatively incompressible, at the pressure differentials which may be encountered in SFE, compressibility must be considered for design. Low clearances for positive displacement pumps may be indicated to ensure desired flow rates.

7.2.2.2 Enthalpy, entropy, specific heat and latent heat. In pumping carbon dioxide with high pressure differentials, the low specific heat and latent heat of the liquid (compared to water, for example) may result in substantial warming of the fluid entering and passing through the pumps, and flashing to vapor may occur causing cavitation. Insulation or cooling of the pump inlet and casing may be required. In most SFE, the fluid from the pump discharge will be warmed to the desired supercritical temperature by a heat exchanger before entering the extractor. Proper consideration of the heat of compression and friction in the pump may reduce the supplemental external heating required before entering the extractor. Temperature rises through the pump of 11–17°C may often be encountered.

7.2.2.3 Viscosity, thermal conductivity and density. These properties are dependent on both pressure and temperature. They are factors in fluid flow, heat transfer and mass transfer phenomena.

7.2.2.4 Diffusivity. Diffusion of carbon dioxide into gaskets, seals, and packings within the high pressure area may cause delamination or bubbling of the elastomer during depressuring cycle steps. Control of depressuring rate is important to prevent this problem. Proper elastomer selection will also minimize this effect.

7.2.3 Materials compatibility

7.2.3.1 Corrosion, erosion and chemistry. Corrosion must be evaluated, since carbon dioxide is much more corrosive in the presence of moisture than when dry. Interaction of carbon dioxide with organic acids or other solutes may also result in corrosion. Erosion is possible if traces of extracted compounds precipitate as a liquid or a solid in wear areas. Also, dry ice will form at 75 psia or less and $-55°C$ if high pressure carbon dioxide leaks through packings to the atmosphere. The solid dry ice may then erode packings, or the low temperature may cause the packings to become brittle and fail. Carbon dioxide also will combine with water to form a solid hydrate at temperatures in the range of -1 to $10°C$. Hydrate formation is a function of both temperature and pressure.

7.3 Process criteria

Criteria for pump selection are also related to the processing conditions. These factors are: flow rate; discharge pressure; differential pressure; net positive suction head; and safety and operability.

7.3.1 Flow rate

The flow rate will have a great influence on the type of pump selected. Positive displacement reciprocating pumps handle low flows and high pressures very well. Centrifugal pumps designed for low capacity at high pressure are not obtainable at reasonable cost.

7.3.2 Discharge pressure

The extraction pressure establishes the discharge pressure required and thus the thickness of the pump housing. In the range of pressures used for extraction, the wall thickness required and the sealing of reciprocating or rotating shafts for these pressures will limit the choice of pump. At high flow, the added cost for special centrifugal pump designs can be justified.

7.3.3 Differential pressure

Differential pressure across the pump also influences the selection of pump

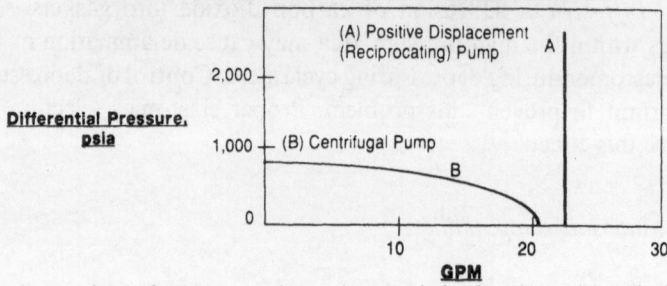

Figure 7.2 Comparison of pressure–volume characteristics for (A) positive displacement (reciprocating) and (B) centrifugal pumps.

type. A low differential will suggest one type, while a high differential will suggest another.

7.3.4   Net positive suction head

Net positive suction head required for the pump must be matched to the suction head available within the system. For centrifugal pumps, high rotary speeds and inlet velocities result in higher suction head requirements to avoid cavitation at the impeller eye. The point of cavitation in a reciprocating pump occurs at the inlet check valve. The velocity past the valve reaches a maximum at the midpoint of the stroke and cavitation will begin at that point. In both cases, static head is converted to velocity head. Reciprocating pump suppliers have a variety of designs and sizes of valves to meet these requirements by minimizing pump inlet velocities. Centrifugal pump suppliers also have inlet designs to minimize the problem.

7.3.5   Safety and operability

System safety and operability must be considered, including failure modes and risks of catastrophic failures. The pumping equipment must be considered in these operability studies. Effects of rapid unplanned depressurization must be reviewed for each process element for hazards to both equipment and personnel.

7.4   Pump types

Pumps can be classified as positive displacement or centrifugal. The variation in flow versus pressure shown in Figure 7.2 graphically demonstrates the functional differences. Positive displacement pumps discharge a constant volume of fluid (proportional to speed), almost independent of the discharge pressure. Centrifugal pumps, however, have a characteristic curve in which the volume delivered increases as the system differential pressure decreases.

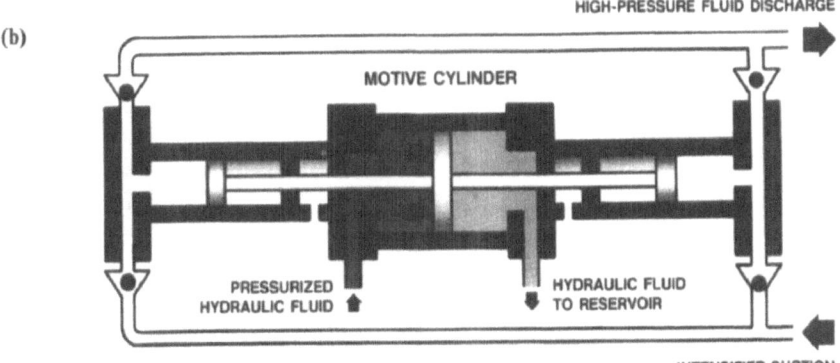

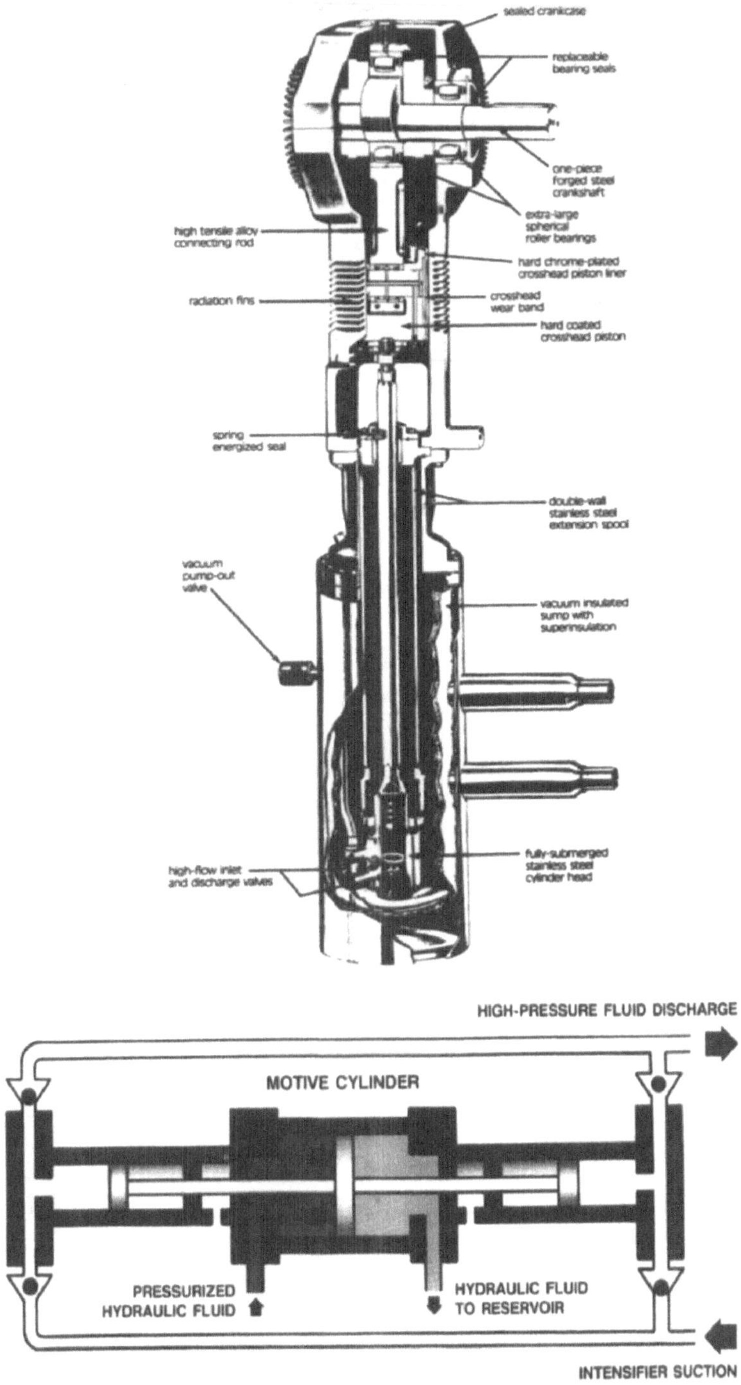

Figure 7.3 (a) Mechanical drive piston pump. (Reprinted with permission. Copyright 1983, CVI Incorporated.) (b) Hydraulic drive piston pump. (Reprinted with permission. Copyright 1988, Hydro-Pac, Inc.)

Figure 7.4 (a) Horizontal packed plunger pump. (Reprinted courtesy of Uraca Pumpenfabrik Urach.) (b) Vertical packed plunger pump. (Reprinted with permission. Copyright 1987, Ingersoll-Rand Company)

7.4.1 Positive displacement reciprocating pumps

Positive displacement pumps may be either reciprocating or rotary. Rotary screw, vane and gear pumps depend upon both very small clearances between rotors and pump housing and rotary shaft seals to achieve volumetric delivery. They cannot develop either the differential pressure or the discharge pressure required for most SCF extraction processes. Reciprocating pumps are of three types: piston pumps; packed plunger pumps; and diaphragm pumps.

7.4.1.1 Piston pumps.   A piston pump depends upon a piston ring seal which moves linearly with the piston and seals against the stationary cylinder wall. Figure 7.3(a) is an example of a piston pump, with rotary motion of the driver mechanically converted to reciprocating motion and speeds may range from 50 to 500 strokes/min. Pistons range in size from 2.5–22.5 cm in bore, with strokes to 20 cm. Very large horsepowers are obtainable by connecting multiple pistons to a common crankshaft and driver. Figure 7.3(b) shows another style, hydraulically driven.

7.4.1.2 Packed plunger pumps.   Packed plungers differ from pistons by the arrangement of the seal, which is stationary. These pumps usually have adjustable packing ring seals, which seal against the reciprocating plunger. These seals can be liquid cooled and may be equipped with lantern rings for flushing with water or another suitable liquid. Packed plunger pumps can range in size from bores of less than 2.5 cm to 22.5 cm. Capacity can be achieved with reasonable plunger speeds (50–500 strokes/min) by ganging the plungers on a common crankshaft at staggered rotation angles. This arrangement limits the pulsations observed at the outlet. A simplex (single plunger) pump has a large pressure and delivery pulsation. Duplex (2 plungers), triplex (3 plungers), quintuplex (5 plungers) and even septuplex (7 plungers) arrangements are common. Figure 7.4(a) shows a horizontal stroke triplex packed plunger pump. Figure 7.4(b) is a cutaway of a vertical stroke septuplex packed plunger pump. The larger the number of plungers, the smaller the variation in discharge volume and pressure.

7.4.1.3 Diaphragm pumps.   The diaphragm pump (Figure 7.5) utilizes a flexible metal diaphragm or flexible elastomer membrane in a chamber of several centimeters diameter to achieve reasonable capacities with very small stroke lengths. For metal diaphragms, a mechanical connection to the crankshaft is possible. For elastomers, as well as metal diaphragms, the stroke can be provided hydraulically, using a packed plunger and hydraulic oil to create the reciprocating motion of the diaphragm. Special features can provide warning of diaphragm failure before catastrophic failure.

7.4.2   Centrifugal pumps

A centrifugal pump may be either single stage or multiple stage.

7.4.2.1 Single stage centrifugal pump.   A single stage centrifugal pump (Figure 7.6) consists of a rotating vaned impeller in a housing, with a rotary shaft seal between the impeller shaft and the housing. The impeller often is direct-connected to the driving motor for speeds of 1200, 1800 or 3600 rev./min with a shaft diameter of 2.5 cm or less, or, as in this example, is gear driven through a speed increasing gear train. Velocity head developed by the impeller is converted to pressure head at the pump discharge. Impeller diameter can vary

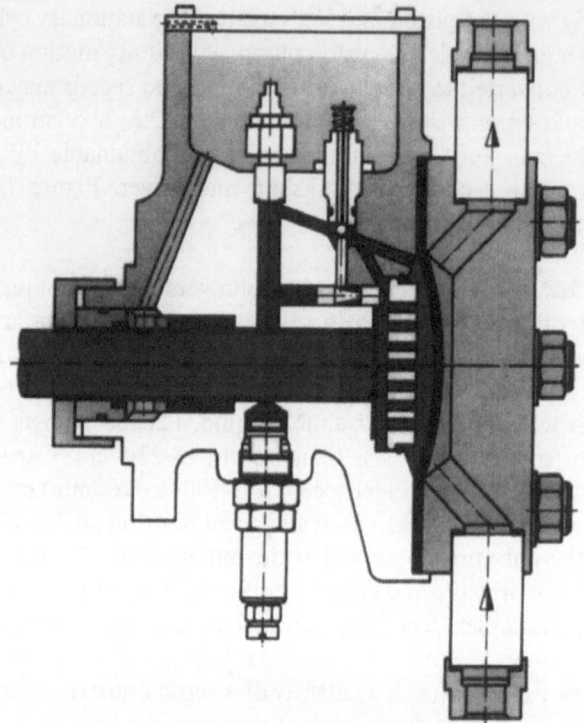

Figure 7.5   Diaphragm metering pump. (Reprinted with permission. Copyright 1990, Bran and Luebbe Inc.)

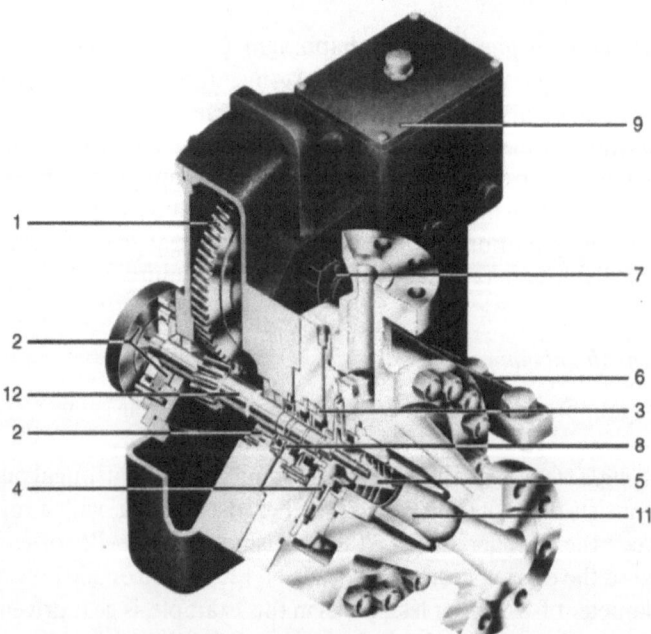

Figure 7.6   Single stage centrifugal pump. (Reprinted with permission. Copyright 1987, Ingersoll-Rand Company.)

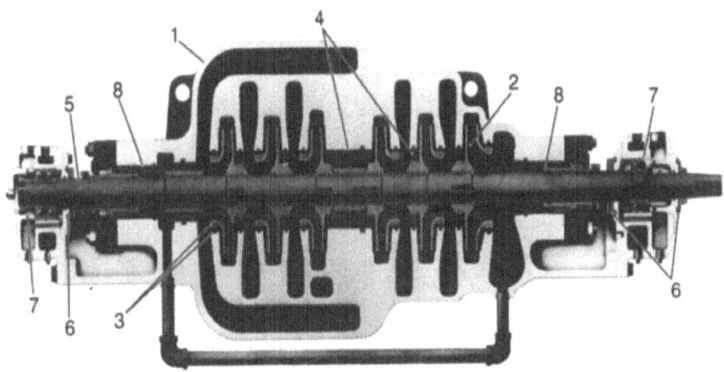

Figure 7.7   Multistage centrifugal pump. (Reprinted with permission. Copyright 1987, Ingersoll-Rand Company.)

from a few centimeters to 30 cm and larger. Width varies from a few mm to 30 cm or more. Housing thickness is determined by discharge pressure at the impeller periphery. Discharge pressure is established by impeller diameter and speed, while capacity is also determined by impeller width.

7.4.2.2   Multiple stage centrifugal pump.   For multistage centrifugal pumps (Figure 7.7), impellers are mounted on a common shaft with separate pumping cavities. The discharge from one impeller is directed to the eye of the second stage impeller, the outlet of the second to the third, etc. Each stage builds upon the discharge pressure of the prior stage. A differential pressure of 800 psi or more per stage can be achieved. This example shows opposing impeller sets, which provide balancing of shaft thrust.

7.5   Applications and selections

Possible pump selections are grouped on the basis of flow rate and differential pressure. All of the applications considered assume high discharge pressure. Selections are tabulated in Figure 7.8. These classifications and selections are not ironclad, but reflect the usage and experience for a number of plants. Sources for reference installations may be found not only in the SFE field, but in the use of carbon dioxide at supercritical conditions for enhanced oil recovery and other high pressure carbon dioxide applications. Valuable experience has been gained on metallurgy, materials for seals, and gasketing from these plants. Selection decisions for a particular installation should be based on both performance experience and equipment cost. The recommendations provided here are for guidance only.

In summary, selection of a pump for an SCF carbon dioxide extraction system requires consideration of certain unique characteristics. Some are related to the solvent fluid properties and others are related to the SFE process requirements. The design engineer should consider these factors for pump

High pressure pump discharge
(1500 to 5000 psia)

(a) Low flow (under 5 gpm)

High differential pressure (over 100 psia)	Low differential pressure (under 100 psia)
1 Diaphragm	1 Diaphragm
2 Packed plunger	2 Packed plunger
3 Piston	3 Centrifugal

(b) Intermediate flow (5 to 50 gpm)

High differential pressure (over 100 psia)	Low differential pressure (under 100 psia)
1 Packed plunger	1 Centrifugal
2 Diaphragm	2 Packed plunger

(c) High flow (over 50 gpm)

High differential pressure (over 100 psia)	Low differential pressure (under 100 psia)
1 Multistage Centrifugal	1 Centrifugal
2 Packed plunger	2 Packed plunger

Figure 7.8 SCF system pump selection matrix.

selection to ensure an efficient, safe and trouble-free plant. As more SCF carbon dioxide extraction plants are put into service, additional experience and information will become available for the design and selection of equipment for this important section of the SFE system.

Bibliography

Anon. (1979) *Horizontal Triplex Pump KD 822*, Uraca Pumpenfabrik Urach, Bad Urach, Germany.
Anon. (1983) *Look inside CVI's New PD-3000 Cryogenic Pump*, CVI Inc., Columbus, OH.
Anon. (1987) *Ingersoll-Rand Pumps*, Ingersoll-Rand Company, Washington, NJ.
Anon. (1988) *Hydro-Pac Product Catalog*, Hydro-Pac, Inc., Fairview, PA, p. C-5.
Anon. (1990) *Metering Pumps and Systems*, Bran+Luebbe Inc., Buffalo Grove, IL, p. 9.

8 Natural antioxidants produced by supercritical extraction

U. NGUYEN, D. A. EVANS, and G. FRAKMAN

Abstract

The variety and activity of synthetic and some natural antioxidants are discussed, and the actions of Carnosic Acid and Carnosol are evaluated. Comparison is made of current (ambient) extraction methods with supercritical extraction with reference to rosemary and sage extracts, and a particular example (Labex™ spice oleoresin) is presented. Supercritical extracts are compared with commercial products and found to be at least equally as effective.

8.1 Introduction

Antioxidants are used in food products containing fats and oils to prevent or retard the development of oxidative rancidity. Oxidation of unsaturated bonds to form hydroperoxides can lead to problems with color, odor and aroma, and once initiated proceeds at an exponential rate dependent on environmental factors such as temperature, atmospheric composition, light, etc. As color, odor and aroma are the main subjective criteria used in assessing product quality, the choice of packaging and storage conditions and the inclusion of antioxidants are factors extending the shelf-life of food products.

Antioxidants act by interfering with the oxidation process, they cannot eliminate oxidative products that have already been produced. Consequently, antioxidants should be added to the product as early in the product's life as possible. Antioxidants do not have any effect on hydrolytic rancidity resulting from hydrolysis of fats and oils to yield free fatty acids. This type of rancidity usually takes place in the presence of moisture and/or lipolytic enzymes and can be controlled by heating to destroy natural lipases and by removing all moisture from the fat/oil.

8.2   Traditional antioxidants

Antioxidants traditionally used in the European and North American food industries consist of two major groups: synthetic (chemical) antioxidants and natural antioxidants.

8.2.1   Synthetic antioxidants

This is the largest group in use today. They consist of the phenolic derivatives:

Butylated hydroxyanisole   (BHA)
Butylated hydroxytoluene   (BHT)
Tertiary butylhydroquinone   (THBQ)
Propyl gallate   (PG).

These antioxidants are manufactured by chemical processes. Their use in the food industry is severely restricted as to both application and level of use. Other problems with this class of compounds include heat sensitivity, discoloration and susceptibility to loss by steam distillation. More recently, researchers in Japan have found links between BHA and cancer in laboratory animals.

Advantages
Cheap
Effective, low usage levels
Very cost-effective
Oil-soluble
Colorless/odorless

Disadvantages
Restricted as to level of use
Restricted as to application
Cannot claim 'ALL NATURAL' or 'NO ARTIFICIAL PRESERVATIVES ADDED' label
Chemical must be identified on label
Consumer aversion
Discoloration and/or volatilization during processing

A further group of chemical antioxidants used in foods comprises the sequestering agents citric acid, ascorbic acid, sodium erythorbate etc.; none of which probably has direct antioxidant activity. They are often used in conjunction with antioxidants.

8.2.2   Natural antioxidants

One of the most important trends in the food industry today is the demand for 'all natural' food ingredients that are free of chemical additives. Natural food antioxidants are extracted from plant materials or by-products. There are currently two major groups.

8.2.2.1 Tocopherols. These are a mixture of alpha, beta, gamma and delta tocopherols obtained as a by-product from the vegetable oil refining industry. Commercial products include Eastman Chemicals Tenox GT-1 (50% mixed tocopherols) and Tenox GT-2 (70% mixed tocopherols) and Henkel Corp. Covi-ox T50 and T70 at identical strengths. They represent the largest group of natural antioxidants currently in use.

Advantages
GRAS for 'Good Manufacturing Practice'
'ALL NATURAL' label claims
Colorless/odorless
Oil soluble

Disadvantages
More expensive
Less effective than synthetics, high usage levels
Low cost-effectiveness
USDA restrictions on applications and level of use
Label declaration of function

8.2.2.2 Spice extracts. Certain botanicals have been long known to exhibit antioxidant properties. Most work in this area has concentrated on the antioxidant activity of natural herbs and spices such as cloves, ginger, mace, nutmeg and the Labiatae herbs rosemary, sage, thyme and oregano. Chipault *et al.* (1952, 1956) have investigated the antioxidant properties of many ground herbs and their ethanol extracts and have shown that rosemary and sage have the highest antioxidant activity.

As the demand for clean labeling and chemical-free all-natural food products has increased, a number of natural antioxidants derived from rosemary and sage are now being marketed. These include Rosemary Deodorized™ (Cal–Pfizer), Spicer Extract AR™ (Nestlé), Herbalox™ (Kalsec) and Flavor Guard™ (OM Ingredients).

Advantages
GRAS for all uses (as a spice extract)
No restrictions on applications or levels of use
ALL NATURAL label claims
Oil and water dispersible forms
No label declaration of function (as spice extract)

Disadvantages
Less effective than synthetics, high usage levels

Expensive
Low cost-effectiveness
Color, aroma and taste residuals

Despite the lower cost-effectiveness of the spice antioxidants their use is increasing due to market pressures.

8.3 Spice antioxidant compounds

Brieskorn *et al.* (1964) isolated a phenolic diterpenic lactone of the ferruginol type from leaves of *Rosmarinus officinalis* and *Salvia officinalis* which was shown to be carnosol. Later however, Wenkert *et al.* (1965) showed that the major terpenic constituent of rosemary leaves was carnosic acid and that carnosol was an artificially produced derivative resulting from the ready oxidative conversion from carnosic acid. Brieskorn and Domling (1969) showed that carnosic acid and carnosol were excellent antioxidants and that the antioxidant property of rosemary and sage was caused by the presence of carnosic acid in the leaves of these plants. The antioxidant activity of carnosic acid and carnosol was shown to be equivalent to the synthetic antioxidant BHT.

Two further minor but structurally related antioxidants have been isolated and identified from rosemary leaves. These are rosmanol (Nakatani and Inatani, 1981; Inatani *et al.*, 1982, 1983) and rosmaridiphenol (Houlihan *et al.*, 1984). Whereas rosmaridiphenol showed antioxidant activity equivalent to the synthetic antioxidants BHA and BHT, rosmanol showed four times that activity.

The chemical structures of carnosic acid, carnosol, rosmanol, rosmaridiphenol and a further compound, the methyl ester of carnosic acid (methyl carnosate), are shown in Figure 8.1.

8.4 Extraction of spice antioxidants

Various processes for obtaining antioxidant extracts from rosemary, for example, have been used commercially. The major problems to overcome are to obtain the extract with sufficient antioxidant activity to allow usage at levels equivalent to the synthetic antioxidants (0.01–0.10% of fat/oil) and to remove flavour, odour and colour components which may be detectable in the treated food product at the usage levels required. Many extraction processes have been employed, which generally invoke the following methods, to obtain antioxidant extracts from the Labiatae family of herbs: solvent extraction (polar and non-polar), (Chang *et al.*, 1973; Kimura *et al.*, 1983; Aesbach *et al.*, 1987; Todd, 1989); aqueous alkaline extraction, (Vinai, 1977); extraction

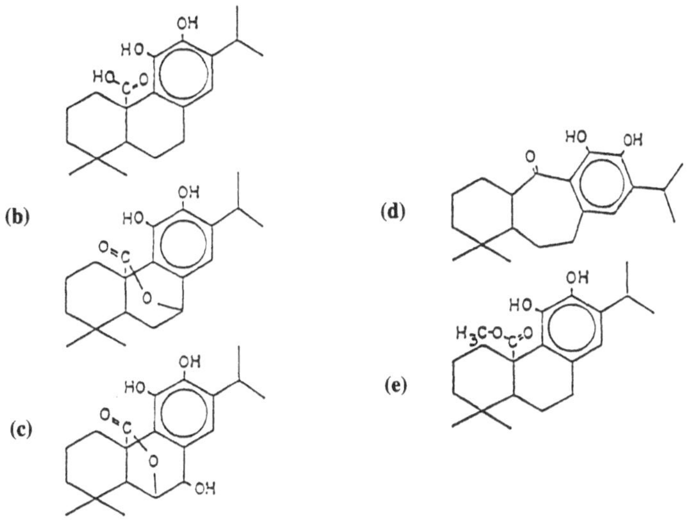

Figure 8.1 Identified phenolic diterpenes: (a) Carnosic acid; (b) Carnosol; (c) Rosmanol; (d) Rosmaridi-phenol; (e) Methyl carnosate.

with vegetable oils and/or mono- and diglycerides, (Berner *et al.*, 1973); steam distillation and molecular distillation. These processes suffer from a number of disadvantages. The solvents used are not effectively selective for the active antioxidant compounds and consequently the resulting extracts are not as strong as the synthetic chemical antioxidants. The solvents used include compounds such as hexane, acetone and methyl chloride which can leave unwanted residues in the food products and which in some instances are prohibited from use in food by regulation. Processes using molecular distillation to concentrate the active fraction and to remove color, aroma and flavor components result in a different type of dilution effect due to the presence of the distillation carrier. The presence of carriers may have a detrimental impact on fat/oil solubility of the extract.

8.5 Supercritical extraction

The application of supercritical carbon dioxide (SC-CO_2) fluid to the extraction and fractionation of lipophilic materials has been reviewed by Stahl *et al.* (1987).

Tateo and Fellin (1988) describe a process in which ground rosemary leaves are extracted with SC-CO_2 at a pressure of 300 bar and 35°C to remove the rosemary oleoresin containing the essential oil. The ground leaf residue remaining after extraction is then re-extracted with ethyl alcohol, filtered, evaporated and dried to obtain the antioxidant principle. Activity of the

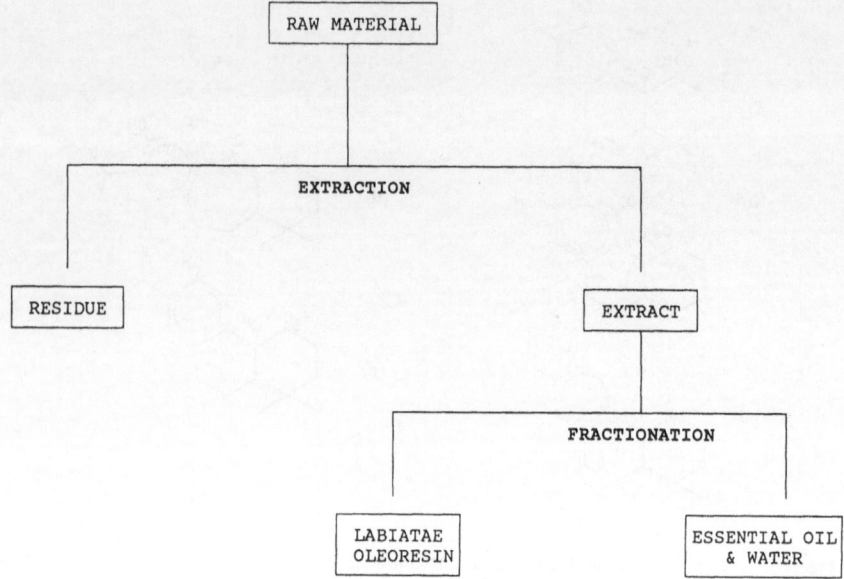

Figure 8.2 Process flow.

antioxidant extract (peroxide value (meq/kg) of prime steam lard at 100°C for 18 hours) was equivalent to a similar ethanol extract of ground leaf residue remaining after steam distillation to remove essential oil, equivalent to a commercial antioxidant extract (Rosemary Extract AR) produced by more complex solvent technology, but less than BHA used at the same level (0.03% fat/oil).

The Pfizer Corporation (since *ca.* 1988) have produced a commercial product (Rosemary Deodorised) in which the oleoresin fraction obtained by extracting rosemary with SC-CO$_2$ is then subjected to molecular distillation to improve color and flavor. The resulting product contains about 80% capric and caprylic triglyceride (the distillation carrier), has poor oil solubility and relatively weak antioxidant activity.

8.6 Labex™ spice oleoresin SC™

Nguyen *et al.* (1991) describe the extraction of the Labiatae herbs rosemary, sage, oregano and thyme with SC-CO$_2$ at pressures in the vicinity of 500 bar and temperatures of 80–100°C (see Table 8.1; Figure 8.2). The extracted oleoresin is precipitated by modification of pressure and temperature into two fractions. The first is a green-brown, oil-soluble, heat-stable, resin containing less than 2% essential oil and exhibiting remarkable antioxidant properties while the second is the essential oil containing in excess of 95 ml steam distillable oil per 100 grams.

Table 8.1 Labiatae oleoresin antioxidants: supercritical extracts – Summary

1. Produced without organic solvents
2. Contain less than 2% essential oil
3. GRAS (as spice oleoresins)
4. Unrestricted as to application and level of use
5. Allow 'all natural' labels
6. Equivalent to BHA/BHT (1:1) at 0.02% (200 ppm) usage
7. 4–5 × stronger than existing labiatae products
8. Oil and ethanol soluble
9. Fully effective in food systems
10. Food industry analytical methods are available

Table 8.2 Antioxidant activity of
Labiatae oleoresin. Modified Schaal
oven testing. 0.03% in prime steam lard
for 18 hours at 100°C.

Antioxidant	PV meq/kg
Base	0.5
Control	33.7
BHA/BHT (1:1)	1.9
Rosemary officinalis extract	1.5
Sage officinalis extract	1.5
Sage triloba extract	2.1
Oregano vulgare extract	3.8
Thyme Vulgaris extract	2.9

The antioxidant resin fraction from rosemary and sage has been named
Labex. The antioxidant activities of a number of such extracts are shown in
Table 8.2. Those from *R. officinalis* and *S. officinalis* are at least as strong as
BHA/BHT (1:1) in prime steam lard at 300 ppm inclusion. Figure 8.3 sum-
marises a large number of similar trials and shows that the Labex extract is
fully effective at a usage of 200–300 ppm while natural mixed tocopherols and
two commercial rosemary extracts require greater than 1000 ppm of usage and
still cannot match the Labex effectiveness. Figures 8.4 and 8.5 show similar
results based on Rancimat testing in both lard and canola oil.

When tested in food applications, Labex also shows exceptional antioxidant
effectiveness. Usage levels are low enough to preclude aroma and flavor effects.
Figure 8.6 shows the effect of Labex, incorporated in a meat binder to give a
usage level of 100 ppm of meat, on oxidative stability of cooked mechanically deboned chicken during storage at 4°C. Satisfactory TBA values were
maintained for 25 days with a product that previously could be stored at
freezer temperatures only. Labex has also proven effective in the reduction of
color value loss from paprika oleoresin due to oxidation of the carotenoid
pigments. Figure 8.7 illustrates the protection from color loss during extended
heating.

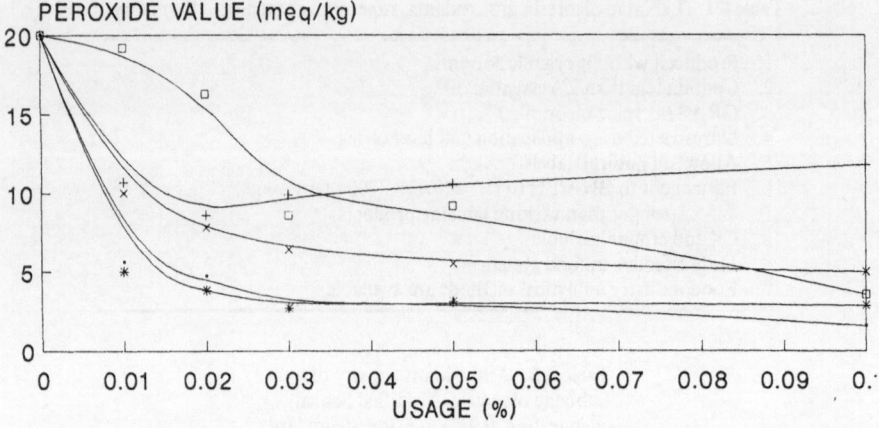

Figure 8.3 Peroxide value estimates of prime steam lard after 18 hours at 100°C, from 23 experiments and 321 observations ●, BHA/BHT (1:1); +, tocopherol; *, Labex; ☐, Com-A; ×, com-B.

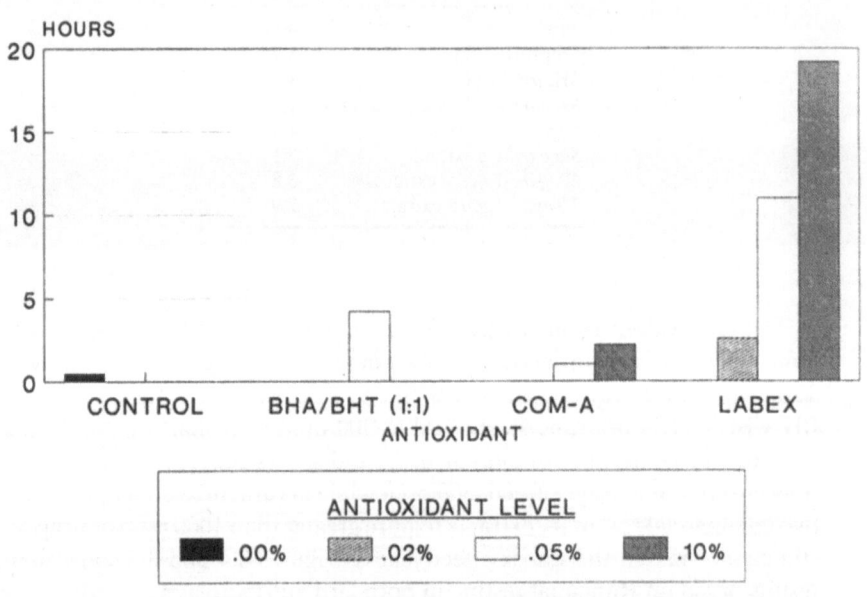

Figure 8.4 Labex spice oleoresin—Rancimat antioxidant comparison induction time (hours) at 120°C using prime steam lard.

8.7 Analytical

The extent of the antioxidant activity displayed by the fractionated supercritical Labiatae extracts has led to an investigation of the active antioxidant ingredients. We have found that $SC\text{-}CO_2$, under the specific extraction conditions,

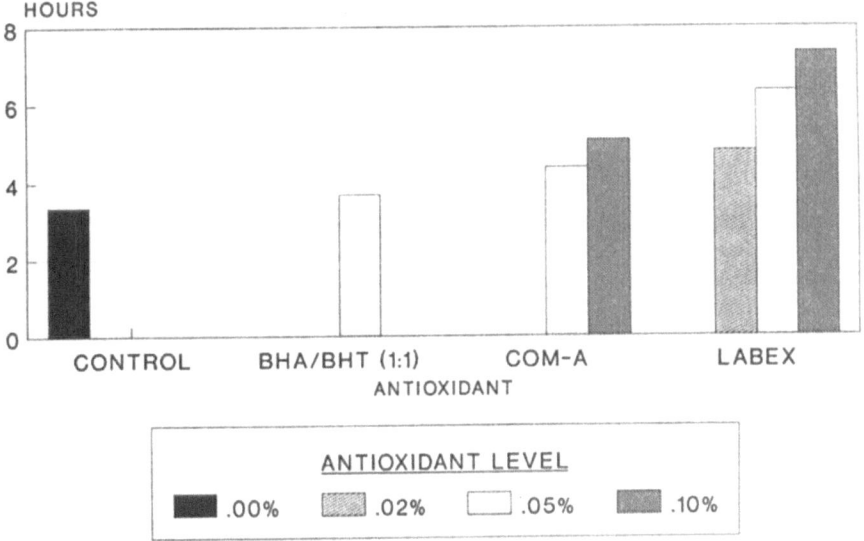

Figure 8.5 Labex spice oleoresin—Rancimat antioxidant comparison induction time (hours) at 120°C using pure refined canola oil.

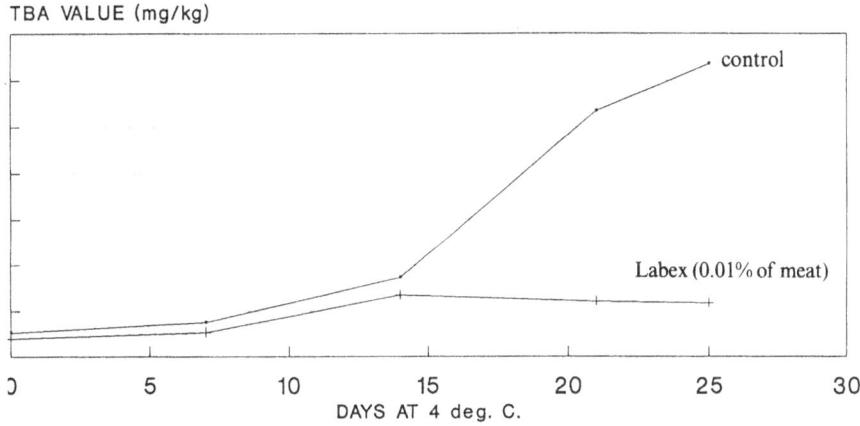

Figure 8.6 Labex spice oleoresin—Cooked chicken storage test of mechanically deboned chicken, from means of 4 replicates (courtesy of Guelph University).

is a very good solvent for carnosic acid. In contrast, organic and/or aqueous solvent extraction of Labiatae herbs results in the oxidative conversion of carnosic acid to carnosol and other unknown by-products. Contrary to earlier papers, carnosol is not the natural antioxidant component of the Labiatae, but is produced by traditional extraction procedures. In contrast, carnosic acid is extremely stable in SC-CO_2, and is fully extracted as such.

Analytical data for *R. officinalis* and *S. officinalis* in comparison with four

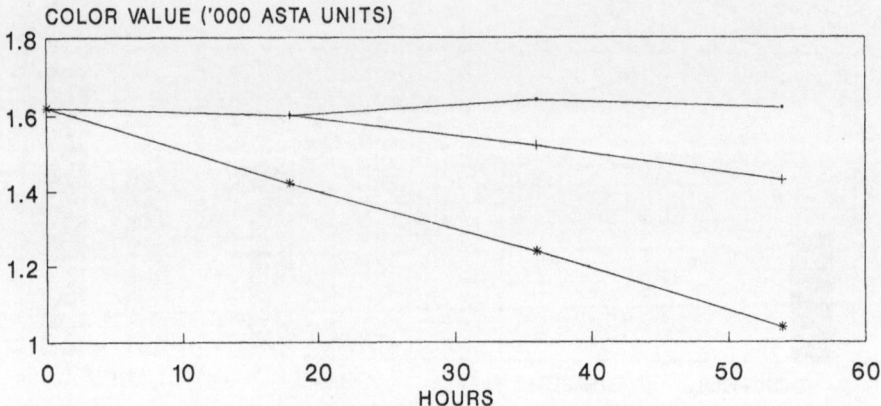

Figure 8.7   Labex spice oleoresin/paprika oleoresin color value test, from means of 3 replicates. ●, control 18°C; * control 100°C; +, Labex 0.03% 100°C.

commercial rosemary extracts are shown in Table 8.3. The supercritical extracts contain some 5–6 times as much total phenolic antioxidant (TPA) as the commercial products and of this, over 80% is in the form of carnosic acid. The distorted carnosol:carnosic acid ratio for two of the commercial products is clearly demonstrated. Consequently, the relative strengths of the two groups of products as established by the peroxide value tests, are fully reflected in their relative phenolic antioxidant contents.

8.8   Summary

Under conditions described by Nguyen *et al.* (1991), supercritical extracts of rosemary and sage have strong antioxidant properties. This is a consequence of their carnosic acid content. During extraction they may be fractionated to contain less than 2% steam volatile essential oil. The essential oil may be recovered separately as a valuable by-product to the extraction process.

At a usage level of 200–300 ppm of fat/oil, their effectiveness is at least equal to that of BHA/BHT (1:1). At this level of usage, residual flavor and aroma effects are not perceptible. They are 4–5 times stronger food antioxidants than existing Labiatae products.

As spice oleoresins, they are GRAS and unrestricted as to application and level of use. They allow the use of 'all natural' labels.

The extracts have the benefit of both oil and ethanol solubility and are easily incorporated into food systems. The active antioxidant compound is stable to cooking temperatures.

Analytical procedures are available to monitor their use and cost-effectiveness in food processes.

Table 8.3 Tentative composition by HPLC against standards for identified phenolic diterpenes. (Frakman *et al.*, in press)

Oleoresin	Carnosic acid %	Methyl carnosate %	Carnosol %	Total %
Rose-SC	26.5	4.2	1.8	32.5
Sage-SC	33.0	2.0	2.3	37.3
Com-A	5.6	0.2	0.8	6.6
Com-B	1.8	0.3	4.5	6.6
Com-C	2.9	0.3	2.6	5.8
Com-D	1.7	0.2	0.9	2.8

References

Aesbach, R. and Philippossion, G. (1987) Swiss Patent 672, 048,A5.

Berner, D. L. and Jacobson, G. A. (1973) US Patent 3,732,111.

Brieskorn, C. H., Fuchs, A., Bredenburg, J. B., McChesney, J. D. and Wankert, E. (1964) *J. Org. Chem.* **29**, 2293.

Brieskorn, C. H. and Domling, H–J., (1969) *Zeitschrift für Lebensmitteluntersuchung und-forschung* **141**(1), 10.

Chang, S. S. (1973) US Patent 3,950,266.

Chipault, J. R., Mizuno, G. R., Hawkins, J. M. and Lundberg, W. O. (1952) *Food Res.* **17**, 46.

Chipault, J. R., Mizuno, G. R. and Lundberg, W. O. (1956) *Food Technol.* **10**(5), 209.

Houlihan, C. M., Ho, C.-T. and Chang, S. S. (1984) *JAOCS* **61**(6), 1036.

Inatani, R., Nakatani, N., Funa, H. and Seto, H. (1982) *Agric. Biol. Chem.* **46**, 1661.

Inatani, R., Nakatani, and Funa, H. (1983) *Agric. Biol. Chem.* **47**(3), 521.

Kimura, Y. and Kanamori, T. (1983) US Patent 4,380,506.

Nakatani, N. and Inatani, R., (1981) *Agric. Biol. Chem.* **45**(10), 2385.

Nguyen, U., Evans, D. A. and Frakman, G. (1991) US Patent 5,017,397.

Pfizer Corporation, (1988) Technical Product Information.

Stahl, E., Quirin, K.-W. and Gerard, D. (1987) *Dense Gases for Extraction and Refining*, Springer-Verlag, Berlin.

Tateo, F., and Fellin, M., (1988) *Perfumer and Flavorist* **13**, 48.

Todd, P. H., (1989) US Patent 4,877,635.

Vinai, (1977) US Patent 4,012,531.

Wenkert, E., Fuchs, A. and McChesney, J. D. (1965) *J. Org. Chem.* **30**, 2931.

9 Separation of ethanol/water solution with supercritical CO_2 in the presence of a membrane

J.-H. HSU and C.-S. TAN

Abstract

The separation of ethanol from an aqueous solution of 6 wt% ethanol with supercritical CO_2 in the presence of a hydrophilic polyamine membrane was investigated. The experiments were performed by means of an apparatus consisting of two chambers divided by the membrane. The aqueous solution and CO_2 flowed concurrently into one of the chambers, while the other was empty. The experimental results indicated that the separation effectiveness, in terms of rejection rate, could be enhanced significantly over the use of membrane alone and the supercritical CO_2 extraction method. The most appropriate operating condition was found to be near the critical point of CO_2 and at a higher ratio of CO_2 flow rate to aqueous solution flow rate.

9.1 Introduction

Recovery or removal of ethanol from dilute aqueous solutions is needed in many industrial processes and in the clean-up of waste streams. Recently, increased use of fermentation to produce ethanol has also led to recovery problems of this kind. Because of the relatively high affinity between ethanol and water, significant energy is required to achieve separation using the conventional evaporation and distillation techniques. Many attempts therefore have been made to develop less energy-intensive separation alternatives which include membrane technology (Mehta, 1982; Leeper and Tsao, 1987), liquid extraction (Arenson *et al.*, 1990), adsorption (Kawabate *et al.*, 1988), and supercritical fluid extraction (Moses and de Filippi, 1984; Brignole *et al.*, 1987). Due to a relatively low distribution ratio and the thermodynamic limitations of each method, the separation effectiveness has not yet reached the industrial standard.

To overcome the limitations of the membrane, some combined technologies, such as pervaporation (Zhu *et al.*, 1989) and perstraction (Acharya *et al.*, 1988) in which a low-pressure vapor or a purge organic liquid is allowed to flow along one interface of a membrane, while the feed flows along the opposite interface, were found to provide better separation of organic solutes from dilute aqueous solutions than the use of membrane alone due to the addition of driving forces for mass transfer. Since the supercritical CO_2 possesses organophilic character and is non-toxic and inexpensive, a method combining supercritical CO_2 and membrane may be another alternative to achieve the above purpose. The operations may depend on whether the membrane used is hydrophobic or hydrophilic. When a hydrophilic membrane is used, both the supercritical CO_2 and the aqueous solution should flow along the same side of the membrane. But when a hydrophobic membrane is used, the supercritical CO_2 should flow along the side opposite to that contacted by the aqueous solution, just as in the perstraction method. The objective of this paper is to investigate the validity of this proposed method in the presence of a hydrophilic membrane.

9.2   Materials and methods

The membrane used in this work was an aromatic polyamine thin-film composite membrane which was purchased from FilmTec Corp. (FT-30). It was stored under conditions specified by the manufacturer and was used without further treatment. A flat sheet membrane was placed in a membrane holder (Milipore Co., 93700) and was supported by a porous sintered stainless steel powder block. The effective contact area between the membrane and aqueous solution was about 9.5 cm^2.

Figure 9.1 illustrates the experimental apparatus. The aqueous solution containing 6 wt% of ethanol was prepared by adding a known amount of ethanol of 99.5% purity into deionized water and mixing thoroughly. The prepared aqueous solution was compressed by a minipump (Millton Roy, 396–89) and was heated by a preheating coil immersed in a constant temperature bath to the desired pressure and temperature. The flow rate of the aqueous solution was fixed at about 100 cm^3/h. Carbon dioxide of 99.8% purity was stored in a high pressure tank. The pressure of the CO_2 stream was adjusted by regulators and metering valves and the temperature was controlled in a constant temperature bath through a preheating coil. The operating pressure varied from 65.3 atm to 88.4 atm and the temperature was fixed at 305 K. Two operations were performed: (1) the CO_2 stream and the aqueous solution were premixed first before they entered into the upper chamber of the membrane holder; (2) these two streams flowed into the upper chamber separately. However, the experimental results show that these two operations provided almost the same separation effectiveness.

The effluent stream of the upper chamber (retentate stream) was depressurized through a back pressure regular (Tescom, TC-06) and passed through a cold trap which collected the condensed liquid solution. One microliter of

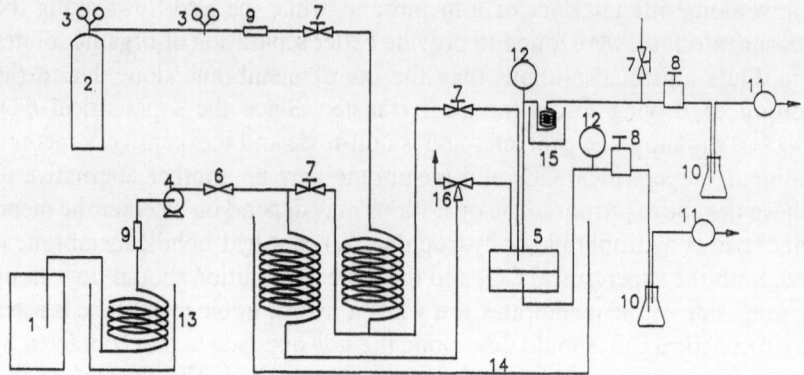

Figure 9.1 Schematic diagram of the experimental apparatus. 1, Ethanol solution tank; 2, CO_2 cylinder; 3, regulator; 4, minipump; 5, membrane holder; 6, check valve; 7, metering valve; 8, back pressure regulator; 9, filter; 10, cold trap; 11, wet test meter; 12, pressure gauge; 13, ice bath; 14, constant temperature bath; 15, oil bath; 16, three-way valve.

the condensed solution was analyzed frequently by a gas chromatograph (Varian 3700) to determine the concentration of ethanol. The amount of CO_2 in the retentate stream was measured by a wet test meter (Yokogawa, W-NK-1B) which was located downstream of the cold trap. The pressure of the lower chamber was maintained at 1 atm. The concentration of ethanol and the amount of CO_2 in the effluent stream of the lower chamber (permeate stream) were determined in a similar way as for the retentate stream.

The reproducibility tests were performed at different pressures and CO_2 flow rates. It was found that the data could be reproduced within 6%. The overall mass of ethanol in the retentate and permeate streams was found to agree well with that in the feed. The deviation was generally less than 2%.

9.3 Results and discussion

Figure 9.2 shows the existence of stable performance of the membrane used for the operations with and without CO_2 flow stream. The latter operation is in fact identical to the conventional membrane operation. The rejection rate on the ordinate of Figure 9.2 is defined as

$$R = (C_o - C_e)/C_o \tag{9.1}$$

where C_o and C_e are the ethanol concentration in the feed and in the retentate stream, respectively. A larger R represents a better separation effectiveness. The flow rate of aqueous solution for both operations was about 100 cm^3/h. It should be mentioned that the rejection rate for the operation without CO_2 flow stream at 60 atm (not shown in Figure 9.2) agreed well with that reported by Mehta (1982).

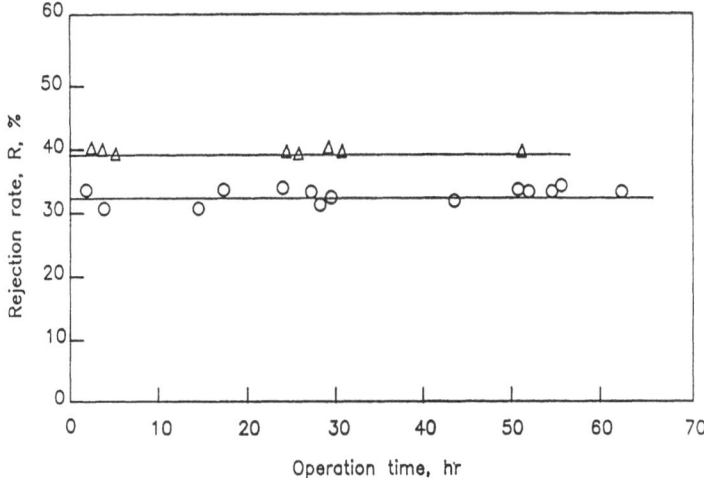

Figure 9.2 Illustration of stable performance of the membrane at 305 K, 74.8 atm. CO_2 flow rate: zero, O, membrane alone; 173 cm^3/h, △, present method.

In order to compare the separation effectiveness between the present method and the use of membrane alone (the conventional membrane method), the relative rejection rate is defined by

$$R.R. = (R - R_o)/R_o \qquad (9.2)$$

where R_o is the rejection rate for the conventional membrane operation at 74.8 atm, at which the highest rejection rate was observed, as compared with the operations at other pressures in the range presently studied. Figure 9.3 shows that better separation could be obtained when supercritical CO_2 stream was introduced. But this is not the case when subcritical CO_2 was introduced. The possible explanation for this observation is that the solubility of ethanol in supercritical CO_2 is higher than that in subcritical CO_2; the resulting affinity of ethanol to CO_2 leads to more ethanol leaving from the polarization layer, if it exists, near the membrane and moving towards the interface between CO_2 and water; consequently, less ethanol passes through the membrane. The pressure effect is especially pronounced at higher CO_2 flow rate, which is shown in Figure 9.3. This may be due to the fact that more contact area between CO_2 and water can be provided under this condition.

It can also be seen from Figure 9.3 that the operation at 74.8 atm, near the critical pressure of CO_2, provided the best separation effectiveness. As pointed out by Eckert *et al.* (1986) and Debenedetti and Kumar (1988), large clusters of approximately 100 solvent molecules per solute molecule may be formed near the critical point. Because of the presence of these clusters, ethanol is believed to be more difficult to pass through the pores of the membrane. This may be the reason that better separation existed near the critical point of CO_2.

The comparison of rejection rates among the proposed method, the use of

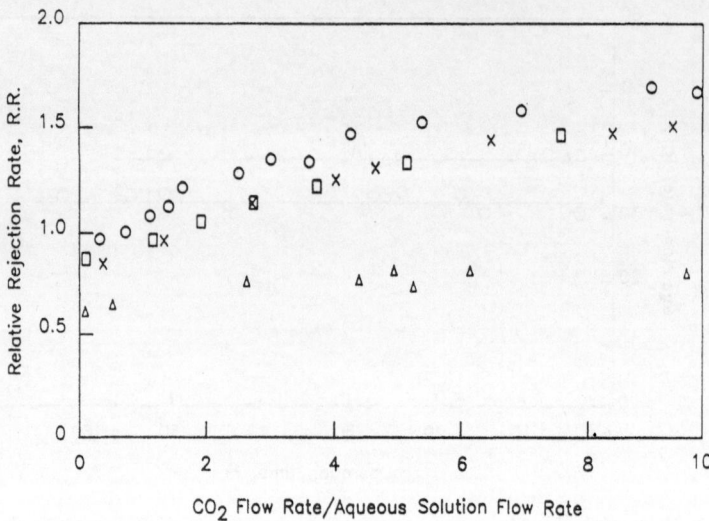

Figure 9.3   Pressure effect on relative rejection rate at 305 K. △, 65.3 atm; ○, 74.8 atm; ×, 81.6 atm; □ 88.4 atm.

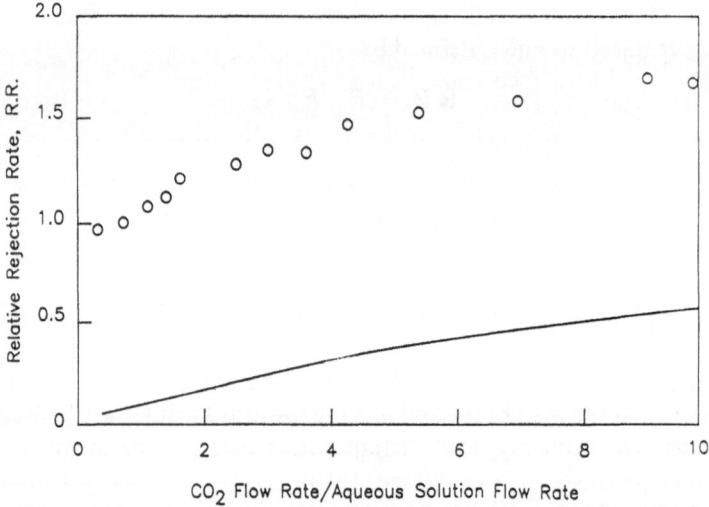

Figure 9.4   Comparison of rejection rate among the proposed method, the conventional membrane method and the SC-CO₂ extraction method, at 305 K, 74.8 atm. ○, Experimental data; —, SCE-CO₂ method.

membrane alone, and the supercritical extraction method at 305 K and 74.8 atm is shown in Figure 9.4. The rejection rate for the supercritical CO_2 extraction method was calculated by performing the flash calculation incorporated with the Peng–Robinson equation of state (Peng and Robinson, 1976) and an unsymmetric mixing rule (Panagiotopoulos and Reid, 1987). It can be

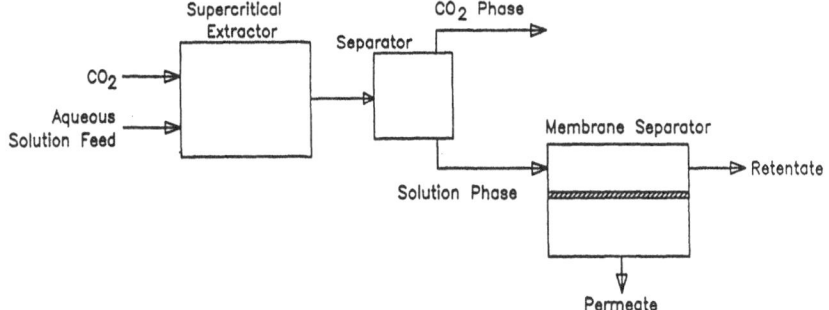

Figure 9.5 Configuration of the proposed model.

clearly seen that the proposed method was superior to the other two methods. The improvement in rejection rate could be 60% more than the use of membrane alone and 180% more than the supercritical CO_2 extraction method at higher CO_2 flow rates.

In order to model the present method, a configuration of the supercritical CO_2 extraction followed by the use of membrane alone was proposed (Figure 9.5). The ethanol content in the effluent liquid solution from the supercritical CO_2 extractor was calculated by the Peng–Robinson equation of state

$$P = RT/(V - b) - a/[V(V + b)+(V - b)] \tag{9.3}$$

where

$$a = \sum_i \sum_j a_{ij} x_i x_j \tag{9.4}$$

$$a_{ij} = [1 - \delta_{ij} + (\delta_{ij} - \delta_{ji})x_i /(x_i + x_j)] (a_i a_j)^{0.5} \tag{9.4}$$

$$b = \sum_i \sum_j b_{ij} x_i x_j \tag{9.5}$$

$$b_{ij} = (1 - \eta_{ij}) (b_i + b_j)/2 \tag{9.6}$$

and the overall mass balance equations

$$Lx_{i,o} + Gy_{i,o} = L'x_i + G'y_i \tag{9.7}$$

$$\sum x_i = \sum y_i = 1 \tag{9.8}$$

The ethanol rejection for the membrane separator was determined by the experimental data for the use of membrane alone. The calculated results shown in Figure 9.6 indicate that this configuration provided adequate prediction for the operations at 81.6 and 88.4 atm, but failed to predict for the operation at 74.8 atm where the prediction was less than the experimental data. There are two possible reasons for the latter situation: (1) the Peng–Robinson equation

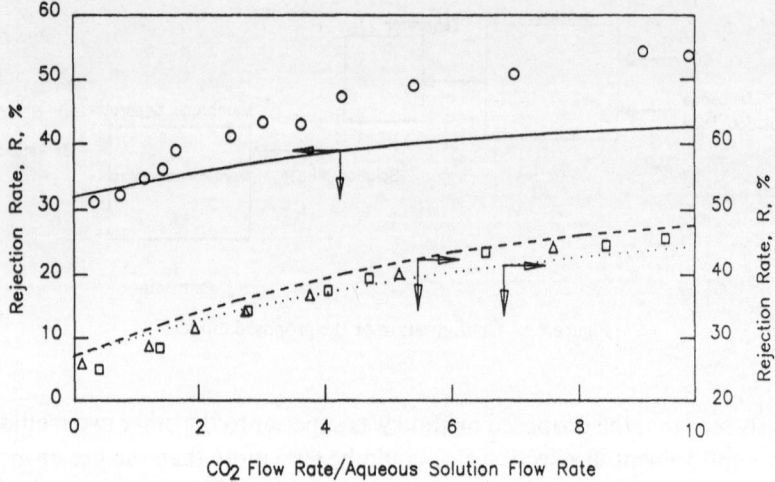

Figure 9.6   Prediction of the rejection rate using the proposed configuration. O, Experimental and —, predicted, at 74.8 atm; □, · · ·, at 81.6 atm; △, – –, at 88.4 atm.

of state may not be accurate enough nearing the critical point; (2) the formation of large clusters which has been discussed above.

Although the proposed method could result in a higher rejection rate compared with the use of membrane alone, the flux was found to reduce. At 305 K and 74.8 atm, the flux of the present method was about 1.0 cm³/h-cm² which is lower than 1.75 cm³/h-cm² for the use of membrane only. This may be due to the adsorption of CO_2 dissolved in aqueous solution by the membrane which consequently hindered the passage of the solution in the membrane. Since the prediction using the proposed configuration indicates that the operations at higher temperature may be more appropriate if the operation is not near the critical point, the temperature effect on the rejection rate was also investigated. However, deterioration of the membrane was observed at temperatures higher than 313 K (at which the manufacturer does not recommend either). Under this situation, the temperature effect could not be verified experimentally.

9.4   Conclusions

A separation method incorporating a hydrophilic polyamine membrane with supercritical CO_2 was proposed to treat an aqueous solution of 6 wt% ethanol. The proposed method was found to be superior to the use of membrane alone and the supercritical CO_2 extraction method regarding the rejection rate. When operated at 305 K, the most appropriate operating pressure was found to be close to the critical pressure of CO_2. This is probably due to the formation of large clusters which are not allowed to pass through the small pores of the membrane.

A configuration, supercritical CO_2 extraction followed by the use of membrane alone, was found to predict the results reasonably well if the operations are not near the critical point of CO_2. This suggests that the present method can be simply regarded as the combination of the advantages of the membrane and supercritical fluid. When the operation is near the critical point of CO_2, the formation of clusters provides an additive separation effect.

Nomenclature

a,b parameters in the Peng–Robinson equation of state
C_e ethanol concentration in permeate, wt%
C_o ethanol concentration in aqueous solution feed, wt%
G feed rate of supercritical CO_2 phase, mol/h
G' effluent rate of supercritical CO_2 phase, mol/h
L feed rate of liquid phase, mol/h
L' effluent rate of liquid phase, mol/h
R ethanol rejection, defined in equation (9.1)
R_o ethanol rejection for the use of membrane alone at 299.5 K and 65.3 atm
R.R.   relative ethanol rejection, defined in equation (9.2)
x_i mole fraction of the ith component in effluent liquid phase
$x_{i,o}$ mole fraction of the ith component in liquid feed
y_i mole fraction of the ith component in effluent supercritical CO_2 phase
$y_{i,o}$ mole fraction of the ith component in supercritical CO_2 feed
δ_{ij}, η_{ij} interaction parameters in the Peng–Robinson equation of state

References

Acharya, H. R., Stern, S. A., Liu, H. R. and Cabasso, I. (1988) Separation of liquid benzene/cyclohexane mixtures by perstraction and pervaporation. *J. Membr. Sci.* **37**, 205.
Arenson, D. R., Kertes, A. S. and King, C. J. (1990) Extraction of ethanol from aqueous solution with phenolic extractants. *Ind. Eng. Chem. Res.* **29**, 607.
Brignole, E. A., Andersen, P. M. and Fredenslund, A. (1987) Supercritical fluid extraction of alcohols from water. *Ind. Eng. Chem. Res.* **26**, 254.
Debenedetti, P. G. and Kumar, S. K. (1988) The molecular basis of temperature effects in supercritical extraction. *AIChE J.* **34**, 645.
Eckert, C. A., Ziger, D. H., Johnston, K. P. and Kim, S. (1986) Solute partial molar volumes in supercritical fluids. *J. Phys. Chem.* **90**, 2738.
Kawabate, N., Sumigama, Y. and Matsuura, N. (1988) Separation of alcohols from water by adsorption on cross-linked polymethacrylic ester containing a pyridinium group. *Ind. Eng. Chem. Res.* **27**, 1882.
Leeper, S. A. and Tsao, G. T. (1987) Membrane separations in ethanol recovery: an analysis of two applications of hyperfiltration. *J. Membr. Sci.* **30**, 289.

Mehta, G. A. (1982) Comparison of membrane processes with distillation for alcohol/water separation. *J. Membr. Sci.* **12,** 1.

Moses, J. M. and de Filippi, R. P. (1984) Critical-fluid extraction of organics from water. Volume I: Engineering analysis, Report to the US Department of Commerce, DE84 012669, Contract No. DE-AC01-79C540258.

Panagiotopoulos, A. Z. and Reid, R. C. (1987) High-pressure equilibria in ternary fluid mixtures with a supercritical component, in *Supercritical Fluids, ACS Symp. Ser. 329,* eds. T. G. Squires and M. E. Paulaitis, ACS, pp. 115–129.

Peng, D. Y. and Robinson, D. B. (1976) A new two-constant equation of state. *Ind. Eng. Chem. Fundam.* **15,** 58.

Zhu, C., Liu, M., Xu, W. and Ji, W. (1989) A study on characteristics and enhancement of pervaporation-membrane separation process. *Desalination* **71,** 1.

10 Supercritical fluid fractionation of butter oil

W. MAJEWSKI, P. MENGAL, M. PERRUT and
J. P. ECALARD

Abstract

Butter oil is extracted with supercritical CO_2 operating at 50–60°C and pressure 100–250 bar in a batchwise apparatus. Fractionation of triglycerides and selectivity of supercritical CO_2 for cholesterol removal are considered. Remarkably, fractionation of triglycerides occurs at relatively low pressure of 100 bar and temperature of 50°C. Higher pressures leading to a high solubility of butter oil decrease the solvent selectivity. Maximum selectivity of CO_2 for cholesterol removal is achieved at 50°C and pressure ranging from 150 to 175 bar. Under these conditions, about 65% reduction in cholesterol content is observed in the residue. Results show that one-stage supercritical extraction of butter is not sufficient to obtain a product with low melting point and low cholesterol content. A multistage fractionation process and/or utilization of polar solids for cholesterol adsorption should be taken into account.

10.1 Introduction

Large scale extraction with supercritical CO_2 has already been used in the food industry (Perrut, 1990), especially for decaffeination of coffee, extraction of hop resins, spices, aromas, etc.

Carbon dioxide under supercritical conditions possesses many desirable features for separation of natural products:

1. low temperature extraction;
2. solvent volatility;
3. non-toxicity;
4. non-flammability;
5. selectivity;
6. low viscosity and high diffusivity.

Modification of butter properties is of great interest due to its poor spreadability at domestic refrigerator temperatures, and its cholesterol content.

Butter oil consists of a large number of components; about 98% are triglycerides with carbon number varying from 26 to 54. The other constituents are present at low or even trace levels (wt%): diglycerides (0.36), monoglycerides (0.027), free fatty acids (0.027), free phospholipids (0.6), cholesterol and its esters (0.31). Variation in molecular weight and unsaturation degree of triglycerides lead to a wide disproportion in the melting properties (−30 to 40°C) of butter oil. In general, fractionation by crystallization gives two fractions: a liquid oleic fraction (18–22°C) where triglycerides of 24–42 carbon atoms are preferentially concentrated and solid stearic fraction containing mainly high molecular weight triglycerides with 44–54 carbon atoms.

A fractionation process including extraction with supercritical CO_2 (Kaufmann et al., 1982; Bradley, 1989; Kankare et al., 1989; Rizvi et al., 1989) is one of the technically possible methods to improve the properties of butter. Cholesterol, a minor constituent in butter, is simultaneously extracted with triglycerides (Arul et al., 1988). Although several methods for the reduction of cholesterol in butter oil are offered (vacuum or molecular distillation (Bracco, 1980), melt crystallization (Norris et al., 1971; Arul et al., 1988) complexation with cyclodextrins (Courregelongue and Maffrand, 1988), biological degradation, no one-step supercritical extraction process (Shishikura et al., 1986; Lim et al., 1991) to obtain butter with low cholesterol content and low melting point is available.

10.2   Materials and methods

10.2.1   Apparatus

A flow diagram of our home-made apparatus (SEPAREX) is shown in Figure 10.1. The high pressure extractor (2), equipped with a special container of 1 l capacity, can operate up to 300 bar and 80°C.

Liquid commercial grade CO_2 was obtained from the cylinder or reservoir (1) and fed, through the heater (8) to the bottom of the extractor. After percolating the autoclave filled with butter, supercritical CO_2 was gradually decompressed in three high performance separators (3, 4, 5) where solute mixture was precipitated. Gaseous CO_2 was then condensed in a heat exchanger (7) and recycled to the extractor through the pump (6).

10.2.2   Analytical methods

The cholesterol content was determined by the method developed by G. Lognay (U.E.R. Chimie Générale et Organique, FSA Gembloux, Belgium). It is based on saponification of the triglycerides and gas chromatography (GC) analysis of the unsaponified part of the sample using betuline as an internal standard.

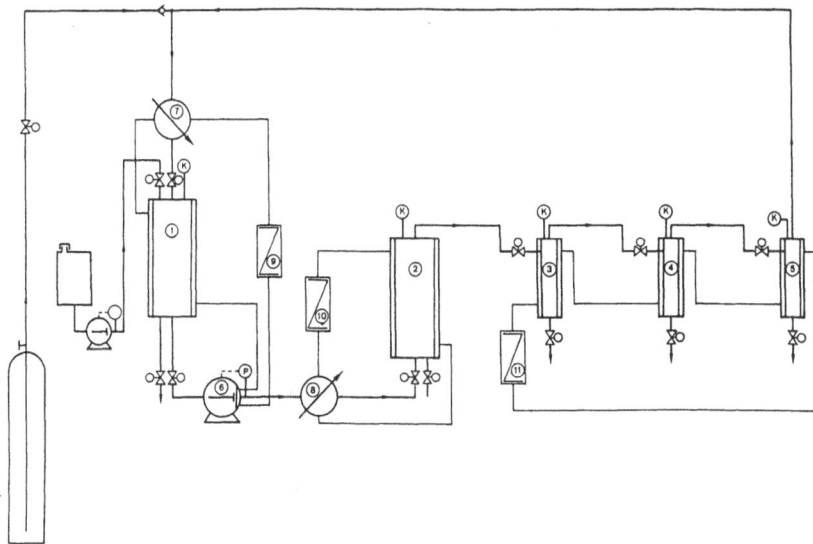

Figure 10.1 Flowsheet of supercritical extraction equipment made by SEPAREX. 1, CO_2 reservoir; 2, autoclave; 3, separator 1; 4, separator 2; 5, separator 3; 6, CO_2 pump; 7, condenser; 8, heater; 9, cooling bath; 10, 11 heating baths.

The composition of fatty acids was determined by GC after conversion to ethyl esters. The triglyceride composition was also determined by GC using a capillary fused-silica column (chemically bonded Carbowax).

The melting properties (slip point) of the extraction products were esti-mated by measurement of the migration temperature of the fat in a capillary tubing (1 mm i.d.) subjected to static pressure of 981 Pa. The heating rate of the water bath was 5°C/min. The melting curves were determined by a differen-tial scanning calorimeter, (Perkin Elmer DSC-7), with a temperature gradient of 10°C/min.

10.3 Results

Extraction of butter with supercritical CO_2 was carried out at pressures varying from 100 to 250 bar and temperatures of 50°C and 60°C. The solvent ratio was in the range of 68–172 kg/kg. These conditions were found to be appropriate through preliminary investigations performed in a batchwise apparatus.

10.3.1 Influence of process parameters on extraction yield

The extraction yield was mainly determined by temperature, pressure or solvent density and duration of extraction. In Figure 10.2, the extraction yield

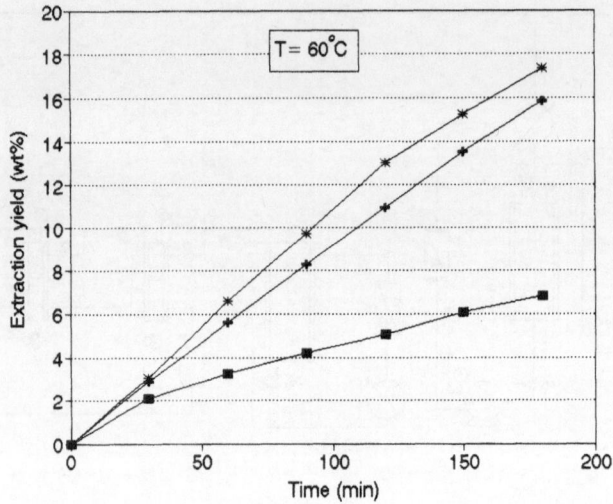

Figure 10.2   Effect of pressure and extraction time on yield of extracts. ■, 150 bar; +, 175 bar; *, 200 bar.

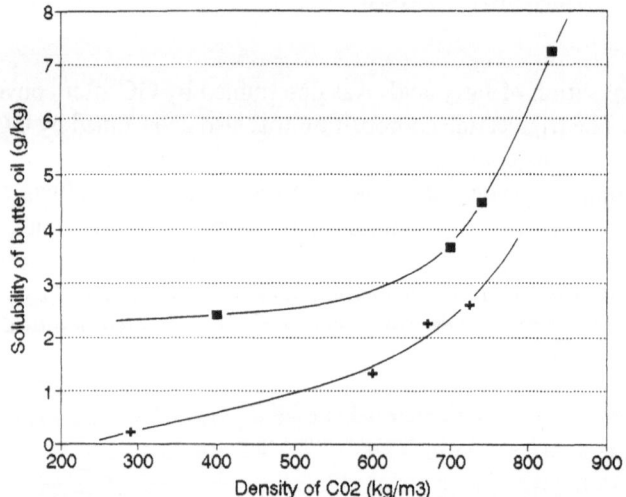

Figure 10.3   Solubility of butter oil in supercritical CO_2 as a function of the solvent density and temperature. ■, 50°C; +, 60°C.

obtained at 60°C and different pressures is shown against the extraction time. As can be anticipated, the extraction yield is enhanced by an increase in pressure under isothermal conditions. For longer extraction times, non-linear profiles can be expected.

The effect of the CO_2 density on total solubility of the butter components is shown in Figure 10.3. It is seen that solubility increases dramatically at higher

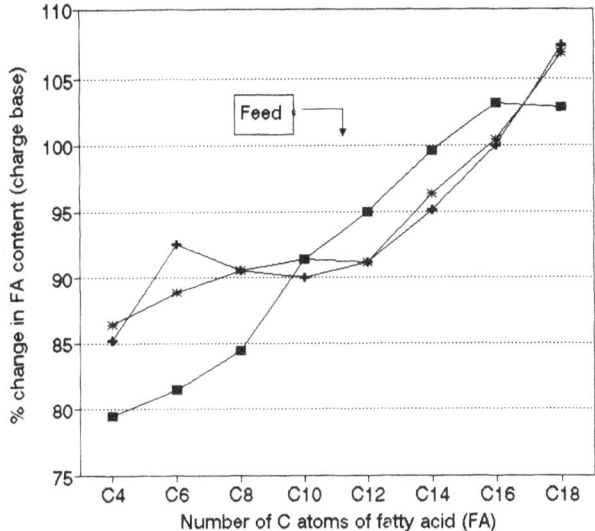

Figure 10.4 Profiles of fatty acid content in the residues as a function of extraction pressure. ■, 100 bar; +, 150 bar; *, 200 bar.

solvent densities. At the same value of density, loading of the solvent is higher for experiments conducted at lower temperature (50°C). This solubility behaviour is similar to what has been obtained by numerous investigators on all types of oils and fats.

10.3.2 Fractionation of triglycerides

The fractionation effect of supercritical extraction of butter is caused by a large disproportion in solubilities of triglycerides in CO_2. Triglycerides with short chains and low polarity are more soluble in supercritical CO_2 and can be easily removed from extracting material. It is no surprise that a higher concentration of low molecular weight compounds in supercritical fluid leads to changes in the proportion of triglycerides in the remaining raffinate.

Figure 10.4 shows some profiles of fatty acid composition in butter extracted at different pressures in comparison with the starting material. Decrease in the concentration of triglycerides containing fatty acids with carbon atoms <16 is visible for all the experimental runs at 50°C. For short-chain fatty acids (C < 8) maximum reduction is observed under pressures of 100 bar. As frequently observed, this confirms that the supercritical fluid selectivity decreases when loading (density, pressure) of the solvent is increased.

The influence of the extraction temperature on fractionation of triglycerides is illustrated in Figure 10.5. Decreasing the extraction temperature from 60 to 50°C causes large differences in the fatty acid (C < 16) composition of residual product. As a result of short-chain fatty acid removal,

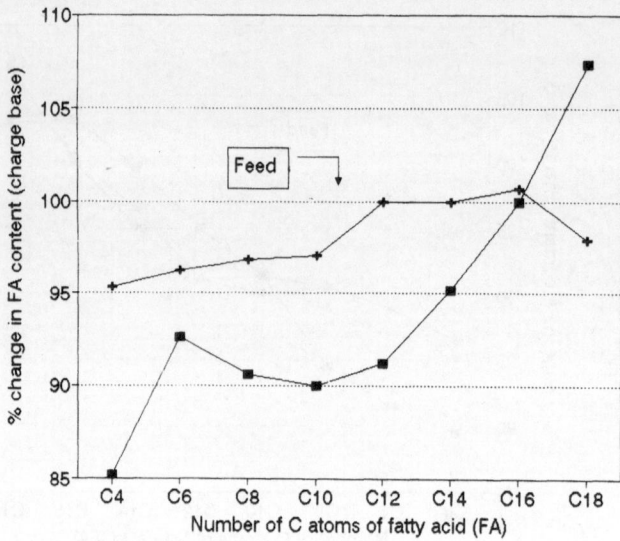

Figure 10.5 Profiles of fatty acid content in the residues as a function of extraction temperature. ■, 50°C, +, 60°C.

concentration of triglycerides containing long-chain fatty acid (C > 16) can be observed.

From these results, the optimum condition of butter fractionation by supercritical CO_2 has been found to be 50°C and 100 bar.

10.3.3   Reduction in cholesterol content

The cholesterol content of the products derived from extraction of butter with supercritical CO_2 is shown in Table 10.1. As seen in the table, the cholesterol

Table 10.1 Cholesterol concentration in residual product of butter extraction with supercritical CO_2.

Run	Pressure (bar)	Temperature (°C)	Extraction yield (wt%)	Cholesterol content (mg/100g)	
				Extract	Residue
Feed	–	–	–	–	280
1	100	50	43	200	160
2	150	50	54	–	130
3	200	50	39	–	170
4	100	60	–	–	–
5	150	60	52	–	130
6	200	60	46	–	150
7	125	50	54	–	130
8	175	50	65	320	100
9	175	60	55	380	120
10	150	50	64	350	100
11	200	50	52	–	130

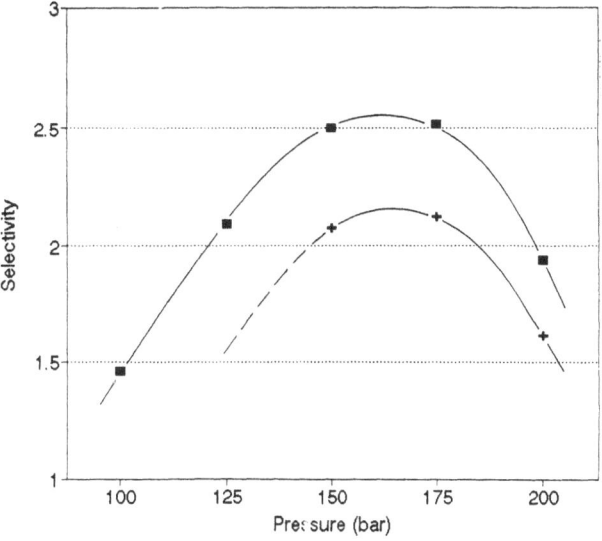

Figure 10.6 Carbon dioxide selectivity for cholesterol removal. ■, 50°C; +, 60°C.

content in the residue can be reduced to about 100 mg/100 g. Compared with the starting material, 65% of cholesterol is co-extracted with triglycerides. Cholesterol, theoretically less soluble than triglycerides in supercritical CO_2, possesses a high affinity for the short- and medium-chain triglycerides. This phenomenon can explain why supercritical extracts are enriched in cholesterol compared to the original butter.

On the basis of cholesterol and fatty acid concentration in feed and residues, the coefficient of solvent selectivity can be defined as

$$\text{Selectivity} = \frac{\text{ratio of cholesterol to fatty acids (C<14) in feed}}{\text{ratio of cholesterol to fatty acids (C<14) in residue}} \quad (10.1)$$

Figure 10.6 shows the selectivity of CO_2 for the removal of cholesterol from butter. For both temperatures investigated, selectivity of supercritical fluid reaches a maximum for pressures between 150 and 175 bar. Since a decrease in temperature of extraction favorably influences selectivity, greater values of selectivity than 2.6 could be expected at temperatures below 50°C; however, as butter viscosity increases sharply when temperature decreases, it is not certain that better results could be obtained.

10.3.4 Melting properties of extraction products

Owing to the fractionating effect of supercritical extraction, the melting properties of the products change. Compared with butter oil having a slip point about 32.3°C, residues melt at higher temperatures between 33.3° and 35.7°C. Melting data for two extracts obtained at 50°C and pressures of 100 and 200 bar are shown in Table 10.2.

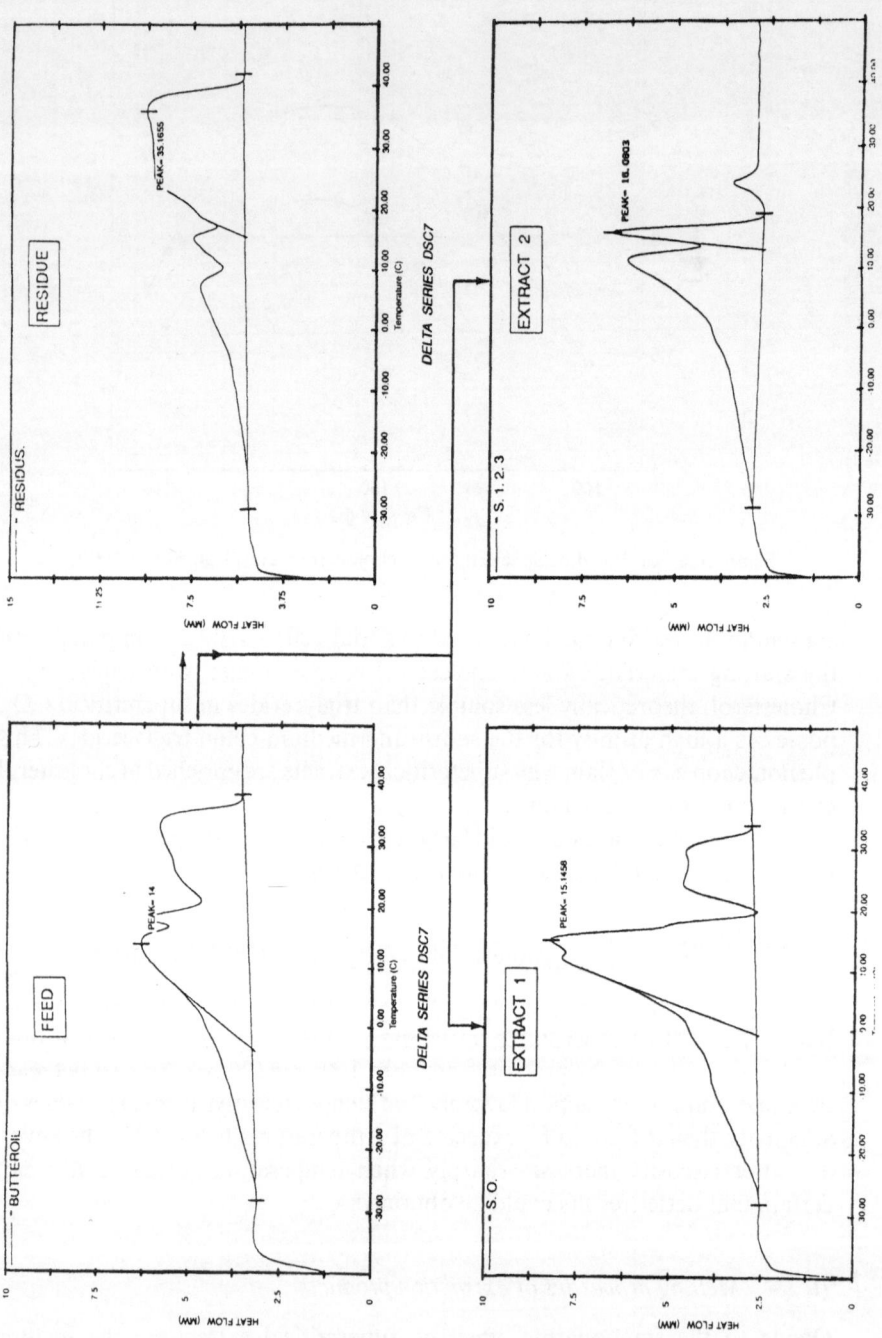

Figure 10.7 Melting curves of the products from butter oil fractionation at 50°C and 250 bar.

Table 10.2 Melting points of extracts from runs at 50°C

Extraction time (min)	Melting point (°C)					
	Separator 1		Separator 2		Separator 3	
	100 bar	200 bar	100 bar	200 bar	100 bar	200 bar
90	24.2	24.5	23.5	22.3	19.7	23.3
150	24.0	23.7	23.5	23.2	20.5	–
210	24.2	24.7	24.5	23.5	20.3	25.0

There are considerable amounts of short-chain triglycerides with low melting point in extracts collected in the last separator (3) for extraction pressure of 100 bar. Although the melting temperatures of extracts are clearly lower than for butter oil or residue, nevertheless this parameter is not good enough to characterize the product as a function of extraction time or separation steps.

More information comes from analysis of melting profiles performed by scanning calorimetry. Figure 10.7 shows the melting curves of original butter oil and products of extraction carried out at 50°C and 250 bar. A large difference in melting properties between extracts precipitated under different pressures (extract 1, 150 bar; extract 2, 70 bar) and residue proves the fractionating capability of supercritical CO_2.

10.4 Conclusion

One-step extraction of butter with supercritical CO_2 improves the properties of the products (increase in amount of low temperature melting triglycerides in extracts, decrease in cholesterol concentration in residue). However, fractionation based mainly on molecular size of triglycerides is not sharp and poor selectivity of CO_2 leads to only modest removal of cholesterol.

Further work on a multistage fractionation column coupled with cholesterol adsorption on polar solids is being undertaken.

References

Arul, J., Boudreau, A., Makhlouf, J., Tardif, R. and Grenier, B. (1988) Distribution of cholesterol in milk fat fractions. *J. Dairy Res.* **55**, 361.

Bracco, U. (1980) Butter-like food product, British Patent 1559064.

Bradley, R. L. Jr., (1989) Removal of cholesterol from milk fat using supercritical carbon dioxide. *J. Dairy Sci.* **72**, 2834.

Courregelongue, J. and Maffrand, J.-P. (1988) Procéde d'élimination du cholestérol contenu dans une m.g. d'origine animale et m.g. appauvrie en cholestérol obtenue, European Patent Office, 0256911.

Kankare, V., Antila, V., Harvala, T. and Komppa, V. (1989) Extraction of milk fat with supercritical carbon dioxide. *Milchwissenschaft* **44**, 407.

Kaufmann, W., Biernoth, G., Frede, E., Merk, W., Precht, D. and Timmen, H. (1982) Fraktionierung von Butterfett durch Extraction mit überkritischem CO_2. *Milchwissenschaft* **37**, 92.

Lim, S., Lim, G.-B. and Rizvi, S. S. H. (1991) Continuous supercritical CO_2 processing of milk fat, *Proc. 8th World Congress on Food Science and Technology*, Boston, MA, p. 292.

Norris, R., Gray, I. K., McDowell, A. K. R. and Dolby, R. M. (1971) The chemical composition and physical properties of milk fat obtained by a commercial fractionation process. *J. Dairy Res.* **38**, 179.

Perrut, M. (1990) Applications en abondance. *Inf. Chim.* **321**, 166.

Rizvi, S. S. H., Lim, S., Nikoopour, H., Singh, M. and Yu, Z. (1989) *Engineering and Food*, Vol. 3, Elsevier Applied Science, London, p. 145.

Shishikura, A., Fujimoto, K., Kaneda, T., Arai, K. and Saito, S. (1986) Modification of butter oil by extraction with supercritical carbon dioxide. *Agric. Biol. Chem.* **50**, 1209.

11 Supercritical carbon dioxide processing of orange juice: effects on pectinesterase, microbiology and quality attributes

A. G. ARREOLA, M. O. BALABAN,
M. R. MARSHALL, C. I. WEI, A. J. PEPLOW and
J. A. CORNELL

Abstract

Single strength Valencia and Pineapple variety orange juices were treated with supercritical CO_2 (SC-CO_2). Effect of pressure (7–34 MPa), temperature (35–60°C), and time (15–180 min) on pectinesterase (PE) activity, cloud stability, microorganisms and sensory attributes were investigated. SC-CO_2 inactivated PE below thermal inactivation temperatures. Activation energy was 97.4 kJ/mol at 31 MPa, and 166.6 kJ/mol at atmospheric pressure. SC-CO_2 treatment did not change pH, °Brix or total acidity, but stabilized and enhanced cloud and color, even in the presence of active PE. Ascorbic acid was better preserved. Sensory evaluations of color and cloudiness of treated juice were better compared to untreated juice. Flavor, aroma and overall acceptability were not different. Analysis of aerobic total microbial plate counts showed that the D value of treated juice decreased as pressure increased. D values at 35, 45 and 60°C of SC-CO_2 treated juice at 33.1 MPa were 28, 22.6 and 12.7 min, respectively. This method can be used to 'minimally treat' orange and other fruit juices.

11.1 Introduction

Supercritical fluid (SCF) technology has many applications in pharmaceutical and food industries. For example, in West Germany and the United States, commercial plants employ SC-CO_2 for the decaffeination of coffee (Novak and Robey, 1988), and in Australia SC-CO_2 is used for the recovery of hops extracts (Hubert and Vitzthum, 1980). There are many other potential applications to foods and biological materials. The motivation to employ SCF technology is based on: (1) tightening environmental regulations on conventional solvent residues; (2) concern over use of chemical solvents in food manufacture; (3) increased demand for higher quality products which traditional processing techniques cannot meet; (4) increased cost of energy. Applications of SCF technology to foods has been reviewed in the literature (Rizvi *et al.*, 1986a, b).

11.1.1   Conventional applications of SC-CO₂ to citrus processing

11.1.1.1   Debittering of citrus juices.   Kimball (1987) used SC-CO₂ to extract bitter triterpenoids such as limonin from orange juice. He found that temperature (30–60°C) had no significant effect on limonin reduction, however, pressure increase from 21.4 to 42.8 MPa was effective. Small amounts of D-limonene were also extracted. A change in the final pH, vitamin C, pulp content, amino acids and percent acid could not be detected.

11.1.1.2   Extraction and/or concentration of citrus essential oils.   Lemon flavors were extracted with SC-CO₂ (Kalra *et al.*, 1987), and phase behavior data on 75% limonene and 25% citral, and lemon oil consisting of 23 components in the C10 to C15 range was reported. Pressure was from 9.31 to 10.69 MPa, and temperatures were from 50 to 80°C. As temperature increased, solubility of CO₂ in the liquid phase decreased, and the fraction of limonene in the vapor phase increased. In addition, as temperature increased, the fraction of *trans*-citral in the vapor phase increased from 61 to 72%. The authors concluded that it was possible to selectively extract components from lemon oil and to fractionate the extracts with SC-CO₂.

Temelli (1987) reported on the concentration of flavor components in cold pressed orange oil (consisting of 96% terpenes, mostly D-limonene) with SC-CO₂. Oxygenated compounds such as aldehydes, alcohols and ketones, accounting for less than 4% of the oil provide much of the characteristic flavor of citrus oil. At 8.3 MPa and 70°C, and at 12.4 MPa and 40°C terpenes were more solubilized in SC-CO₂ than oxygenated materials. They are non-polar, have smaller molecular weight and higher vapor pressure, and are more soluble in SC-CO₂. Therefore, most of the extract was D-limonene. Temelli *et al.* (1988) reported that it was possible to concentrate the flavor portion of citrus oils with SC-CO₂, and gave 70°C and 8.3 MPa as the conditions for minimum amount of flavors lost with the extract, which also resulted in low extraction yields.

11.1.2   Unconventional applications of SC-CO₂ to citrus processing

11.1.2.1   Effect on the enzyme pectinesterase.   The primary goal of heat treating single strength orange juice (SSOJ) is to inactivate pectinesterase (PE) which causes loss of juice cloud (Rouse and Atkins, 1952). The current practice is a treatment of 90°C for up to 1 min, which also kills most spoilage organisms. However, this causes undesirable aroma and flavor changes (Varsel, 1980). There are non-thermal methods to stabilize cloud in SSOJ, including the use of enzymes (Baker and Bruemmer, 1972), clouding agents (Crandall *et al.*, 1983), oligo-galacturonic acids (Termote *et al.*, 1977), membrane operations (Dziezak, 1989), and low pH (Baker, 1979; Owusu-Yaw *et al.*, 1988). However, all those methods have disadvantages. Addition of foreign materials such as enzymes to orange juice is not allowed. Addition of HCl, or use of cation exchange resins to lower pH results in unacceptable quality from a sensory point of view.

Solution of high pressure CO$_2$ in water causes carbonic acid formation, temporarily lowering pH, and affecting enzymes and microorganisms. When water content is low in the treated medium, enzymes such as alpha-amylase, glucose oxidase, lipase and catalase retained their activities (Taniguchi *et al.*, 1987). SC-CO$_2$ treatment can be used for citrus juice cloud stabilization. When pressure is returned to atmospheric, CO$_2$ separates from the juice, restoring the original pH. This low temperature method circumvents previous processing disadvantages since there is no net addition or removal of materials from SSOJ. Haas *et al.* (1989) reported PE was not inactivated when orange juice was treated with SC-CO$_2$ at 6.2 MPa. This pressure may have been too low for inactivation.

11.1.2.2 Effect on quality attributes and microorganisms.   Quality of SSOJ also depends on other attributes such as pH, stability, total soluble solids (°Brix), total acidity, color, ascorbic acid content, flavor, aroma, and general appearance. In addition, prevention of undesirable flavors caused by microorganisms growing under low pH conditions is important. Microbial counts of fresh orange juice are between 10^4 and 10^5 colony forming units (CFU) per milliliter (Faville *et al.*, 1951). pH has a significant effect on microbial growth. Rushing *et al.* (1956) have shown that the growth rate of *Lactobacillus brevis* and *Leuconostoc mesenteroides* in orange juice at 21°C decreased by 200% when pH was reduced from 4.0 to 3.4. Kamihira *et al.* (1987) found a sterilizing effect of SC-CO$_2$ on various microorganisms at 20.3 MPa and 35°C (water content 70–90%). Dry cells were not sterilized when treated under the same conditions. Haas *et al.* (1989) sterilized fresh herbs and spices with pressurized CO$_2$. Molin (1983) found that 100% CO$_2$ reduced the growth rate of several microorganisms including *Lactobacillus* spp. The inhibitory effect of CO$_2$ on different types of bacteria increases as temperature decreases, possibly due to increased CO$_2$ solubility in water at lower temperatures (Molin, 1983). Wei *et al.* (1991) studied the effects of high pressure CO$_2$ on *Listeria* and *Salmonella*. Effects of SC-CO$_2$ treatment on microorganisms involves not only pressure and pH, but also possible chemical interactions between the organisms and CO$_2$. Therefore, SC-CO$_2$ treatment of orange juice may have the added benefit of reducing microbial numbers.

The objectives of this study were: (1) to determine the effect of SC-CO$_2$ treatment conditions such as time, temperature and pressure on degree of inactivation of pectinesterase in SSOJ; (2) to evaluate the effect of SC-CO$_2$

Table 11.1 Proximate analyses of freshly squeezed Valencia orange juice[a]

Parameter	Range	Average
°Brix[b]	11.23–11.43	11.27
% Acid[c]	0.65–0.67	0.67
% Oil (v/v)	0.012–0.0132	0.0123
Pulp (% volume)[d]	12.00	12.00
% Trans[e]	8–9	8.4
Pulp (g/l) 20 mesh[f]	Trace–0.8	0.6
Pulp (g/l) 60 mesh[f]	6.0–11.6	10.0
Color score	39.9–40.2	40.1
pH	3.78–3.81	3.79

[a] Performed at FMC laboratory, Lakeland, Florida.
[b] Corrected for acid.
[c] Citric.
[d] Centrifuged.
[e] Light transmission.
[f] Screened.

treatment on the total plate count (TPC) of the microbial loads in SSOJ; and (3) to determine the effects of SC-CO_2 treatment on pH, cloud stability, °Brix, total acidity, color, ascorbic acid content, flavor, aroma and general appearance of SSOJ. Kinetic parameters for enzymatic and microbial reduction at different conditions were also determined.

11.2 Materials and methods

11.2.1 Orange juice

Freshly squeezed orange juice (FSOJ) from Valencia oranges, quickly frozen with liquid N_2 in 1.4-l cans was provided by FMC Corporation (Lakeland, FL), and stored at -28.8°C. Table 11.1 shows the proximate analysis of the juice. In addition, FSOJ from Pineapple variety oranges were placed in aluminum pouches, flushed with N_2, and quickly frozen by Citrus World, Inc. (Lake Wales, FL). Both samples were shipped to the laboratory in refrigerated trucks.

The initial TPC of the FMC orange juice used in microbiological experiments on orange serum agar was found to be low (54 CFU/ml). Therefore, a commercial freshly squeezed non-pasteurized orange juice (Black Welder Groves Inc., Haines City, FL) purchased from a local supermarket was left at room temperature for 2 days to allow multiplication of the natural flora (stock juice). The resulting count was determined to be 2.9×10^6 CFU/ml. This was used to spike the FMC orange juice in the ratio of 14:1 (original juicestock juice).

11.2.2   Experimental design

A central composite design was used to determine the levels of three experimental process variables: pressure (P), temperature (T), and time (t). Minimum P was chosen as 6.9 MPa, near the critical P for CO$_2$. Maximum P was set by the operating limit of the supercritical equipment (34.4 MPa). Minimum T (35°C) was slightly above the critical T for CO$_2$, while maximum T of 60°C was selected because at least one of the multiple forms of PE is thermally inactivated (Versteeg et al., 1980) at that T. The time levels were chosen arbitrarily between 15 and 180 min.

11.2.3   Supercritical equipment

Two different supercritical systems were used: (1) Milton Roy Supercritical X-10 System (Milton Roy, Riviera Beach, FL); and (2) a system with custom made parts by Baydatco Inc. (Ithaca, NY), and designated in this study as the custom-made supercritical (CMSC) system. The Milton Roy system had a 150-ml vessel, and CO$_2$ could not be recirculated, so phase equilibrium was reached slowly. A new system was designed to allow continuous recirculation of CO$_2$ from the top of the vessel through a recirculating pump (Model 12286, Baydatco USA, Inc.) back to the bottom. A metering pump (Lewa, Inc., Holliston, MA) was used to pump CO$_2$ up to 11 l/h at a maximum discharge pressure of 34.5 MPa. It had a 2-l vessel. Figure 11.1 shows a diagram of the CMSC system. For both systems, a copper-constantan thermocouple (0.F. Ecklund, Cape Coral, FL) was placed inside the vessel to read the temperature of SSOJ. A heating jacket was placed around the vessel and another thermo-couple was placed between the outer surface of the vessel and the inner surface of the jacket to maintain a constant temperature using a controller. SSOJ was placed into the vessel, flushed and pressurized by CO$_2$ and brought to the desired pressure and temperature. For the CMSC system, CO$_2$ was recirculated once the experimental pressure was reached. Samples were withdrawn periodically from a bottom valve, and various parameters measured.

11.2.4   Thermal inactivation of PE

Orange juice was thawed and divided into three portions. The first portion was used for initial determinations of pH, PE, Brix, ascorbic acid, acidity, and microbial loads. The second was temperature control (TC), placed in a water bath and held at the same temperature and time as the SC-CO$_2$ treatment. The third was used for SC-CO$_2$ treatment. After SC-CO$_2$ treatment and depress-urization, small amounts of CO$_2$ remaining in the juice quickly diffused out.

Residual PE activity of the TC samples was used to calculate thermal inactivation kinetics. We assumed the following first order rate and Arrhenius relationship:

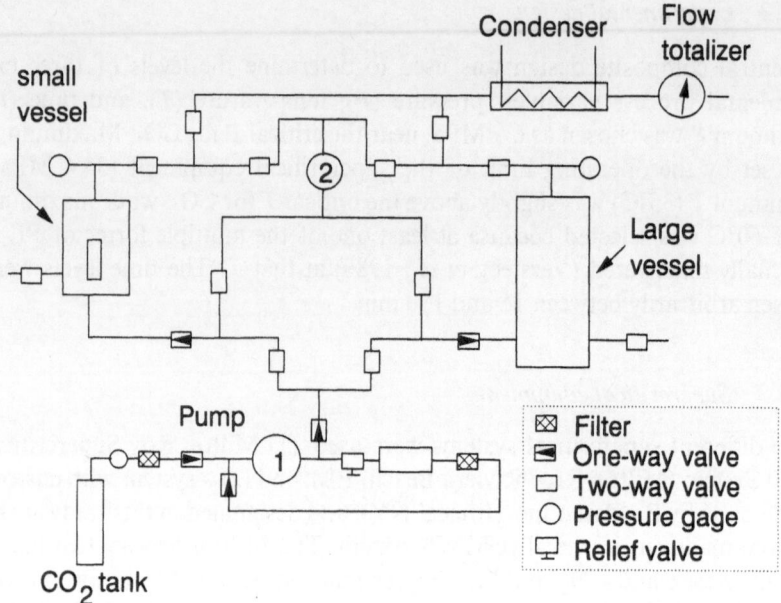

Figure 11.1 Schematic representation of the custom made supercritical system; 2, recirculation pump.

$$N = N_o \exp[-A \exp(-E_a/RT)\, t] = N_o \exp[-k\, t] \qquad (11.1)$$

where N is the PE activity of the TC sample after time t, N_o is the PE of initial juice, A is the frequency factor (l/min), E_a is the energy of activation (J/mol), R is the gas constant 8.31 J/mol K, T is the absolute temperature (K), and t is the experimental time (min). A plot of $\ln(k)$ versus $1/T$ resulted in a straight line. The slope allowed calculation of E_a, the energy of activation for thermal degradation.

11.2.5   Combined thermal and supercritical inactivation of PE

Experiments were conducted to calculate the kinetics of combined thermal and SC-CO_2 inactivation of PE. The pressure was set at 31 MPa. Temperatures were set to 40, 55 and 60°C. Samples were taken at different time intervals, and PE activity was measured. Inactivation kinetics (D and z values) of both the TC samples kept at atmospheric pressure and the SC-CO_2 treated samples were calculated.

11.2.6   Pectinesterase activity determination

PE activity was determined by automatic titration (Rouse and Atkins, 1955).

A 1% high methoxyl pectin (Kodak, Rochester, NY) solution in water, with 0.15 M NaCl and 1 mM sodium azide was used as substrate. A 25-ml aliquot of pectin solution (initial pH 3.9) was adjusted to pH 7.50 with NaOH, then 1 ml of the juice sample was added to the solution. The pH was re-adjusted back to 7.5 using NaOH. The amount of NaOH (0.05 N) used versus time at 27°C was measured. PE activity was calculated by the following formula:

$$PE\ units/ml = \frac{(ml\ NaOH)\ (Normality\ of\ NaOH)\ (1000)}{(time)\ (ml\ sample)} \qquad (11.2)$$

11.2.7 Effect of storage on PE and cloud of SC-CO₂ treated orange juice

In the Milton Roy system, a sample was treated at 28.95 MPa, 45°C for 4 h. Samples were removed every 30 min to monitor PE activity and cloud. All samples were kept in amber jars to measure PE activity and cloud after 1, 2, 3, 6 and 9 weeks storage at 4.4°C. This experiment was repeated with the CMSC system. Treatment conditions were 28.95 MPa, 45°C for 2 h.

11.2.8 pH and °Brix measurement

pH of orange juice before and after treatments was measured by a pH meter (EA 920, Orion Inc., Cambridge, MA). A 10-ml sample was placed in a beaker containing a stirring bar. The pH was recorded when stable.

°Brix was measured with an ABBE refractometer with temperature correction (Warner-Lambert Inc., Buffalo, NY). The sample was adjusted to 20°C and mixed well before reading. Appropriate acid correction to Brix was done after determining titratable acidity (Redd et al., 1986).

11.2.9 Cloud measurement

The method of Versteeg et al. (1980) was used. Samples of SSOJ were placed in an Eppendorf centrifuge (Brinkman Instr. Inc., Westbury, NY) and centrifuged at $320 \times g$ for 10 min. The supernatant was poured into 3-ml cuvettes, placed in a spectrophotometer (Beckman Inc., Fullerton, CA) calibrated with distilled water, and the absorbance at 660 nm was read.

11.2.10 Ascorbic acid and total acidity measurement

The titration method by Redd et al. (1986) was used for ascorbic acid determination. A 10-ml sample was mixed with 90 ml of distilled water. Three drops of 1% phenolphthalein solution in isopropyl alcohol were added, and the sample was titrated with 0.1 N NaOH. The g acid/100 ml juice was calculated from the following formula:

$$g\ citric\ acid/100\ ml\ juice = (ml\ NaOH)\ (N\ NaOH)\ (0.064)\ (100) \qquad (11.3)$$

Titratable acidity as percent citric acid was measured according to Redd *et al.* (1986).

11.2.11   Color measurement

A Hunterlab Color/Difference meter (Model D25, Hunter Inc., Fairfax, VA) was used to determine L (lightness), a (redness), and b (yellowness) values. The colorimeter was calibrated with a standard white tile. A 25-ml sample of SSOJ was placed in a glass cup and covered with the same standard tile used for calibration.

11.2.12   Microbiological studies

Orange serum agar (Bacto OSA dehydrated, Difco Laboratories, Detroit, MI) was prepared on the day of the experiments. The media was autoclaved and placed in a water bath at 50°C. Butterfield's buffer solution (pH 7.2) was used for serial dilutions of the samples (Speck, 1984). Before each experiment, the Milton Roy system was cleaned with sterile distilled water and alcohol. The vessel was washed with sterile distilled water and placed overnight in an oven at 100°C.

A can of orange juice was thawed, 100 ml of the stock juice was added and mixed thoroughly. A sample was taken to determine the initial microbial count. The spiked juice was stored at 4.4°C. Samples were taken after 120 and 240 min of storage to monitor any change in microbial numbers during the day. Three samples of 100 ml each were taken from the spiked juice. Sample 1 was left at room temperature and aliquots were taken every 15 min up to 1 h to determine the microbial growth. Sample 2 was placed in a water bath at the experimental T as atmospheric pressure (AP) sample. When the juice reached experimental T, CO_2 was bubbled through it. A sample was immediately taken and was considered as the 'time zero sample' for AP. Sample 3 was placed into the supercritical vessel, and the system was allowed to reach experimental T and P. Then a sample was taken and considered as 'time zero sample' for SC-CO_2 treatments. Samples were taken from controls simultaneously with the SC-CO_2 treated juice. Twelve experiments involving four levels of pressure (atmospheric, 8.3, 20.7, and 33.1 MPa) and temperature (room temperature, 35, 45, and 60°C) were performed. All experiments were duplicated or triplicated. Each sample was placed in a sterile test tube and put on ice. After dilution, 0.1 ml of each dilution was placed in four petri dishes and mixed with orange serum agar. Plates were inverted and incubated at room temperature for 3 days before colonies were counted. No attempt was made to differentiate microbial types. Results were analyzed statistically. Kinetic parameters for microbial destruction were evaluated. D values were obtained by plotting the log CFU/ml versus time. z values were calculated by plotting the log of D values against temperature, at each pressure.

11.2.13   Sensory evaluations

About 6 l of SSOJ were thawed, and divided into two portions. The 'untreated' portion was stored at 4°C until served. The second portion was SC-CO₂ treated at 34 MPa and 33°C for 1 h, in two batches. The treated batches were mixed, and brought to 4°C before serving. In addition, a commercial non-pasteurized SSOJ was obtained from a local supermarket, and kept at 4°C until served. Coded juice samples were presented to 30 panelists. Unsalted crackers were provided. Panelists ranked the juices according to preference (best, second, least) based on color, cloud, flavor, aroma, and overall acceptability. The taste test was duplicated with the same panelists, with different coded numbers.

11.3   Results and discussion

11.3.1   Thermal inactivation of PE

An Arrhenius analysis of temperature control data (Figure 11.2) resulted in a calculated frequency factor A of 1.7×10^{24} min^{-1}, and energy of activation E_a of 166.6 kJ/mol. Results confirmed that thermal inactivation of PE followed a first order kinetics. Versteeg *et al.* (1978) reported three isoenzymes of PE. The z values for these were 6.5°C for PE-I and PE-III, and 11°C for PE-II. Wicker and Temelli (1988) also reported values of 6.5°C and 10.5°C for those isoenzymes. The z value calculated in our study was 8.8°C, which is between reported values of the heat stable and heat labile forms of PE. Since no attempt was made to treat the isoenzymes separately in our study, the reported value of 8.8°C represented a 'lump-sum' value of z.

11.3.2   Combined thermal and supercritical inactivation of PE

Figure 11.2 also shows measured PE activity versus time for SC-CO₂ treated juice. The TC samples at low temperature (40°C) had very little PE inactivation, while at the same temperature, the SC-CO₂ treated sample had greater PE inactivation, ranging from 43.2% to 100% beyond temperature control levels. D values at each temperature were calculated for both TC and SC-CO₂ treated samples. When temperature was combined with pressure, D values decreased, from 2673 min at atmospheric pressure to 104.6 min at 31 MPa and 40°C, and from 56.6 min at atmospheric pressure to 10.6 min at 31 MPa and 60°C. The combined effect of supercritical pressure and temperature decreased the time required to inactivate the enzyme at any T. z values were also calculated and found to be 5.2°C with supercritical pressure and 8.8°C at atmospheric pressure. These results suggest that PE is more sensitive to temperature increases under pressure. k values for inactivation increased as pressure was increased from atmospheric to 31 MPa. At atmospheric pressure,

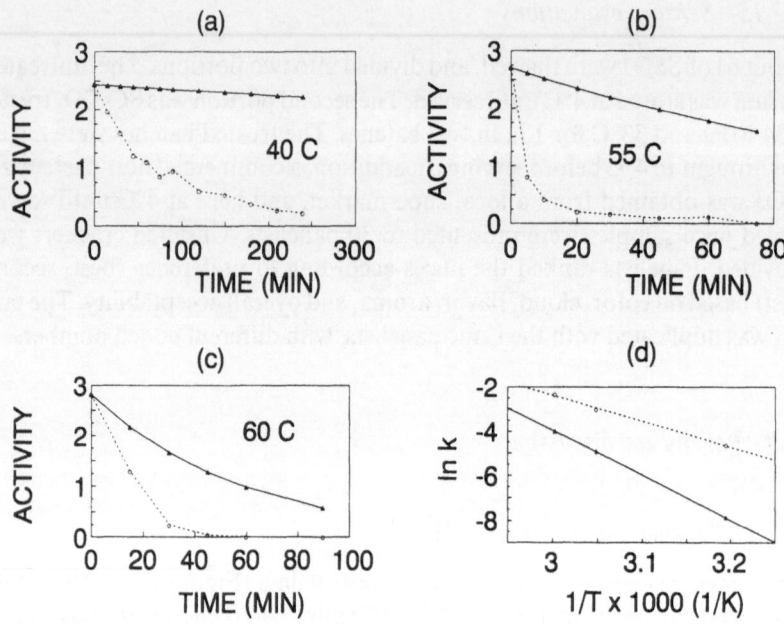

Figure 11.2 Inactivation of pectinesterase in supercritical CO_2 treated Valencia orange juice at 31 MPa and different temperatures, and temperature control samples.

$k = 1.77 \times 10^{-2}$ min^{-1}, at 13.8 MPa, $k = 8.26 \times 10^{-2}$ min^{-1}, and at 31 MPa $k = 9.45 \times 10^{-2}$ min^{-1}. The energy of activation at 31 MPa was 97.4 kJ/mol, suggesting that increasing the pressure at a given temperature would increase the rate of inactivation of PE. Recirculation of CO_2 reduced the time required for inactivation.

11.3.3   Effect of SC-CO_2 on PE activity and cloud stability during storage at 4.4°C

Figures 11.3 and 11.4 show measured PE activity and cloud of TC and SC-CO_2 treated samples during 66 days of storage, for the Milton Roy system. Super-critical conditions of 29 MPa and 50°C for 4 h resulted in complete inactivation of PE. Cloud was retained and increased as much as fourfold. TC samples had some thermal inactivation of PE, however, during storage PE activity increased. This caused the cloud to decrease rapidly to near zero after 66 days of storage. SC-CO_2 treated samples started with zero PE activity and showed some activity after 15 days. This activity increased at a very slow rate during storage. The cause of this increase is not known. The cloud of the SC-CO_2 treated samples was not affected during storage and remained high. These results suggest that cloud is retained, enhanced and stabilized, and that cloud stability is not only related to the activity of PE but also to a possible

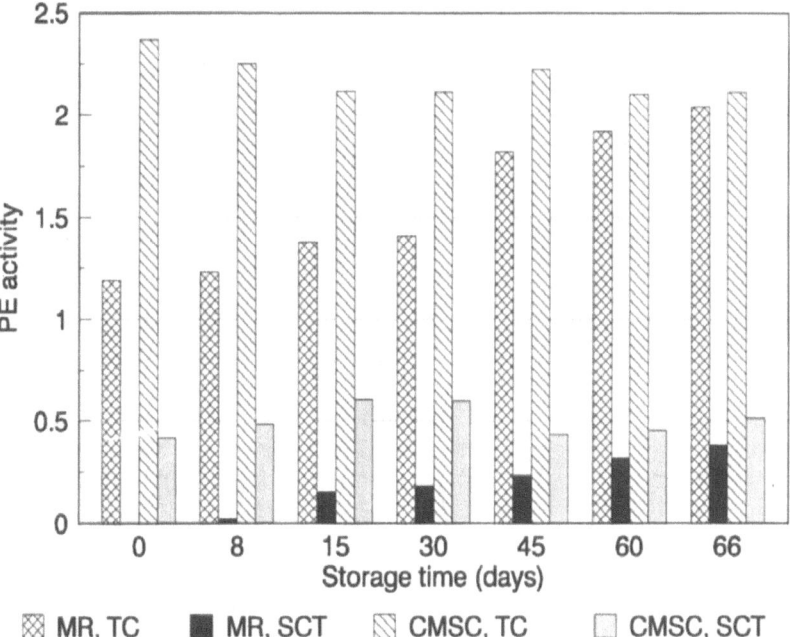

Figure 11.3 PE activity (units/ml) of SC-CO₂ treated (SCT) and temperature control (TC) samples during storage. MR, Milton Roy system: 29 MPa, 50°C, 4 h. CMSC, custom made supercritical system: 29 MPa, 45°C, 2 h.

modification of pectin. During storage, TC samples lost their cloud quickly. SC-CO₂ treated samples retained their enhanced cloud. On day 45, the TC samples gelled while the SC-CO₂ treated samples remained fluid.

Figures 11.3 and 11.4 also show the PE activities and cloud of similar samples treated with the CMSC system. Supercritical conditions were 29 MPa at 45°C for 2 h. After SC-CO₂ treatment, PE was inactivated by 84%, and the cloud was not only retained but increased fourfold. PE activity showed some increase in SC-CO₂ treated samples after 8 days of storage, however, cloud was not affected and remained high. TC samples gelled at 45 days, while the SC-CO₂ treated sample remained fluid at day 66.

11.3.4   Effect of SC-CO₂ on pH and total soluble solids (°Brix)

Table 11.2 shows the measured pH values for the Valencia juice before and after SC-CO₂ treatment, using the CMSC system. There were no significant ($p<0.01$) differences between pH values before and after SC-CO₂ treatment. It was shown that the pH of the juice at 31 MPa and 35°C was lowered by about 0.7 pH units (Balaban *et al.*, 1991). Therefore, although the pH of the juice under pressure decreases (Arreola, 1990), as soon as pressure is released, CO₂ separates from the juice and the pH returns to its original value. However, Owusu-Yaw *et al.* (1988) showed that pH must be lowered to 2.4 for substantial

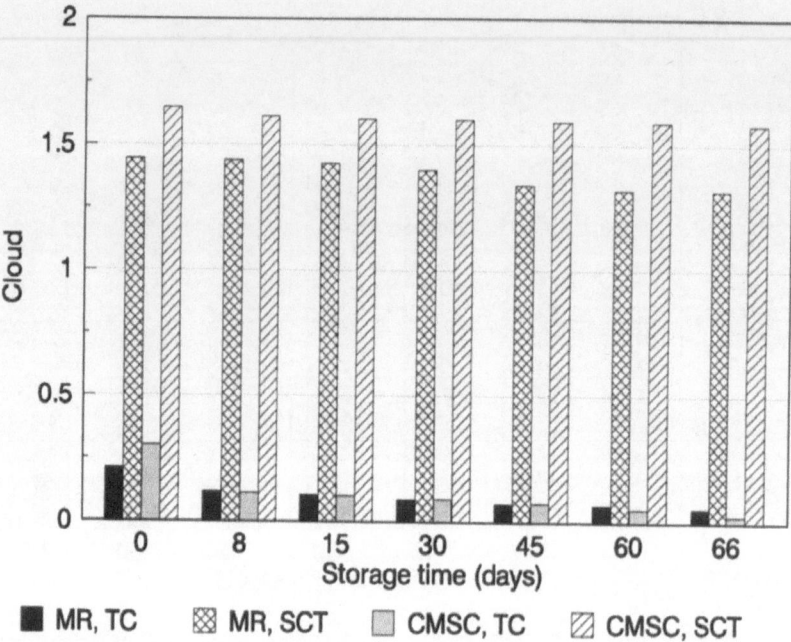

Figure 11.4 Cloud of SC-CO₂ treated (SCT) and temperature control (TC) samples during storage. MR, Milton Roy system: 29 MPa, 50°C, 4 h. CMSC, custom made supercritical system: 29 MPa, 45°C, 2 h. Cloud is absorbance at 660 nm following centrifugation of orange juice at $320 \times g$ for 10 min.

PE inactivation. Thus, combined effects of P, T, pH reduction, possible shear during sample gathering, and time are suggested as the cause of PE inactivation. Table 11.2 also shows that the °Brix values of Valencia SSOJ before and after SC-CO₂ treatment are not significantly different ($p<0.01$). The results of the CCD experiments using the Milton Roy system regarding pH and °Brix were similar.

Table 11.3 shows various attributes of Pineapple SSOJ before and after treatment with SC-CO₂ under different conditions. In this case, a vacuum was applied to the juice to remove residual CO₂ immediately after the experiment. It is apparent that pH and °Brix are not significantly changed due to SC-CO₂ treatment.

11.3.5   Effect of SC-CO₂ on cloud

Table 11.2 shows the measured cloud of the Valencia juice before and after SC-CO₂ treatment for the CMSC system. TC cloud at high T (55°C) was retained up to 60% due to the partial inactivation of PE by heat. However at low T, cloud was retained poorly and in some cases, it was lost completely at the end of experiments. Paired t-tests confirmed that the average difference between cloud before and after SC-CO₂ treatment was large and significant

Table 11.2 Effects of SC-CO$_2$ on pH, °Brix, ascorbic acid and total acidity of orange juice using the CMSC system

Exp	Pressure (MPa)	T (°C)	Time (min)	pH before SC-CO$_2$	pH after SC-CO$_2$	°Brix before SC-CO$_2$	°Brix after SC-CO$_2$	% AA retention of TC[a]	% AA retention of SC-CO$_2$	Total acidity SC-CO$_2$[b]	Total acidity SC-CO$_2$[b]	Cloud before SC-CO$_2$[c]	Cloud after SC-CO$_2$
1	13.1	44	50.0	4.17	4.16	11.3	10.1	71.1	83.8	0.69	0.62	0.301	0.887
2	26.9	44	50.0	4.23	4.16	12.0	10.7	71.1	86.4	0.69	0.63	0.310	0.854
3	13.1	56	50.0	4.16	4.16	12.0	11.6	76.4	94.0	0.92	0.71	0.244	0.820
4	26.9	56	50.0	4.15	4.15	11.4	11.1	76.4	76.4	0.88	0.70	0.308	0.832
5	13.1	44	145.0	4.17	4.17	11.3	10.2	64.2	81.6	0.69	0.55	0.301	0.682
6	26.9	44	145.0	4.23	4.16	12.0	10.8	64.1	85.7	0.69	0.66	0.310	0.395
7	13.1	56	145.0	4.16	4.16	12.0	12.2	78.7	93.5	0.94	0.67	0.244	0.902
8	26.9	56	145.0	4.15	4.15	11.4	11.2	62.9	73.0	0.87	0.60	0.308	0.972
9	8.3	50	97.50	4.19	4.17	12.0	11.7	73.5	76.3	0.72	0.63	0.284	0.689
10	31.7	50	97.50	4.18	4.16	11.9	11.7	62.4	81.8	0.71	0.65	0.262	0.638
11	20.0	40	97.50	4.20	4.19	11.0	11.7	83.0	90.7	0.71	0.70	0.203	0.684
12	20.0	60	97.5	4.20	4.18	12.1	11.3	70.3	70.6	0.70	0.62	0.252	0.523
13	20.0	50	15.0	4.15	4.15	11.7	11.6	78.8	95.2	0.89	0.60	0.276	0.797
14	20.0	50	180.0	4.18	4.17	12.2	12.2	67.5	92.0	0.70	0.67	0.249	0.999
15	20.0	50	97.5	4.18	4.16	11.8	12.1	68.6	91.3	0.60	0.63	0.249	0.973

[a] Temperature control (TC) samples were held at the same experimental T as the SC-CO$_2$ treated samples for the same time period.
[b] Total acidity was expressed in equivalent g citric acid/100 ml juice.
[c] Cloud was measured as absorbance at 660 nm after centrifugation at 320 × g for 10 min.

Table 11.3   Various attributes of Pineapple orange juice before and after SC-CO_2 treatments[a]

	Control	Treatment 1 (35°C, 5.8 MPa, 30 min)	Treatment 2 (40°C, 5.8 MPa, 60 min)
Cloud	0.368	0.858	
pH	3.786	3.77	
Brix	9.067	9.18	
Corrected Brix	9.193	9.310	
Acidity	0.630	0.635	
Color			
L	51.25	51.3	
a	− 6.1	− 6.62	
b	24.46	23.62	
Cloud	0.391	1.110	1.121
pH	3.735	3.742	3.753
Brix	10.07	10.2	10.17
Corrected Brix	10.21	10.35	10.31
Acidity	0.702	0.720	0.718
Color			
L	52.43	53.07	53.72
a	− 5.75	− 6.58	− 6.35
b	25.53	25.1	25.83
Cloud	0.511	1.038	1.076
pH	3.74	3.723	3.72
Brix	10.1	10.11	10.22
Corrected Brix	10.24	10.27	10.36
Acidity	0.702	0.738	0.715
Color			
L	52.5	53.55	53.93
a	− 5.87	− 6.5	− 6.77
b	25.55	25.98	25.82
Cloud	0.321	0.964	1.068
pH	3.68	3.68	3.678
Brix	10.1	10.15	10.23
Corrected Brix	10.19	10.29	10.38
Acidity	0.710	0.715	0.721
Color			
L	52.73	53.45	53.5
a	− 5.87	− 6.37	− 6.4
b	26.02	25.92	25.95
Cloud	0.286	1.095	1.071
pH	3.73	3.75	3.758
Brix	10.0	10.02	10.23
Corrected Brix	10.14	10.16	10.38
Acidity	0.710	0.701	0.720
Color			
L	52.78	53.97	53.07
a	− 5.67	− 6.08	− 6.32
b	26.3	26.32	25.55

[a] Orange juice was thawed under cold running tap water, while in the bag. First sample stirred 10 min, vacuumed 15 min at 500 mm/Hg after treatment. Second sample stirred 10 min, vacuumed 20 min at 700 mm/Hg after treatment. Brix is corrected for acidity.

($p<0.01$). Cloud was retained and enhanced from 1.3 to 4 times by SC-CO_2 treatment regardless of the temperature and time of treatment. There was less cloud enhancement in orange juice drained after the system was depressurized compared to the cloud of samples taken while the system was under pressure.

Cloud of Pineapple SSOJ before and after treatment with SC-CO_2 at different conditions is presented in Table 11.3. There was significant cloud increase due to SC-CO_2 treatment.

11.3.6 Effect of SC-CO_2 on total acidity and ascorbic acid

Table 11.2 shows the total acidity of Valencia orange juice (expressed as g citric acid/100 ml juice) before and after SC-CO_2 treatment, using the CMSC system. Similar results were obtained for the Milton Roy CCD experiments. Significant differences in total acidity occurred following SC-CO_2 treatment as determined by t-test ($p<0.01$). A trend towards a decrease in total acidity after SC-CO_2 treatment was observed. The reason for this was not clear, and this seemed to contradict Kimball's data on total acidity. Further experiments with Pineapple variety showed that there was no significant change in total acidity (Table 11.3).

Table 11.2 shows the changes in ascorbic acid (AA) in Valencia orange juice after SC-CO_2 treatment for the CMSC system. Similar results were obtained for the Milton Roy CCD experiments. The retention of AA in the samples after SC-CO_2 treatment is 71–98%. The percent retention of AA in TC samples was significantly ($p<0.01$) lower than in SC-CO_2 treated samples, as shown by t-tests. This is due to the presence of oxygen and the long exposure to warm T in the TC samples (Trammel et al., 1986). Higher retention of AA in SC-CO_2 treated samples is attributed to the oxygen-free environment during SC-CO_2 treatment. In addition, since ascorbic acid is not soluble in SC-CO_2, it is not extracted.

11.3.7 Effect of SC-CO_2 on color

Table 11.4 shows the measured lightness (L), redness (a) and yellowness (b) values of Valencia orange juice before and after SC-CO_2 treatment, and those of the TC samples. The L value of the TC sample increased as T increased. The a value of the TC sample decreased with low T and remained almost the same with high T. The b value of the TC samples increased with T. The L value of orange juice after SC-CO_2 treatment increased as P was increased. This increase was larger than the increase in lightness for TC samples. The a value of the juice after SC-CO_2 treatment decreased as P increased. Pressure did not have an effect on the yellowness of the SC-CO_2 treated Valencia juice. The yellowness of the SC-CO_2 treated juice decreased with low T.

The effect of SC-CO_2 treatment on color parameters of Pineapple variety

Table 11.4 L, a, and b values of Valencia orange juice before SC-CO$_2$ treatment, after SC-CO$_2$ treatment and temperature control, in the CMSC system.

Exp.	Pressure (MPa)	T (°C)	Time (min)	Before SC-CO$_2$ treatment			Temperature control			After SC-CO$_2$ treatment		
				L	a	b	L	a	b	L	a	b
1	13.1	44	50.0	47.3	−17.1	24.2	47.6	−18.4	35.6	47.9	−20.3	25.8
2	26.9	44	50.0	47.4	−17.2	24.1	47.8	−18.2	25.9	47.0	−21.8	24.3
3	13.1	56	50.0	47.4	−17.2	24.2	48.4	−17.7	26.2	46.1	−23.0	24.0
4	26.9	56	50.0	47.4	−17.0	24.3	48.5	−17.3	26.5	49.2	−19.4	27.0
5	13.1	44	145.0	47.2	−17.0	24.1	48.5	−17.1	25.5	45.7	−23.0	23.5
6	26.9	44	145.0	47.5	−17.0	24.3	47.8	−18.2	25.9	46.7	−22.3	24.6
7	13.1	56	145.0	47.5	−17.4	24.1	49.0	−15.9	26.8	48.5	−18.6	26.3
8	26.9	56	145.0	47.1	−17.1	24.3	48.7	−16.8	26.6	49.3	−16.9	27.1
9	8.3	50	97.50	47.1	−17.1	24.3	48.1	−17.9	26.1	45.9	−22.7	23.5
10	31.7	50	97.50	47.1	−17.0	24.2	48.1	−17.9	26.1	48.0	−19.5	25.8
11	20.0	40	97.50	47.0	−17.1	24.1	47.6	−18.4	25.7	45.8	−23.1	23.8
12	20.0	60	97.5	47.5	−17.3	24.0	49.5	−16.3	27.1	48.3	−22.3	24.0
13	20.0	50	15.0	47.5	−16.9	24.2	48.4	−17.7	26.1	49.5	−18.2	27.1
14	20.0	50	180.0	47.3	−16.7	24.3	48.4	−17.3	26.4	48.1	−19.7	25.9
15	20.0	50	97.5	47.3	−17.0	24.2	48.2	−17.6	26.2	48.8	−19.4	26.6

juice is shown in Table 11.3. A trend of increased L value, and decreased a and b values as a result of SC-CO_2 treatment was observed.

11.3.8   Effects on total microbial plate count

Tables 11.5–11.7 show the TPC of the experiments at 35, 45 and 60°C, respectively. The microbial numbers of the room temperature, AP and SC-CO_2 treated samples from the same batch at 'time zero' were different, since it took different times for each one of the samples to reach the experimental conditions. In addition, there were differences in the initial numbers of microorganisms between experiments performed on different days. The data were analyzed statistically using the following equation:

$$Y = (\log_{10} X_c - \log_{10} X_i)/\log_{10} X_c \qquad (11.4)$$

where X_c is the TPC of the AP sample, and X_i are the TPC of the SC-CO_2 treated samples at different sampling times. The quantity Y represents the decrease in the log of the TPC of the SC-CO_2 treated sample relative to the log of the TPC of the AP sample. A straight line $Y = \beta_o + \beta_1$ (time) was fitted to the data after confirming the absence of a second order relation. An analysis of variance was also performed. Results showed that the slopes of the lines were significantly different from zero, meaning that there was a significant reduction in CFU/ml in treated samples compared to TC.

A t-test was performed to compare the difference between the slopes of the resulting lines. It was apparent that at 35°C, the slopes of the equations for 33.1 and 20.7 MPa were not significantly different ($\alpha = 0.005$). The slopes of the equations for 33.1 and 8.3 MPa were not significantly different at the same level of significance, meaning that at 35°C, pressure does not make a difference in the reduction of the TPC. Numbers were reduced by using SC-CO_2 at 35°C compared to AP, although most lactic acid bacteria grow best between 20 and 37°C (Parish, 1988).

At 45°C, the slopes of the equations for 33.1 and 20.7 MPa were found to be not significantly different, but the slopes of the equations at 33.1 and 8.3 MPa were significantly different ($p<0.005$), meaning that at the 45°C level, the decrease in the log of the TPC is greater at 33.1 MPa than at 8.3 MPa.

At 60°C, the slopes of the equations for 33.1 and 8.3 MPa, and those of 20.7 and 8.3 MPa were found to be not significantly different. However, slopes of equations for 33.1 and 20.7 MPa were significantly different ($p<0.01$), meaning that the decrease in the log of the TPC at 60°C is greater at 33.1 MPa than it is at 20.7 MPa.

The D value of the AP at 45°C was large (241 min) while the D value at 60°C was very low (14.7 min). This shows the lethal effect of T on the microflora of the juice. The D values at 35, 45 and 60°C of the SC-CO_2 treated samples at 33.1 MPa were 28, 22.6, and 12.7 min, respectively. These results showed that even at low T, significantly shorter times are needed to reduce

Table 11.5 Effects of SC-CO_2 on total plate count (colony forming units/milliliters) at 35°C at different pressures

Time (min)	Room temperature sample	AP[a]	33.1 MPa	20.7 MPa	8.3 MPa
0	2.6×10^4	4.2×10^4	2×10^4	1×10^5	1.9×10^5
	4.0×10^4*	5.1×10^4			
	4.1×10^4*	5.6×10^4	6.8×10^3	1.3×10^4	3.4×10^4
15	1.8×10^4	3.9×10^4	1.1×10^4	3.2×10^4	6.9×10^4
	3.1×10^4*	4.0×10^4			
	5.6×10^4*	5.8×10^4	5.6×10^3	1.2×10^3	1.3×10^4
30	4.2×10^4	1.5×10^5	2.5×10^3	1.4×10^4	1.1×10^4
	4.4×10^4*	3.7×10^4			
	3.9×10^4*	5.1×10^4	580	850	8.4×10^3
45	9.6×10^4	1.6×10^5	550	660	8.6×10^4
	5.6×10^4*	4.5×10^4			
	3.7×10^4*	4.4×10^4	490	570	620
60	2.5×10^5	1.9×10^5	110	105	1.2×10^3
	4.0×10^4*	2.3×10^4			
	5.4×10^4*	2.5×10^4	108	260	320
120	5.6×10^4*				
	2.4×10^4*				
240	4.1×10^4*				
	4.8×10^4*				
D (min)		242.7	28	40.5	27.9

[a] Atmospheric pressure CO_2 treatment.
*Duplicate experiments performed on different days

Table 11.6 Effects of SC-CO_2 on total plate count (colony forming units/milliliters) at 45°C at different pressures

Time (min)	Room temperature sample	AP[a]	33.1 MPa	20.7 MPa	8.3 MPa
0	3.2×10^4	2.5×10^4	8.6×10^3	3.5×10^4	3.3×10^4
	4.0×10^4*	4.2×10^4	1.6×10^4	1.4×10^4	2.8×10^4
15	3.1×10^4	2.6×10^4	2.9×10^3	2.8×10^4	2.4×10^4
	3.1×10^4*	3.2×10^4	3.0×10^3	2.7×10^3	3.2×10^4
30	3.8×10^4	2.7×10^4	170	2.4×10^4	6.5×10^3
	4.4×10^4*	3.0×10^4	750	380	1.0×10^4
45	4.1×10^4	1.7×10^5	27	1.9×10^4	6.5×10^3
	5.6×10^4*	2.2×10^4	95	37	1.3×10^3
60	6.4×10^5	1.9×10^4	15	1.6×10^4	1.0×10^4
	4.0×10^4*	2.1×10^4	66	23	380
120	3.9×10^4				
	5.6×10^4*				
240	3.9×10^4				
	4.1×10^4*				
D (min)		241	22.6	20.2	29.3

[a] Atmospheric pressure CO_2 treatment.
*Duplicate experiments performed on different days

Table 11.7 Effects of SC-CO$_2$ on total plate count (colony forming units/milliliters) at 60°C at different pressures

Time (min)	Room temperature sample	AP[a]	33.1 MPa	20.7 MPa	8.3 MPa
0	5.7×10^4 $6.4 \times 10^{4*}$	2.5×10^3 110	7.5×10^3 1.5×10^3	2.0×10^4 1.2×10^4	2.8×10^4 6.6×10^4
15	5.6×10^4 $4.4 \times 10^{4*}$	27 0	89 50	380 1.4×10^3	700 2.1×10^4
30	6.3×10^4 $5.7 \times 10^{4*}$	0 0	2 24	190 860	60 960
45	6.6×10^4 $5.3 \times 10^{4*}$	0 0	0 0	130 730	34 250
60	7.0×10^4 $5.9 \times 10^{4*}$	0 0	0 0	110 560	35 160
120	8.2×10^4 $6.4 \times 10^{4*}$				
240	1.0×10^5 $6.2 \times 10^{4*}$				
D (min)		14.7	12.7	27	20.7

a Atmospheric pressure CO$_2$ treatment.
*Duplicate experiments performed on different days.

TPC of microorganisms compared to the AP at the same temperature. The D values at 35, 45, and 60°C of the SC-CO$_2$ treated samples at 20.7 MPa were 40.5, 20.2, and 27 min, respectively. The value of D at 60°C is unexpected, because higher T should reduce the value of D compared to that at 45°C. It is suspected that this discrepancy is due to the mixed microbial population in the juice. The combination of different pressure, pH and temperature levels could affect various microorganisms differently. The D values at 35, 45, and 60°C of the SC-CO$_2$ treated samples at 8.3 MPa were 27.9, 29.3, and 20.7 min, respectively. Again, the D at 45°C was expected to be between 27.9 and 20.7 min. The z value at 8.3 MPa was calculated as 180°C, the z value at 20.7 MPa was 167°C, and that at 33.1 MPa was 72°C.

The growth rate of *Lactobacillus brevis* and *Leuconostoc mesenteroides* in orange juice at 21°C decreases by 200% when the pH is changed from 4.0 to 3.4, resulting in significant increases in generation times (Parish, 1988). Since high pressure CO$_2$ causes a pH decrease in orange juice, this may partially explain the changes in TPC at different pressures. The decrease in the z values with increasing pressure indicates that pressure is 'sensitizing' the general microflora of orange juice to the effects of temperature. The calculated D and z values could be used in the prediction of the total microbial reduction in SC-CO$_2$ treated orange juice.

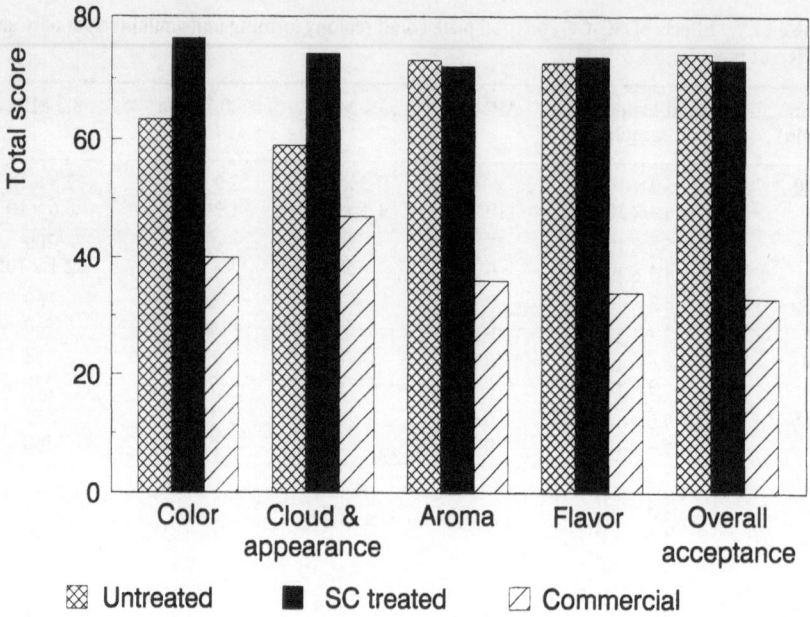

Figure 11.5   Results of sensory evaluation tests. Total score is the sum of 30 evaluation results (3, best; 2, second, 1, least liked). Average of two evaluations.

11.3.9   Sensory evaluation tests

Figure 11.5 shows the results of the sensory evaluation tests. Data were analyzed by Duncan's multiple range test at the 5% level. Judges preferred the color and cloudiness of the $SC\text{-}CO_2$ treated juice over that of the untreated juice, and that of the commercial juice. There was no significant ($p < 0.01$) difference between the treated and untreated juices for flavor, aroma and overall acceptability. The commercial juice was the least liked in all tests. This suggests that even a $SC\text{-}CO_2$ treatment of 1 h at 34 MPa and 33°C does not seem to have an adverse effect on the sensory attributes of SSOJ.

11.4   Summary and conclusions

$SC\text{-}CO_2$ treatment of orange juice temporarily reduces the pH of the juice. This and other possible factors such as pressure, shear, and effects of CO_2 inactivates PE. The extent of inactivation depended on P, T and time. Higher P, T and time resulted in more inactivation. High pressure inactivation was possible at temperatures where thermal inactivation was insignificant. The activation energy for the 'reaction' was reduced significantly when the pressure was increased to 31 MPa from atmospheric pressure. At a given T, increasing P resulted in decrease in the D value for inactivation.

After depressurization, pH was restored to its original value. There was no change in the °Brix before and after treatment. More ascorbic acid was retained during SC-CO_2 treatment than temperature controls at the same conditions. Total acidity was also preserved. Cloud was enhanced and stabilized, even in the presence of residual PE, by SC-CO_2 treating and shearing the juice at high pressure through small openings during depressurization. Sensory quality of the juice is not adversely affected by the SC-CO_2 treatment, and color and cloud scores are actually improved. As an added benefit, SC-CO_2 treatment of single strength orange juice resulted in reduction of microflora of the juice even at low T. A combination of high pressure, shear to which the juice is subjected during depressurization, and lower pH due to the formation of carbonic acid could have inhibitory effects on the normal flora of orange juice. D values for microbial reduction at the same T were generally lower at higher pressures. z values decreased as pressure increased.

This method offers many potentially beneficial processing avenues for orange juice and for other juices, especially in the area of minimally processed, limited shelf-stable products.

Acknowledgements

Financial support and raw materials for this study were provided by the FMC Corporation, Santa Clara, California, and by Citrus World, Inc., Lake Wales, Florida. Their assistance is greatly appreciated.

References

Arreola, A. G. (1990) Effects of supercritical carbon dioxide on some quality attributes of single strength orange juice, M.Sc. Thesis, University of Florida, Gainesville, FL.

Baker, R. A. (1979) Effect of pH on clarification of citrus juices by low methoxyl pectins. *Proc. Fla. State Hortic. Soc.* **92**, 156.

Baker, R. A. and Bruemmer, J. H. (1972) Pectinase stabilization of orange juice cloud. *J. Agric. Food Chem.* **20**, 1169.

Balaban, M. O., Arreola, A. G., Marshall, M. R., Peplow, A., Wei, C. I. and Cornell, J. (1991) Inactivation of pectinesterase in orange juice by supercritical CO_2. *J. Food Sci.* **56**, 743.

Crandall, P. G., Matthews, R. F. and Baker, R. A. (1983). Citrus beverage clouding agents—review status. *Food Technol.* **37**, (12), 106.

Dziezak, I. D. (1989) New process concentrates juices, preserving "fresh notes". *Food Technol.* **43**(10), 148.

Faville, L. W., Hill, E. C. and Parish, E. C. (1951) Survival of microorganisms in concentrated orange juice. *Food Technol.* **5**, 33.

Haas, G. J., Prescott, H. E., Dudley, E., Dik, R., Hintlian, C. and Keane, L. (1989) Inactivation of microorganisms by carbon dioxide under pressure. *J. Food Safety* **9**, 253.

Hubert, P. and O. G. Vitzthum (1980) Fluid extraction of hops, spices, and tobacco with supercritical gases, in *Extraction with Supercritical Gases*, eds. G. M. Schneider, E. Stahl and G. Wilke, Verlag Chemie Deerfield Beach, FL.

Kalra, H., Chung, S. Y. K. and Chen, C. J. (1987) Phase equilibrium data for SC extraction of lemon flavors and palm oils with CO_2. *Fluid Phase Equil.* **36**, 263.

Kamihira, M., Taniguchi, M. and Kobayashi, T. (1987) Sterilization of microorganisms with supercritical carbon dioxide. *Agric. Biol. Chem.* **2**, 407.

Kimball, D. A. (1987) Debittering of Citrus juices using S.C. CO2. *J. Food Sci.* **52,** 481.

Molin, G. (1983) The resistance to carbon dioxide of some food related bacteria. *Eur. J. Appl. Microbiol. Biotechnol.* **18,** 214.

Novak, R. A. and Robey, R. J. (1988) Supercritical fluid extraction of flavoring materials: design and economics, presented at the AIChE 1988 annual meeting in Washington, DC.

Owusu-Yaw, J., Marshall, M. R., Koburger, J. A. and Wei, C. I. (1988) Low pH inactivation of pectinesterase in single strength orange juice. *J. Food Sci.* **53,** 504.

Parish, M. E. (1988) Microbiological aspects of fresh squeezed citrus juice, in *Ready to Serve Citrus Juices and Juice Added Beverages,* ed. R. F. Matthews, Food Industry Short Course Proc. IFT Fla. Section, p. 79.

Redd, J. B., Hendrix Jr., C. M. and Hendrix, D. L. (1986) *Quality Control for Citrus Processing Plants,* Intercit, Inc., Safety Harbor, FL.

Rizvi, S. S. H., Benado, A. L., Zollweg, J. A. and Daniels, J. A. (1986a). Supercritical fluid extraction: fundamental principles and modeling methods. *Food Technol.* **40**(6), 55.

Rizvi, S. S. H., Daniels, J. A., Benado, A. L. and Zollweg, J. A. (1986b) Supercritical fluid extraction: operating principles and food applications. *Food Technol.* **40**(7) 57.

Rouse, A. H. and Atkins, C. D. (1952) Heat inactivation of pectinesterase in citrus juices. *Food Technol.* **6**(8), 291.

Rouse, A. H. and Atkins, C. D. (1955) Pectinesterase and pectin in commercial citrus juices as determined by methods used at the Citrus Experiment Station, *Univ. Fla. Agric. Exp. Stn. Bull.* 570.

Rushing, N. B., Veldhuis, M. K. and Senn, V. J. (1956). Growth rates of *Lactobacillus* and *Leuconostoc* species in orange juice as affected by pH and juice concentration. *Appl. Microbiol.* **4,** 97.

Speck, M. L. (1984) *Compendium of Methods for the Microbiological Examination of Foods,* Am. Publ. Health Assoc. Washington, DC.

Taniguchi, M., Kamihira, M., and Kobayashi, T. (1987) Effect of treatment with supercritical carbon dioxide on enzymatic activity. *Agric. Biol. Chem.* **2,** 593.

Temelli, F. (1987) $SC-CO_2$ extraction of terpenes from cold pressed valencia orange oil, Ph.D. dissertation, University of Florida, Gainesville.

Temelli, F., Chen, C. S. and Braddock, R. J. (1988) $SC-CO_2$ extraction in citrus oil processing. *Food Technol.* **8,** 145.

Termote, F., Rombouts, F. M. and Pilnik, W. (1977) Stabilization of cloud in pectinesterase active orange juice by pectic acid hydrolysates. *J. Food Biochem.* **1,** 15.

Trammel, D. J., Dalsis, D. E. and Malone, C. T. (1986) Effect of oxygen on taste, ascorbic acid loss and browning for HTST-pasteurized, single strength orange juice. *J. Food Sci.* **51,** 1021.

Varsel, C. (1980) Citrus juice processing as related to quality and nutrition, in *Citrus Nutrition and Quality,* eds. S. Nagy and J. A. Attaway, ACS Symp. series, No. 143, ACS, Washington, DC.

Versteeg, C., Rombouts, F. M., and Pilnik, W. (1978) Purification and some characteristics of two pectin esterase isoenzymes from orange. Lebensm.-Wiss. Technol. **11,** 267.

Versteeg, C., Rombouts, F. M., Spaansen, C. H. and Pilnik, W. (1980) Thermostability and orange juice cloud destabilizing properties of multiple pectinesterases from orange. *J. Food Sci.* **45,** 969.

Wei, C. I., Balaban, M. O., Fernando, S. and Peplow, A. J. (1991) Bacterial effect of high pressure CO_2 treatment on foods spiked with *Listeria* or *Salmonella. J. Food Protection* **54,** 189.

Wicker, L. and Temelli, F. (1988) Heat inactivation of pectinesterase in orange juice pulp. *J. Food Sci.* **53,** 162.

12 Supercritical fluid carbon dioxide technology for extraction of spices and other high value bio-active compounds

K. UDAYA SANKAR

Abstract

Supercritical fluid (SCF) carbon dioxide was used for extraction of pepper (*Piper nigrum* L.) and ginger (*Zingibere officinale* R.) at a cross-section of pressures from 16 to 32 MPa and temperatures from 40 to 70°C. Experiments were conducted as per the Box–Behnken model with three variables: pressure, temperature and flow rate at three different levels. Multiple linear regression equations for the overall yield of oleoresin, piperine and non-volatile extract were obtained. In the case of ginger, a multilinear regression equation was obtained for the yield of oleoresin with temperature and pressure at constant flow rate. The significance of pressure and temperature were discussed in relation to the yield of oleoresin. The quality of the oleoresin was found to be superior by physico-chemical and sensory characteristics. It was found that by suitable variation of pressure and temperature, the essential oils of pepper and ginger can be obtained without much contamination of non-volatile matter. The essential oils obtained at 8, 10, 12 MPa and at temperatures of 40, 50, 60°C were examined and found to contain higher amount of high boiling volatile components compared to steam distilled oil. Jasmine absolutes were produced from jasmine concrete by carbon dioxide extraction. The absolutes thus obtained were clear and compare well organoleptically with those produced using alcohol. Extraction of saponified rice (*Oryza sativa*) wax using supercritical carbon dioxide was found to result in the fractionation of fatty alcohols. The removal of undesirable sterol from the dry mouldy bran (DMB) in the solid state fermentation (SSF) process for purification of gibberellin A_3 is indicated.

12.1 Introduction

Supercritical fluids as 'solvents' or 'magic fluids' have been in the forefront of research in the last two decades. The attention they have been receiving is reflected in the number of symposia held (Wilke *et al.*, 1981; Paulaitis *et al.*, 1983; Penninger *et al.*, 1985; Squires and Paulaitis, 1987), books written (Stahl *et al.*, 1987; Perrut, 1988), the innumerable patents filed and scores of articles published. The earlier promise of SCF technology as an 'energy conservation' alternative and an 'exciting technology to take off' (Paul and Wise, 1971), did not take the industry by storm. Nevertheless, its rise to prominence as a separation technology in the food industry is evident from the establishment of a number of plants for extraction of hops (Hopfen Extraction GmbH, Munchenmuster, 50,000 Mtons/year; HVG Barth, Raiser & Co, 20 000 Mtons/year; Carlton & United Breweries, Melbourne; Hops Extraction Co., Yakuma, Washington; Wohlnzach, Pfizer Inc., New York); for decaffeination of coffee (HAG Bremen, Germany; Krafts General Foods, Texas; Supercritical Processing Inc., Allentown), for isolation of flavours and pyrethrin (Pauls & White, Reigate; Camilli & Loule, Grasse; SKW Trostberg, Germany; Botanical Resources, Emeryville) (Parkinson and Johnson, 1989).

The use of carbon dioxide as solvent is the main attraction in this technology. Carbon dioxide does not leave any harmful solvent residues, meets the higher product quality demands, creates no environmental hazards and finally has lower energy and operating costs. These advantages could overcome the high capital investment and inherent risk in the use of high pressures. The high capital investment cost has directed the efforts of research into the areas where the product costs are advantageous such as spices, high value bio-active compounds, etc.,

Use of SCF or liquid carbon dioxide for flavours and spices has been well reviewed (Hubert and Vitzthum, 1978; Calame and Steiner, 1981; Caragay, 1981; Coenen and Kriegel, 1984; Stahl and Gerard, 1985; Rizvi *et al.*, 1986a, b; Moyler and Heath, 1988; Richard *et al.*, 1989). Liquid carbon dioxide extracted ginger oils, vanillin absolute and clove bud oil have been found to be rich in flavour and find use in specialised applications (Moyler and Heath, 1988). Systematic study has been reported on extraction of cinnamon (Tateo and Chizzini, 1989), vanilla beans (Vidal *et al.*, 1989a), pepper (Vidal and Richard, 1987), rosemary (Tateo *et al.*, 1988) and clove buds (Tateo *et al.*, 1988) using liquid or SCF carbon dioxide with or without entrainers.

Realising the importance of the SCF extraction technology for foods with carbon dioxide, a systematic study on the extraction of spices and other bio-active compounds was undertaken at the Central Food Technological Research Institute (CFTRI), Mysore. This chapter highlights the results of some of these experiments carried out at CFTRI.

12.2 Pepper oil

Pepper oil, generally produced by steam distillation of coarse powder of pepper corns (*Piper nigrum* L.), is used in pharmaceutical preparations and

Table 12.1 Physical and chemical characteristics of pepper oil obtained by steam distillation and SC-CO₂

	Steam distillation	SC-CO₂ 10 MPa, 40°C	SC-CO₂ 10 MPa, 60°C
Yield of oil (%)	2.42	2.83	2.62
Refractive index	1.479	1.483	1.483
Specific gravity	0.8834	0.8859	0.8759
Optical rotation	– 6.74	+0.13	+0.35
Total terpene hydrocarbons (%)	84.32	82.72	78.6
Oxygenated hydrocarbons (%)	3.24	6.69	7.90
Monoterpenes (%)	73.19	61.01	51.45
Sesquiterpenes (%)	26.7	38.99	48.55

flavourings. The oil, rich in top note, is very much preferred in the trade and commerce (Verghese, 1989). The use of SC carbon dioxide (SC-CO₂) for obtaining essential oil raised doubts on the recovery of the volatile components and contamination with the non-volatile constituents. But by employing suitable pressure and temperature at specific conditions, it has been possible to obtain an oil rich in sesquiterpene hydrocarbon fraction, suitable for tailoring the oleoresin. The recovery of the oil is complete and its physicochemical characteristics are found to be superior to the oil obtained by steam distillation. The former has a higher amount of sesquiterpenes, oxygenated fraction and better sensory qualities (Sankar, 1989). The physical constants and the chemical constituents of the oil obtained by the two methods are given in Table 12.1.

12.3 Pepper oleoresin

Black pepper oleoresin is obtained by homogenisation of the pepper oil with the solvent extracted resin. In the two-stage process of oleoresin manufacture, degradation of piperine, an active principle of the oleoresin, has been found to occur (Verghese, 1989) during distillation of the oil and subsequent removal of solvent from the solvent extract. The problem of formation of mesityl oxide during use of acetone for spice extraction and the resultant catty odour in meat formulations has narrowed down the choice of solvent to either ethanol or carbon dioxide (Paginton, 1983). Pepper oleoresin obtained by SC-CO₂ is golden yellow in colour and rich in aroma of pepper. The recovery of piperine has been found to be 99%. The piperine content of the final oleoresin has been more than 50%, showing a complete recovery of piperine with little loss in extraction.

Table 12.2 Parameters of the quadratic polynomial coefficients

	A_0	A_1	A_2	A_3	A_{11}
Oleoresin	85.503	− 2.794	19.582	3.426	− 3.875
Volatile oil	98.989	0.273	2.828	1.809	0.356
Non-volatile extract	77.153	− 4.431	26.813	4.055	− 5.423
Piperine	89.735	− 5.262	26.766	4.281	− 6.322

	A_{22}	A_{33}	A_{12}	A_{13}	A_{23}
Oleoresin	− 18.774	− 5.555	3.489	− 2.232	2.748
Volatile oil	− 2.212	− 1.999	0.387	0.362	− 2.77
Non-volatile extract	− 25.139	− 5.707	4.499	− 3.545	5.246
Piperine	− 29.456	− 11.201	7.196	− 4.83	6.995

Table 12.3 Experimental and theoretical predicted yields for oleoresin, volatile oil, non-volatile extract and piperine for various experimental conditions

Experimental conditions			Oleoresin	
Temperature (°C)	Pressure (MPa)	Flow rate (cm/min)	Exp. (%)	Theor. (%)
40	16	2.4	59.74	47.85
40	24	1.2	66.1	68.64
40	24	3.6	73.0	82.19
40	32	2.4	76.4	76.55
55	16	1.2	30.3	40.91
55	24	2.4	87.2	85.5
55	32	3.6	95.0	86.93
55	16	3.6	40.0	42.27
55	32	1.2	77.7	74.58
70	16	2.4	30.0	29.0
70	24	1.2	77.0	66.96
70	32	2.4	64.2	78.63
70	32	3.6	79.3	75.9
Correlation coefficient			0.9067	

The oleoresin yield $(OY) = 85.503 - 2.794T + 19.582P + 3.426F - 3.875T^2 - 18.774P^2 - 5.555F^2 + 3.489PT - 2.232/TF + 2.748PF$ (where $T = T^* - 55)/10°C$; $P = (P^* - 24)/8$ MPa; $F = (F^* - 2.4)/1.2$ cm/min).

To optimise the process conditions, a statistical method based on the Box–Behncan model, has been chosen (Box and Behncan, 1983). A multilinear regression equation has been derived for the yield of oleoresin, non-volatile extract and piperine with a correlation coefficient greater than 0.95 for second

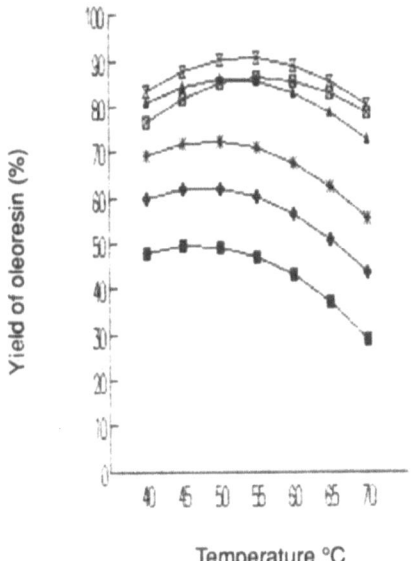

Figure 12.1 Extraction of oleoresin with temperature (linear velocity 2.4 cm/min). ■, 16 MPa; +, 18 MPa; *, 20 MPa; ▲, 24 MPa; Ⴟ, 28 MPa; ☐, 32 MPa.

degree interaction of pressure, temperature and flow rate (Table 12.2). These calculations are based on the experimental values obtained at pressures of 16, 24 and 32 MPa, temperatures of 40, 55 and 70°C and at interstitial velocities of 1.2, 2.4 and 3.6 cm/min (Sankar, 1991). The experimental and theoretical values are presented in Table 12.3. Figure 12.1 gives the extraction yield of oleoresin for different temperatures at the interstitial velocity of 2.4 cm/min of carbon dioxide. Based on the above studies, the optimum conditions of extraction are found to be 28 MPa and 55°C with an interstitial velocity of 2.4 cm/min. The crystals of oleoresin are of micrometre size and can be easily mixed homogenously. Figure 12.2 gives comparative sizes of the crystals in solvent and carbon dioxide extracted oleoresin.

12.4 Ginger oil

Conventionally, ginger oil is obtained by steam distillation from rhizomes of *Zingiber officinale* Roscoe (Mathew, 1973). The distillation time for recovery of ginger oil is longer than in the case of cardamom and pepper due to the presence of a large proportion of sesquiterpenes (Lawrence, 1984). The liquid carbon dioxide extraction technique reduces the adverse effect due to moist heat damage (Tuley, 1985; Moyler and Heath, 1988), but the extract contains up to 66% of non-volatile matter (Chen and Ho, 1988). Using relatively mild conditions of SC-CO$_2$ at 8, 10, 12 MPa of pressure and 40, 50 and 60°C, it is possible to extract the volatile oil of ginger without much contamination of

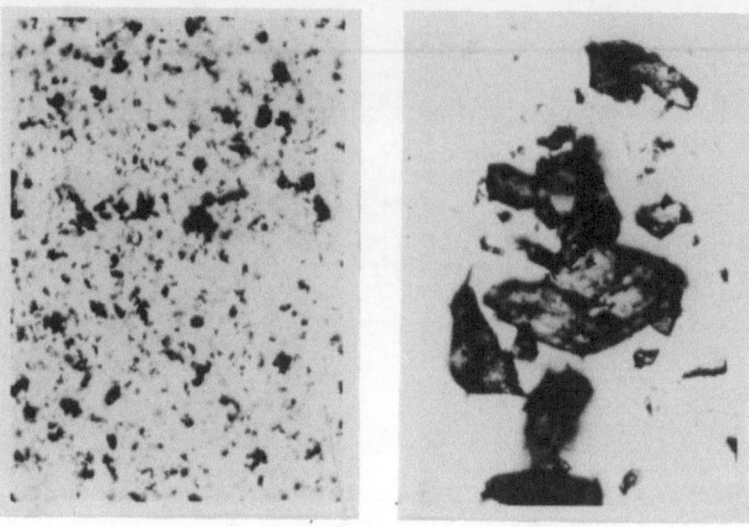

Figure 12.2   The crystals of piperine obtained by (A) carbon dioxide, (B) solvent extraction.

non-volatile pungent principles. The physico-chemical characteristics compare favourably with ginger oil obtained by steam distillation (Sankar, 1993). The monoterpene hydrocarbon content is less than in steam distilled oils. The $(-)$-α-zingiberene alone accounts for more than 24.36% of the total oil, while the sesquiterpene alcohols and the oxygenated fraction form 21–30%. In comparison, the steam distilled oil contains the $(-)$-α-zingiberene and oxygenated fraction at 21.77% and 14.8%, respectively. The ginger oil obtained at 10 MPa and 60°C of carbon dioxide is found to be superior by chemical characteristics such as a higher amount of sesquiterpene hydrocarbons, lesser percentage of monoterpene hydrocarbons and a higher quantity of oxygenated fraction and better sensory characteristics (Table 12.4).

Table 12.4   Physical constants of ginger oil obtained by SC-CO_2 and steam distillation

	Steam distillation	SC-CO_2 10 MPa, 60°C	SC-CO_2 12 MPa, 60°C
Yield of oil (%)	1.86	1.71	2.02
Refractive index	1.480	1.497	1.496
Specific gravity	0.8806	0.8904	0.8972
Optical rotation	-28.4	-43.50	-40.50
Evaporation residue (%)	30.23	34.07	32.18
Total terpene hydrocarbons (%)	79.66	69.62	68.22
Oxygenated hydrocarbons (%)	6.37	8.10	10.39
Esters and unknown alcohols (%)	8.50	16.63	18.67
Monoterpenes (%)	19.20	11.13	6.82
Sesquiterpenes (%)	60.38	58.48	61.40

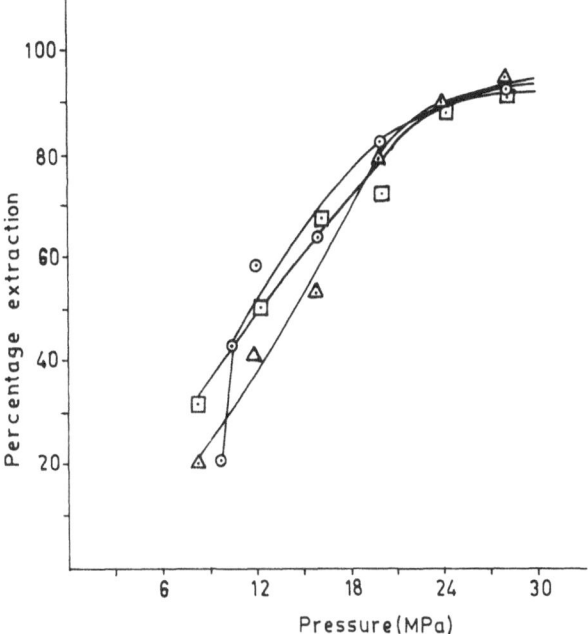

Figure 12.3 Extraction yield of ginger oleoresin with temperature. ○, 40°C; △, 50°C; □, 60°C.

12.5 Ginger oleoresin

Ginger oleoresin contains the active hot principles known as 6- or 8- or 10-gingerols. These organic compounds are thermally labile (McHale *et al.*, 1989) and transform themselves into shogaols. In liquid carbon dioxide extraction of ginger, the percentage of gingerols in the oleoresin has been reported as 15.95% with traces of shogaols (Chen *et al.*, 1986), whereas the SC-CO$_2$ extracts showed more than 18–24% of gingerols, depending on the process time. Extraction of more than 98% of gingerols was feasible from ginger. The oleoresin is a dark yellow viscous liquid with highly characteristic natural ginger flavour. The oil obtained from the oleoresin found to have an optical rotation of − 48, contains more than 60% sesquiterpene hydrocarbons and the oxygenated fraction around 22%. The yield of oleoresin has been found to fit a second order equation at constant mass flow rate conditions (pressure (*P*) in bar and temperature (*T*) in °C). The multiple regression equation for the overall yield (OY) of oleoresin is

$$OY = 129.53 + 0.657P - 5.75T - 0.001P^2 + 0.52T^2 + 0.001PT \quad (12.1)$$

at mass flow rate of 6 kg/h (Sankar, 1993). Figure 12.3 presents the yield of oleoresin versus temperature at various pressures. Increasing the pressure of extraction resulted in increased yields. Increasing temperature at lower pressures resulted in higher yields but at higher pressures the increase is not substantial.

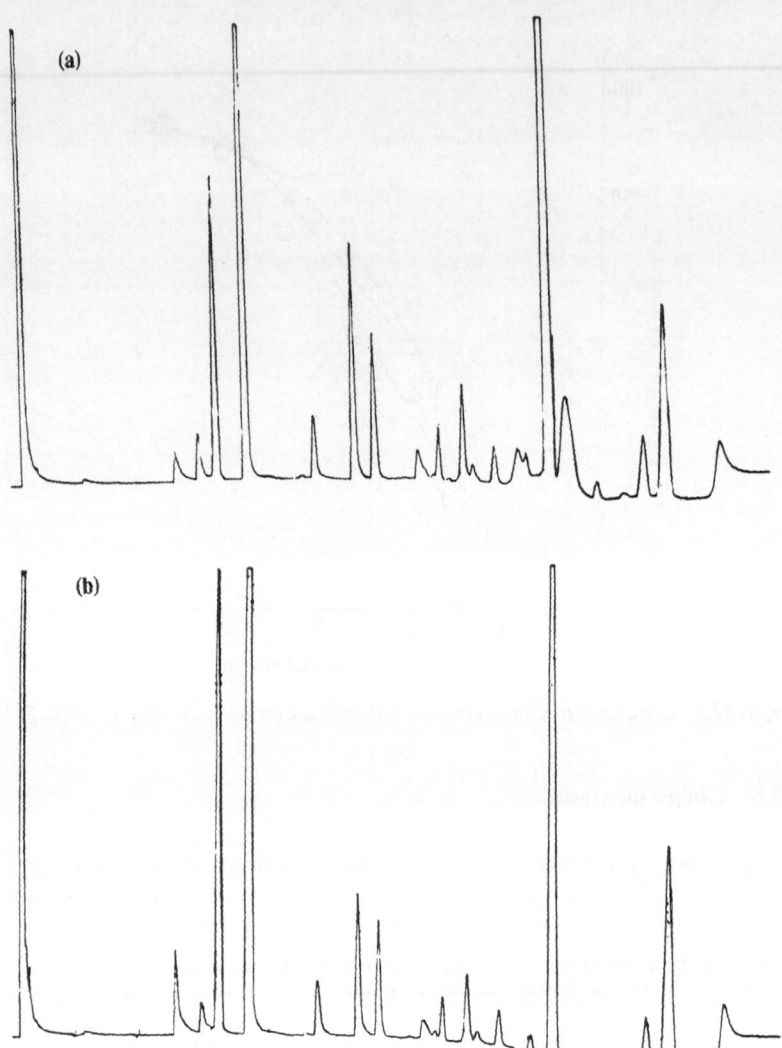

Figure 12.4 Gas chromatograms of the jasmine absolute obtained by (a) carbon dioxide and (b) alcohol.

12.6   Jasmine absolute

The jasmine absolute is undoubtedly one of the most important floral ingredients used in the perfumery industry. Different techniques such as enfleurage
and solvent extraction are used for making 'pomades' or 'concretes' rich in
jasmine odour. Then the plant waxes are separated from the concrete by
extraction with absolute ethanol followed by chilling to obtain 'absolute' of
jasmine (Muller, 1965).

The jasmine absolute is obtained by extraction of the concrete with SC-CO$_2$

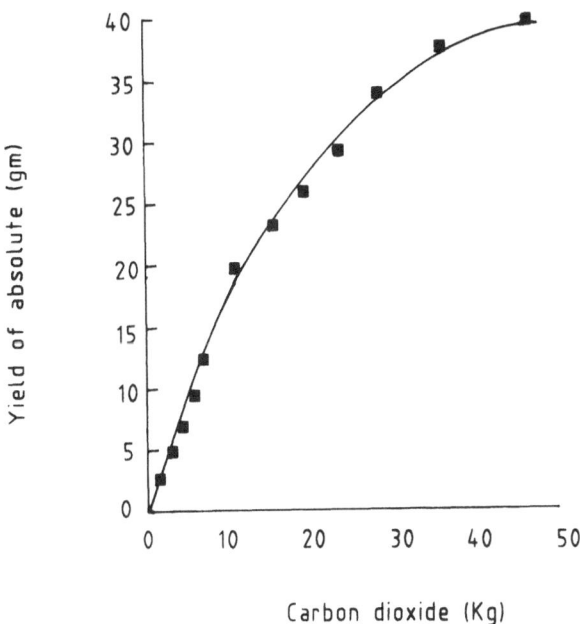

Figure 12.5   Extraction yield of jasmine absolute with carbon dioxide.

at 10 MPa and 50°C. The absolute thus obtained is rich in top notes of jasmine. The gas chromatographs of the absolutes obtained by the two methods are presented in Figure 12.4. The chromatograms indicate that the oils are qualitatively similar although the concentration of the individual components differs. The chemical composition of the jasmine absolute has been studied and found to contain 130 different compounds (Polak, 1973; Lawrence, 1977). The effect of quantity of the carbon dioxide used on the yield of jasmine absolute is depicted in Figure 12.5.

12.7   Bio-active compounds

12.7.1   Fatty alcohols from saponified rice bran wax

Newer products are being tried as plant growth regulators (Louis, 1978). N-Triacontanol, isolated from tea wax (Rao et al., 1987), has been shown to be a plant growth regulator for lucerne (alpha-alpha), tomato, cucumber, lettuce, rice and maize (Ries and Houtz, 1983). There are several problems in isolating the long chain fatty alcohols in pure form. The waxes are saponified after deoiling and subjected to SC-CO$_2$ extraction yielding relatively pure fatty alcohols. The chromatogram of the acetate derivatives of the alcohols presented show that it is devoid of C-28 alcohol (Figure 12.6). The effect of

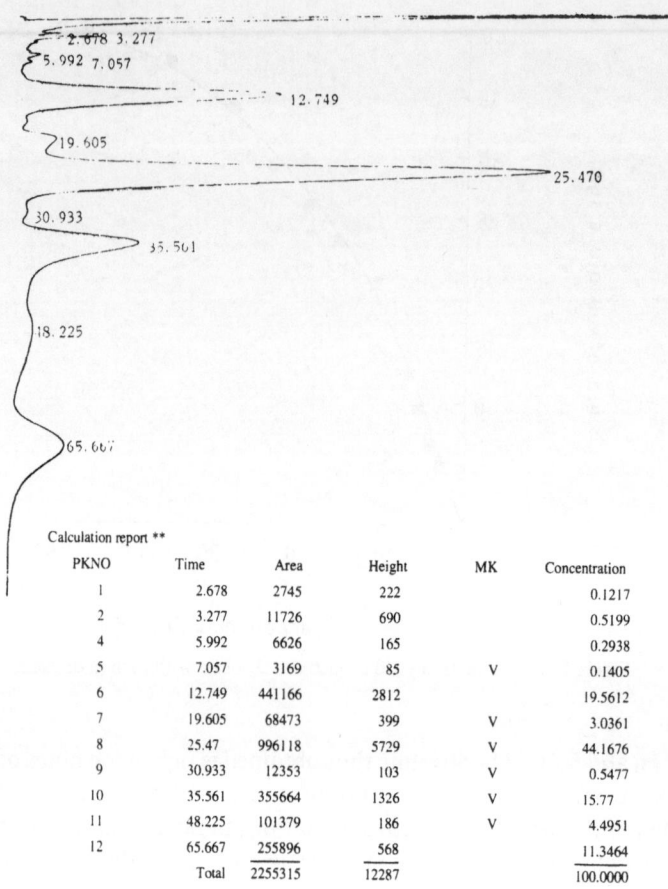

Calculation report **

PKNO	Time	Area	Height	MK	Concentration
1	2.678	2745	222		0.1217
2	3.277	11726	690		0.5199
4	5.992	6626	165		0.2938
5	7.057	3169	85	V	0.1405
6	12.749	441166	2812		19.5612
7	19.605	68473	399	V	3.0361
8	25.47	996118	5729	V	44.1676
9	30.933	12353	103	V	0.5477
10	35.561	355664	1326	V	15.77
11	48.225	101379	186	V	4.4951
12	65.667	255896	568		11.3464
	Total	2255315	12287		100.0000

Figure 12.6 Gas chromatograms of the acetates of fatty alcohols obtained by carbon dioxide extraction.

triacontanol on sprouting of potatoes and the harmful effect of C-28 alcohol (hexacosanol) is mentioned (Narasimham, personal communication) by Shashirekha and Narasimhan (1990). The mixture of fatty alcohols can be an effective raw material for a bio-active formulation.

12.7.2   Down stream processing in bio-technological applications

SC-CO_2 is also of interest to biological process industries (Wilson, 1985). The SC-CO_2 has been recognised as a valuable tool in bio-technology in down stream processing (Wong and Johnston, 1986) and deactivation of specific enzymes (Randolph *et al.*, 1985; Taniguchi *et al.*, 1987a). The importance of carbon dioxide extraction in the recovery of dry mouldy bran (DMB) produced under solid state fermentation (SSF) process has been stressed by

Mudget (1986). However, no information is available on the use of SC-CO$_2$ except in the killing of microorganisms in defatted rice koji with simultaneous extraction of lipids (Tokuda *et al.*, 1986), and killing of organisms alone. (Kamihira *et al.*, 1987; Taniguchi *et al.*, 1987b).

In SSF processes where gibberellic acid (GA$_3$) is one of the high value products, it has been found that the presence of an undesirable sterol disturbs the defencive mechanism in rice seedlings. A process has been developed for efficient extraction of GA$_3$ from DMB by ethanolic extraction under a counter current process (Kumar and Lonsane, 1987). Under these conditions, the undesirable sterol is also extracted and is present in the final product. The preferential extraction of the undesirable sterol and the other eleven fluorescent compounds is achieved with SC-CO$_2$ at 35 MPa and 40°C entrained with ethanol at a flow rate of 4.019 kg/h of carbon dioxide (Kumar *et al.*, 1991).

12.7.3 Vanilla absolute

The natural absolute made by conventional extraction finds special preference in certain specific flavouring applications. Considerable expertise is essential to obtain product with the full, rounded and well balanced profile of vanilla beans. The vanilla absolute is generally dark brown in semi-solid form, and used in high quality flavouring applications. Liquid carbon dioxide absolutes have already been successfully tried in high quality ice creams and desserts (Moyler and Heath, 1988). Vanilla beans extracted at 8.5–10 MPa at 20°C with 2.4% ethanol gave an absolute with 25% vanillin with an extraction yield of 92.8% (Vidal *et al.*, 1989b). SC-CO$_2$ entrained with alcohol at 28 MPa and 50°C gave extracts rich in natural vanilla aroma and fine crystals of vanillin were found to separate out on standing. Vanillin is extracted up to 98% from the beans as determined by spectroscopic estimation of the spent vanilla beans.

12.8 Summary

The use of SC-CO$_2$ for production of essential oils or oleoresin from spices is possible by suitable selection of pressure and temperature of extraction. The oils extracted are found to be superior in chemical composition containing a higher percentage of sesquiterpene compounds. In the case of ginger, the oxygenated fraction is much higher than in steam distilled oils. The gingerols in oleoresin of ginger are extracted without any decomposition. In the case of oleoresin of pepper, the piperine is extracted with insignificant loss with longer process times. The jasmine and vanilla absolutes from this process have desirable top notes with rounded flavour and are superior to those obtained by conventional processes. Fatty alcohols from rice bran wax can be obtained in pure form to be used as bio-active formulation. In down stream processing of products in bio-technology SC-CO$_2$ extraction is possible to obtain relatively purer products. As an example, gibberellin (GA_3) processing in a SSF process is cited.

References

Box, G. E. P. and Behncan, D. W. T. (1983) *Technometrics* **2**, 29.

Calame, J. P. and Steiner, R. (1981) Carbon dioxide extraction in the flavour and perfumer industries. *Chem. Ind.*, 399.

Caragay, B. A. (1981) Supercritical fluids for extraction of flavours and fragrances from natural products. *Perfumer Flavorist* **6**, 43.

Chen, C. C. and Ho, C. T. (1988) Gas chromatographic analysis of volatile components of ginger oil (*Zingiber officinale* R.) extracted with liquid carbon dioxide. *J. Agric. Food Chem.* **36**, 322.

Chen, C. C., Kuo, M. C., Wu, C. M. and Ho, C. T. (1986) Pungent compounds of ginger (*Zingiber officinale* R.) extracted with liquid carbon dioxide. *J. Agric. Food Chem.* **34**, 477.

Coenen, H. and Kriegel, E. (1984) Applications of supercritical gas extraction processes in the food industry. *German Chem. Eng.* **7**, 335.

Gopalakrishnan, N., Shanti, P. P. and Narayan, C. S. (1990) Composition of clove (*Sizygium aromaticum*) bud oil extracted using carbon dioxide. *J. Sci. Food Agric.* **50**, 111.

Hubert, P. and Vitzthum, O. G. (1978) Fluid extraction of hops, spices and tobacco with supercritical gases. *Angew. Chem. Int. Ed. Engl.* **17**, 710.

Kamihira, M., Taniguchi, M. and Kobayashi, T. (1987) Sterilization of micro-organisms with supercritical carbon dioxide. *Agric. Biol. Chem.* **51**, 407.

Kumar, P. K. R. and Lonsane, B. K. (1987) Extraction of gibberellic acid from dry mouldy bran produced under solid state fermentation. *Process Biochem.* **22**, 139.

Kumar, P. K. R., Sankar, K. U. and Lonsane, B. (1991) Supercritical fluid extraction of sterol from dry mouldy bran in production of gibberellic acid under solid state fermentation. *J. Chem. Eng. Chem. Eng. J.* **46**, 1353.

Lawrence, B. M. (1977) Recent progress in jasmine research. *Cosmetics Perfumery* **88**, 46.

Lawrence, B. M. (1984) Botanical and chemical aspects of ginger. *Perfumer Flavorist* **5**, 1.

Louis, G. L. (1978) Plant growth regulators, controlling biological behaviour with chemicals. *Chem. Eng. News.* **56**, 18.

Mathew, A. G., Krishnamurthy, N., Nambudiri, E. S. and Lewis, Y. S. (1973) Oil of ginger. *The Flavour Ind.* **4**, 226.

Mc Hale, D., Lawurie, W. H. and Sheridan, J. B. (1989) Transformation of the pungent principles in extracts of ginger. *Flavour Fragrance J.* **4**, 9.

Moyler, D. A. and Heath, H. B. (1988) Liquid carbon dioxide extraction of essential oils, in *Developments of Food Science Flavours and Fragrances, a World Perspective*, eds. B. M. Lawrence, B. D. Mookherjee and B. J. Wills, Elsevier, Amsterdam, pp. 41–64.

Mudget, R. E. (1986) Solid state fermentation, in *Manual of Industrial Microbiology and Biotechnology*, eds. A. L. Demain and N. A. Solomon, American Society of Microbiologists, Washington, DC, pp. 66–83.

Muller, P. A. (1965) The jasmine and the jasmine oil. *Perfumer Essential Oil Record* **36** (1965) 658–663.

Paginton, J. S. (1983) A review of oleoresin black pepper and its extraction solvents. *Perfumer Flavorist* **8**, 29.

Parkinson, G. and Johnson, E. (1989) Supercritical gas extraction win CPI acceptance. *Chem. Engl.* **96**(7), 36.

Paul, P. F. M. and Wise, W. S. (1971) in *The Principles of Gas Extraction*, ed. J. G. Cook, Mills and Boon, London.

Paulaitis, M. E., Penninger, J. M. L., Gray, Jr. R. D. and Davidson, P. (1983) *Chemical Engineering at Supercritical Conditions*, Ann Arbor Science, Ann Arbor, MI.

Penninger, J. M. L., Radsoz, M., McHugh, M. A., Krukonis and V. J. (1985) *Supercritical Fluid Technology. Process Technology Proceedings*, Vol. 3, Elsevier, Amsterdam.

Perrut, M. (1988) *Proceedings of the International Symposium on Supercritical Fluids*, Vols. 1 and 2, Vandoeuvre, INPL.

Polak, E. (1973) Recent progress in jasmine research. *Cosmetics Perfumery* **88**, 46.

Randolph, T. W., Blanch, H. W., Prausnitz, J. M. and Wilke, C. R. (1985) Enzymatic catalysis in a supercritical fluid. *Biotechnology Lett.* **7**, 325.

Rao, J. M., Natarajan, C. P. and Seshadri, R. (1987) A study on the occurrence of n-triacontanol, a plant growth regulator in tea. *J. Sci. Food Agric.* **39**, 95.

Richard, H., Loo, A. and Morin, P. (1989) Extraction of aromas using carbon dioxide. *Ind. Aliment. Agric.* **106**, 383.

Ries, S. K. and Houtz, R. (1983) Triacontanol as plant growth regulator. *Hort. Sci.* **18**, 654.

Rizvi, S. S. H., Daniels, J. A., Benado, A. L. and Zollweg, J. A. (1986a) Supercritical extraction: fundamental principles and modelling methods. *Food Technol.* **40**(6), 55.

Rizvi, S. S. H., Daniels, J. A., Benado, A. L. and Zollweg, J. A. (1986b) Supercritical fluid extraction: operating principles and food applications. *Food Technol.* **40**(7), 57.

Sankar, K. U. (1989) Studies on physico-chemical characteristics of volatile oil from pepper (*Piper nigrum* L.) extracted by supercritical carbon dioxide. *J. Sci. Food Agric.* **48**, 483.

Sankar, K. U. (1991) Ph.D. Thesis, University of Mysore.

Sankar, K. U. (1993) Studies on the quality aspects of the volatile oil of ginger (*Zingiber officinale* R.) extracted by supercritical carbon dioxide. *J. Sci. Food Agric.*, submitted.

Shashirekha, M. N. and Narasimhan, P. (1990) Effect of triacontanol on sprout in potatoes (*Solanum tuberosum*). *J. Sci. Food Agric.* **50**, 1.

Squires, T. G. and Paulaitis, M. E. (1987) *Supercritical Fluids Chemical Engineering Principles and Applications*, ACS Symposium Series, 329, American Chemical Society, Washington, DC.

Stahl, E. and Gerard, D. (1985) Solubility behaviour and fractionation of essential oils in dense carbon dioxide. *Perfumer Flavorist* **10**, 29.

Stahl, E., Quirin, K. W. and Gerard, D. (1987) *Verdichtite Gases Zur Extraktion und Raffination*, Springer-Verlag, Berlin.

Taniguchi, M., Kamihira, M. and Kobayashi, T. (1987a) Effect of treatment with supercritical carbon dioxide on enzyme activity. *Agric. Biol. Chem.* **51**, 593.

Taniguchi, M., Suzuku, H., Sato, M. and Kobayashi, T. (1987b) Sterilization of plasma powders by treatment with supercritical carbon dioxide. *Agric. Biol. Chem.* **51**, 3425.

Tateo, F. and Chizzini, F. (1989) The composition and quality of supercritical carbon dioxide extracted cinnamon. *J. Essential Oil Res.* **1**(4), 165.

Tateo, F., Fellin, M. and Verderio, E. (1988) Production of rosemary oleoresin using supercritical carbon dioxide. *Perfumer Flavorist* **13**, 27.

Tokuda, K., Ito, K., Imamura, T., Taniguchi, M., Kobayashi, T., Aki, T. and Hara, S. (1986) *J. Brew. Soc. Jpn.* **81**, 194.

Tuley, L. (1985) Flavour developments: carbon dioxide solvent extraction. *Food Flavourings, Ingredients and Processing,* **7**(4), 29.

Verghese, J. (1989) On the isolation of oleoresin black pepper by steam distillation cum solvent extraction and tailoring of oleoresin. *Perfumer Flavorist* **14**, 33.

Vidal, J. P. and Richard, H. (1987) Production of a black pepper oleoresin by dense carbon dioxide ethanol extraction. *Sci. Alimentes,* **7**, 481.

Vidal, J. P., Fort, J. J., Gaultier, P. and Richard, H. (1989a) Vanilla aroma extraction by dense carbon dioxide. *Sci. Alimentes* **9**(1), 89.

Vidal, J. P., Fort, J. J., Gaultier, P. and Richard, H. (1989b) Vanilla aroma extraction by dense carbon dioxide. *Sci. Alimentes* **9**, 89.

Wilke, A., Stahl, E. and Zosel, K. (1981) *Extraction with Supercritical Gases*, Verlag Chemie, Berlin.

Wilson, R. C. (1985) Supercritical fluid extraction, *Comprehensive Bio-technology. 2. The principles of Biotechnology: Engineering Considerations* eds. C. L. Cooney and A. E. Humphrey, Pergamon Press, Oxford, pp. 567–574.

Wong, J. M. and Johnston, K. P. (1986) Solubilisation of biomolecules in carbon dioxide based supercritical fluids. *Biotechnology Prog.* **22**, 29.

13 Extraction of oil from evening primrose seed with supercritical carbon dioxide

B.-C. LEE, J.-D. KIM, K.-Y. HWANG and Y. Y. LEE

Abstract

Extraction of evening primrose oil (EPO) from the seed with supercritical carbon dioxide (SC-CO_2) was conducted, and solubilities of EPO in compressed carbon dioxide were measured at temperatures of 20–50°C and pressures of 100–300 bar. The average loadings of oil in SC-CO_2 increased with the residence time, and the mass transfer parameter, $k_c a$, was found to increase linearly with the superficial velocity of CO_2 in the extractor. The quality of EPO extracted with SC-CO_2 was compared with that of EPO extracted with n-hexane.

13.1 Introduction

Recently, supercritical fluid extraction technology has been of considerable research interest in food industries, particularly in the extraction of oil from oil-bearing seeds (Stahl *et al.*, 1980; de Filippi, 1982; Friedrich and List, 1982; Eggers *et al.*, 1985). CO_2 as a supercritical fluid has many potential advantages over other organic solvents such as n-hexane, because it is non-toxic, non-flammable, non-corrosive, cheap, readily available and easily removable from the extracted solute. Furthermore, the critical temperature of CO_2 is so low (31.1°C) that the thermal degradation of extracts can be prevented. By controlling the temperature and pressure of CO_2, a wide range of selectivity can be obtained (McHugh and Krukonis, 1986).

SC-CO_2 is unique in that it can have liquid-like density and possess transport properties (e.g. viscosity and diffusivity) intermediate between those typical of a liquid and a gas. Thus, with seed oils, SC-CO_2 extraction is considered to be a promising alternative to the conventional extraction method using n-hexane as a solvent.

This work describes the extraction of oil from evening primrose seeds with compressed CO_2 and compares it with n-hexane extraction.

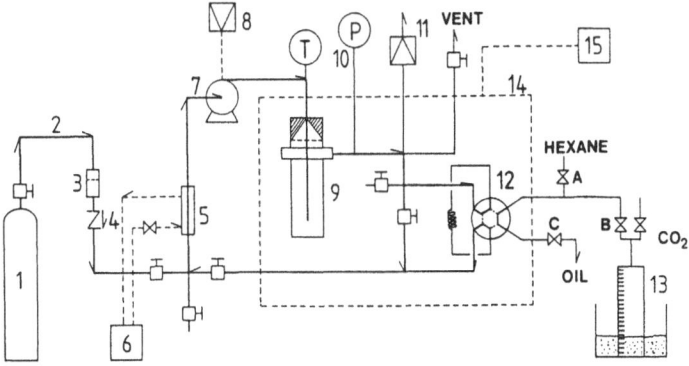

Figure 13.1 Schematic diagram of experimental apparatus for measuring solubility of evening primrose oil in compressed carbon dioxide. 1, CO_2 cylinder; 2, flexible hose; 3, filter; 4, check valve; 5, cooling jacket; 6, refrigerated ciculator; 7, high pressure pump; 8, air compressor; 9, equilibrium cell; 10, pressure gauge; 11, rupture; 12, sampling valve; 13, cylinder, graduated; 14, air bath; 15, temperature controller.

13.2 Experiments

13.2.1 Materials

Evening primrose seeds were pretreated before use as follows: the seeds were ground to a particle size of 25–40 mesh, and then dried for 2–3 h in a vacuum oven at 105°C and 300 mmHg. Commercial grade carbon dioxide was used as the extractant.

13.2.2 Apparatus and procedures for measuring solubility

A schematic diagram of the experimental apparatus used for measuring the solubilities of evening primrose oil (EPO) in compressed CO_2 is shown in Figure 13.1. All equipment was made of stainless steel 316 which could be used up to 1000 bar. The apparatus consisted of a high pressure pump (Haskel model MCP-110-C), a high pressure vessel (40 mm internal diameter, 200 cm³ volume), a sample valve (Valco model C6U) and a pressure gauge (Heise model CMM-42982), in a recirculation layout.

The pump, driven by air from a compressor (ITT pneumotive model LGH-210), was used both to raise the system pressure and to recirculate extractant. The vessel, sampling valve and connecting parts were enclosed in a temperature-controlled, forced convection air bath to keep the system temperature constant. All the equipment was protected by a 5000 psi rupture disc (PPI model H-11U-60K).

The ground seeds were filled in the vessel, and the temperature of the air bath was increased to the desired temperature; CO_2 from a cylinder was introduced into the vessel through a line filter and sub-cooler and then

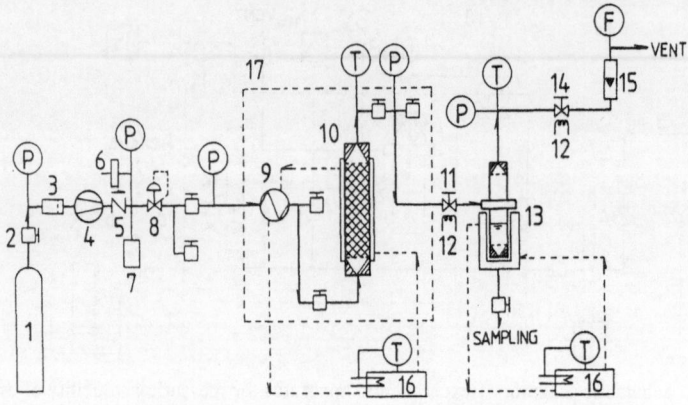

Figure 13.2 Schematic diagram of experimental apparatus for the extraction of evening primrose oil by compressed carbon dioxide. 1, CO_2 cylinder; 2, high pressure valve; 3, filter; 4, compressor; 5, check valve; 6, rupture; 7, damper; 8, pressure regulator; 9, preheater; 10, extraction vessel; 11, pressure control valve; 12, heater; 13, separator;   14, back pressure regulator; 15, rotameter; 16, circulator; 17, air bath.

compressed to the desired pressure. The CO_2 and oil extract were sufficiently recirculated by the pump through the sampling valve with a loop of 1 ml for about 3 h until the equilibrium was reached.

After reaching equilibrium, the contents of the sample loop were analyzed in the following manner. When the loop was switched out of the system, the sample expanded into the transfer lines through valves A, B and C. As a result of this expansion into the transfer lines, the oil component precipitated in the lines. The amount of CO_2 in the sample loop could be ascertained by slowly opening valve B to vent the CO_2 into a 500-ml graduated cylinder filled with CO_2-saturated water at a known temperature. As the CO_2 was being vented, care had to be exercised to avoid entrainment of the oil in the CO_2 gas phase. The volume of CO_2 vented from the sample loop was determined by the volume of the water displaced corrected for the vapor pressure of water and the volume of CO_2 remaining in the sample loop, and it was converted to a weight basis.

The oil component that had precipitated in the sample loop and transfer lines could be removed by flushing a large amount of n-hexane through the system at valve A. To determine the amount of oil component in the n-hexane solution, the absorbance of the solution was measured at 269 nm wavelength using a spectrophotometer (Shimadzu model UV-260) and the amount of oil was obtained from the calibration curve between the absorbances and concentrations of standard n-hexane solutions.

13.2.3 *Apparatus and procedures for extraction*

A schematic diagram of the one-pass flow type apparatus for the extraction of EPO with compressed CO$_2$ is shown in Figure 13.2. Ground seeds (about 100 g) were placed in a cylindrical extraction vessel (40 mm i.d., 200 cm^3 volume). CO$_2$ from the cylinder was charged into an electrically driven double diaphragm compressor through a line filter, and then compressed into the extraction vessel to the desired pressure. It was heated to the desired temperature by a double pipe heat exchanger, a jacket outside extractor and forced-convection air bath.

The pressure in the extractor was controlled by the pressure regulator at the discharge of the compressor. When the desired pressure was reached, the CO$_2$ and oil extract were expanded into the separator through a heated micro-metering valve, which regulated the flow rate of the fluid. The oil dissolved in the compressed CO$_2$ was separated from the CO$_2$ by pressure reduction and collected in the separator. The separator was equipped with a cooling jacket. The oil-free CO$_2$ was passed through a back pressure regulator, which main-tained the pressure of the separator at 30 bar, and then through a rotameter and a flow totalizer before being exhausted.

The amount of CO$_2$ consumed in the experiment was recorded and the oil collected in the separator was withdrawn and weighed. Finally, the oil products were analyzed according to AOCS analytical methods (Mehlenbacher *et al.*, 1973) to observe their properties.

13.3 Results and discussion

13.3.1 *Solubility of evening primrose oil in compressed carbon dioxide*

The solubilities of EPO in compressed CO$_2$ were measured at temperatures of 20–50°C and pressures of 100–300 bar. Figures 13.3 and 13.4 show the effect of temperature and pressure on the solubilities of EPO in compressed CO$_2$. The solubility of EPO in CO$_2$ was found to increase with pressure at a constant temperature. When the pressure was kept constant, the solubility decreased with increase in temperature at pressures below 250 bar, while it increased at pressures above 250 bar.

Although this behavior is too complex to be interpreted exactly, it may be regarded as due to the competing effects between the solvent density and solute volatility (Friedrich *et al.*, 1982; Friedrich and Pryde, 1984; King *et al.*, 1987). As temperature increases, solvent density decreases and solute volatility in-creases. The decrease in the solvent density decreases the probability of a given solute molecule in the extractant phase interacting with a solvent molecule, tending to decrease solubility. Increasing solute volatility, on the other hand, increases the escaping tendency of the solute from the condensed phase, thus tending to increase solubility.

Above 250 bar, as CO$_2$ density is less sensitive to temperature and oil volatility effects dominate, the solubility of EPO increases with a higher

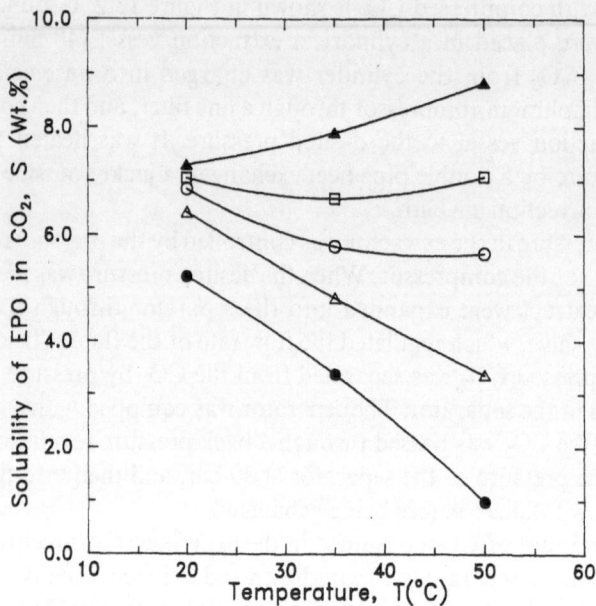

Figure 13.3 Solubility of evening primrose oil in compressed CO_2 as a function of temperature at various pressures. ●, 100 (bar); △, 150; O, 200; □, 250; ▲, 300.

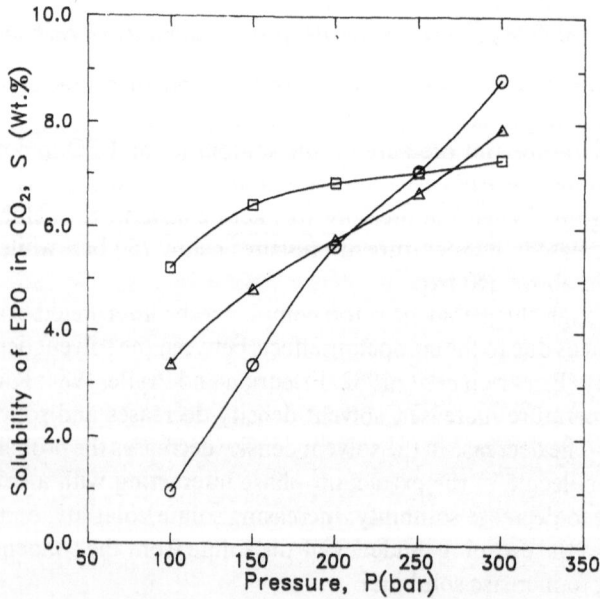

Figure 13.4 Solubility of evening primrose oil in compressed CO_2 as a function of pressure at various temperatures. □, 20 (°C); △, 35; O, 50.

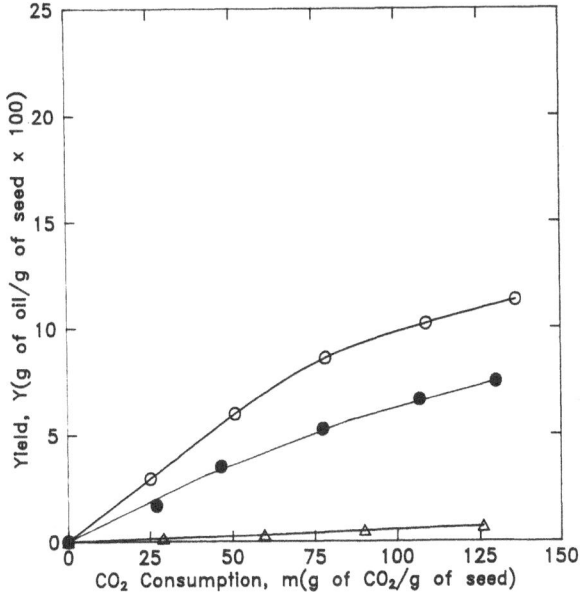

Figure 13.5   Extraction yield as function of the amount of compressed CO$_2$ consumed for various temperatures at 100 bar.O, 20 (°C); ●, 35; △, 50.

temperature at a given pressure. But below 250 bar, due to the rapid decrease of CO$_2$ density, it decreases with a higher temperature. At about 250 bar, the competing effects balance each other, and the solubility remains relatively constant with increasing temperature.

13.3.2   Extraction of oil from evening primrose seed

The oil component was extracted from ground evening primrose seeds with compressed CO$_2$ in the operating range of 20–50°C and 100–300 bar. The CO$_2$ was passed through the extraction vessel at a flow rate of 3.5–4.0 l/min (at 20°C, 1 atm).

Figures 13.5, 13.6 and 13.7 show the effect of temperature and pressure of CO$_2$ on the extraction yield of oil. The extraction yield of oil, Y, is defined as the percentage of the mass of oil extracted per unit mass of seed. The CO$_2$ consumption, m, is the mass of CO$_2$ consumed per unit mass of seed. The increase in the pressure improved the extraction rate at constant temperature. At higher temperatures, the pressure had more influence on the extraction rate. This may be attributed to the greater dependence of the solubility on pressure with increasing temperature. With increasing temperature, the extraction rate decreased at 100 bar, but it increased at 300 bar. It was interesting that the effect of the temperature on the extraction rate was almost negligible at 200 bar.

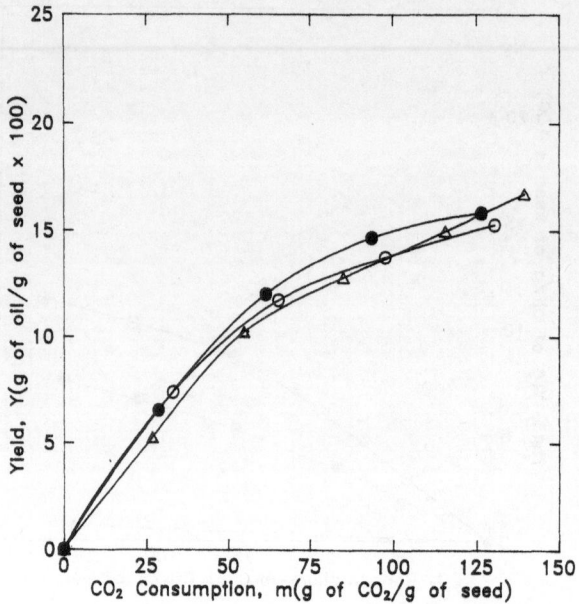

Figure 13.6 Extraction yield as function of the amount of compressed CO_2 consumed for various temperatures at 200 bar. O, 20 (°C); ●, 35; △, 50.

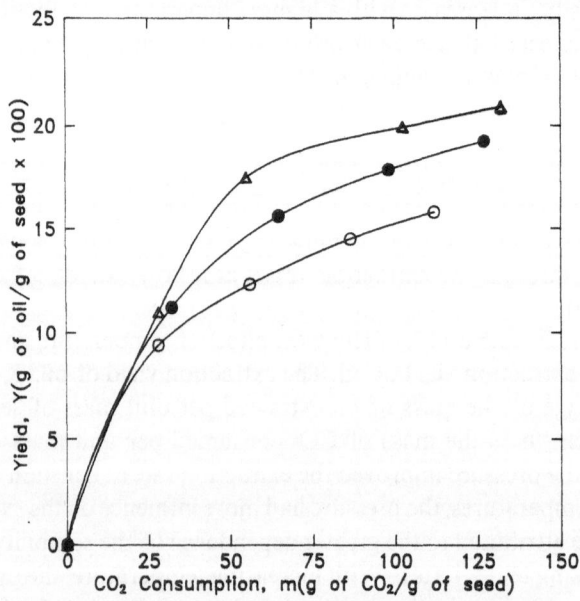

Figure 13.7 Extraction yield as function of the amount of compressed CO_2 consumed for various temperatures at 300 bar. O, 20 (°C); ●, 35; △, 50.

Consequently, the effects of temperature and pressure of CO$_2$ on the extraction yield of oil coincide with the trend in solubility. Under the range covered in this study, the highest yield of extraction, which was about 21 wt% on a seed basis, was obtained at 50°C and 300 bar.

13.3.3   Prediction of extraction rate

To assess the effective extraction rate, it will probably be necessary to predict the mass transfer rate of oil from the seeds into the solvent. An overall mass transfer coefficient can be assumed to be used because the diffusion through the pores inside seed particles can be negligible and a mean oil concentration may be assigned (King *et al.*, 1987). Finally, the equation of the mass transfer rate from seeds to CO$_2$ in the extractor can be expressed as

$$\frac{dC}{dt} = k_c a \, (C^* - C) \tag{13.1}$$

where C is the concentration of oil in the CO$_2$ phase, C^* is the equilibrium value of C, a is the surface area per unit volume of seed, and k_c is the overall mass transfer coefficient. Equilibrium concentration, C^*, which could be obtained from the solubility of oil in CO$_2$ at 50°C and 300 bar, was 84.42 kg of oil/m^3 of CO$_2$.

Integrating equation (13.1) with $C = 0$ at $t = 0$ and $C = C_o$ at $t = t_R$ gives

$$\frac{C_o}{C^*} = 1 - \exp(-k_c a \, t_R) \tag{13.2}$$

$$k_c a = \frac{-\ln(1 - C_o/C^*)}{t_R} \tag{13.3}$$

where C_o is the loading, i.e. the concentration of oil in the CO$_2$ stream leaving the extractor and t_R is the residence time of CO$_2$ in the extractor.

Although the mechanism of mass transfer from the bed of seeds is probably complex, involving the diffusion through pores in seed as well as across a boundary film of solvent, it may be empirically valid to predict the mass transfer rate using equation (13.3).

In the extraction of evening primrose oil with SC-CO$_2$ at 50°C and 300 bar, the mass transfer rate of the oil from the seeds to the CO$_2$ phase was predicted by varying the residence time of CO$_2$ in the extractor. The residence time of CO$_2$, t_R, was varied by carrying out tests at a series of CO$_2$ flow rates in the extractor of a fixed volume.

Figure 13.8 shows the effect of residence time of CO$_2$ on the extraction yield of oil at 50°C, 300 bar. Figure 13.9 shows the loading of oil in the CO$_2$ stream leaving the extractor as the percentage of oil in the seeds. These loadings, C_o, are average values over the interval between the percentage of extraction shown and that corresponding to the previous point in the series. The percent-

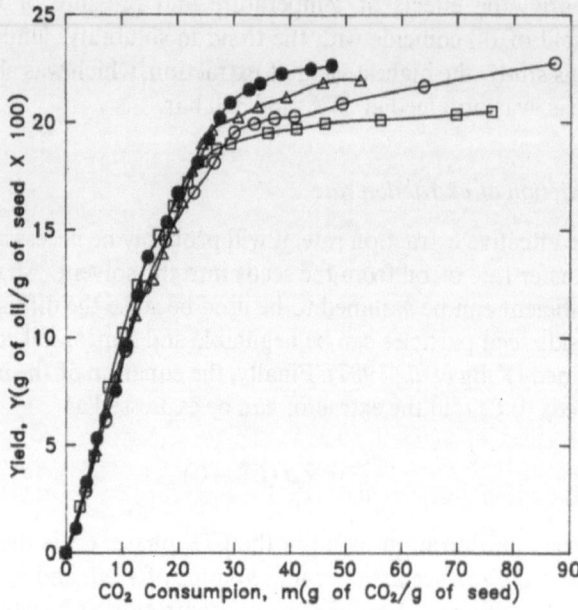

Figure 13.8 The effect of residence time of CO_2 on the extraction yield of oil at 50°C, 300 bar. O, 14.7 (min); □, 20.7; △, 30.1; ●, 54.7.

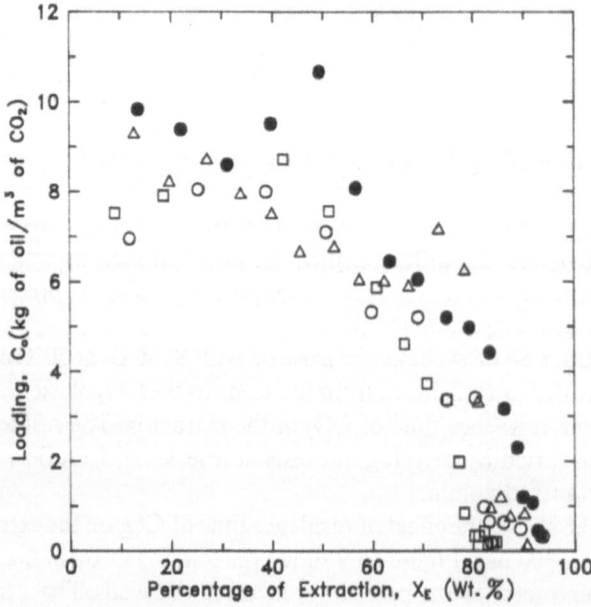

Figure 13.9 The effect of residence time of CO_2 on the extraction loading of oil at 50°C, 300 bar. O, 14.7 (min); □, 20.7; △, 30.1; ●, 54.7.

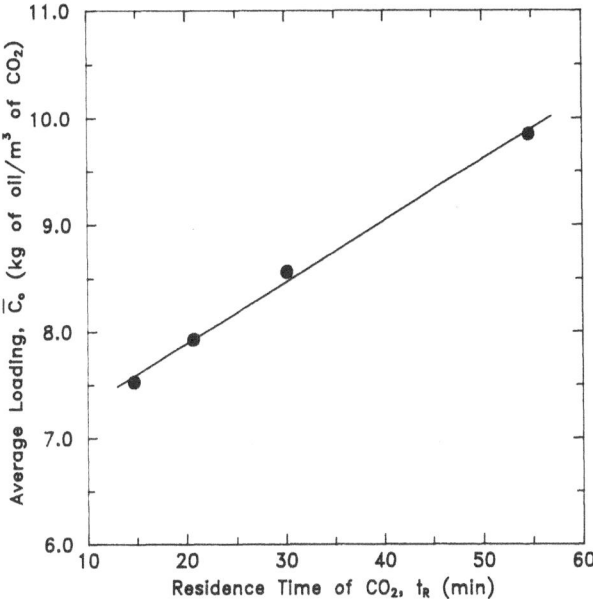

Figure 13.10   The effect of residence time of CO$_2$ on average loading of oil at 50°C, 300 bar.

age of extraction, X_E, is the amount of oil extract accumulated at that instance divided by the total available oil content of the seed (24.3 wt% for evening primrose seed).

It is shown that the loadings remain substantially constant until the percentage of extraction exceeds about 50 wt%, and then a sharp fall takes place in the loadings. This may be regarded as due to a progressive increase in the lengths of the diffusion paths by which the oil reaches the surface of the seed particles as extraction proceeds (King *et al.*, 1987). Figure 13.10 shows the average loadings as a function of the residence time of CO$_2$. The average loading, C_o, at a given residence time is the average value of the loadings until the percentage of extraction is about 50 wt%. As the residence time of CO$_2$ increased, the average loading increased linearly.

Table 13.1   Extraction rate data for the extraction of evening primrose oil using SC-CO$_2$ at 50°C, 300 bar

Residence time t_R (min)	Superficial velocity u_S (cm/min)	Average loading C_o (kg of oil/ m^3 of CO$_2$)	Mass transfer parameter $k_c a \times 10^3$ (min^{-1})
14.7	1.08	7.53	6.35
20.7	0.77	7.93	4.77
30.1	0.53	8.56	3.55
54.7	0.29	9.85	2.27

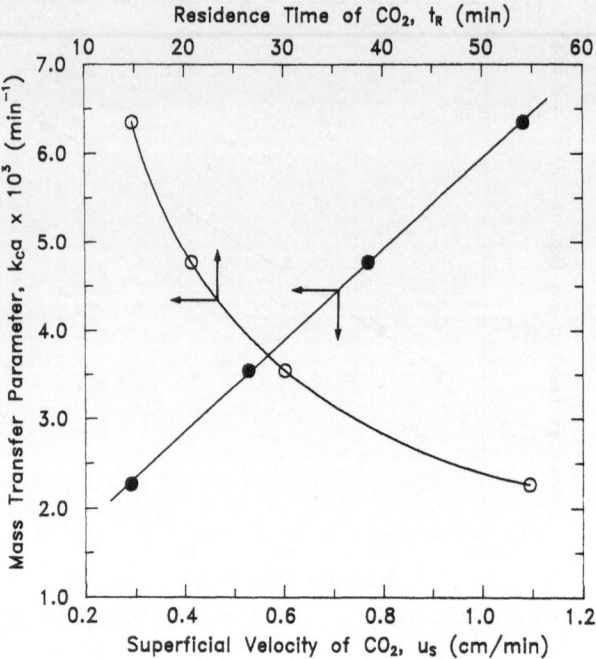

Figure 13.11 The effect of superficial velocity and residence time of CO_2 on mass transfer parameter at 50°C, 300 bar.

Mass transfer parameter, $k_c a$, can be obtained from equation (13.3), as shown in Table 13.1. Figure 13.11 shows the effect of the residence time and superficial velocity of CO_2 on the mass transfer parameter. $k_c a$ decreased with increase in the residence time of CO_2 in the extractor, while it increased linearly with increase in the superficial velocity of CO_2.

13.3.4 Comparison with n-hexane extraction

The physicochemical characteristics of oil extracted with SC-CO_2 were compared with that of the oil extracted with *n*-hexane in a Soxhlet extractor, as shown in Table 13.2. The properties of oils extracted by SC-CO_2 and *n*-hexane were analyzed according to the AOCS methods. Individual methods are referenced in Table 13.2.

The overall yield obtained with SC-CO_2 of 50°C and 300 bar was about 21.0 wt%, which was higher than that extracted with *n*-hexane for about 10 h.

To obtain the composition of their fatty acids, oils were saponified and esterified to the corresponding methyl esters, which were then analyzed by gas chromatography. A gas chromatograph (Hewlett Packard 5890 Series II) fitted with a flame ionization detector was used with nitrogen as a carrier gas. A

Table 13.2 Properties of evening primrose oils extracted with supercritical carbon dioxide and *n*-hexane

Properties	AOCS method	SC-CO$_2$[a]	*n*-Hexane[b]
Acid value	Cd 3a-63	4.59	1.89
Saponification value	Cd 3-25	195.47	195.50
Peroxide value	Cd 8-53	6.32	4.25
Iodine value	Cd 1-25	147.59	154.64
Unsaponifiable matter (%)	Ca 6a-40	1.93	1.88
Phosphorus content (ppm)	Ca 12-55	4.40	30.40
Fatty acid (wt%)			
16:0	Ce 2-66	7.73	7.85
18:0		2.10	2.04
18:1ω9		6.00	6.27
18:2ω6		74.74	74.41
18:3ω6		9.44	9.43
Color[c]			
Red	Cc 13b-45	1.7	7.2
Yellow		40.0	40.5
Blue		0.1	5.1

[a] Oils extracted with SC-CO$_2$ of 50°C and 300 bar.
[b] Oils extracted by *n*-hexane Soxhlet extraction for about 10 h.
[c] Lovibond Tintometer color unit.

column, 3.2 mm × 3.0 m, packed with 10% Silar 7CP on Chromosorb W-HP 100/120 mesh was used. The flow rate of nitrogen was 25 ml/min and the column was operated isothermally at 180°C. The injector and detector temperatures were maintained at 250°C. The composition of the fatty acids was almost the same regardless of the extraction method. Particularly, the composition of γ-linolenic acid was about 10.0 wt%.

The oil extracted with SC-CO$_2$ showed significantly less phosphorus content than *n*-hexane oil. SC-CO$_2$-extracted oil was lighter in color than hexane-extracted oil. In free fatty acids, peroxides and unsaponifiables, there is no significant difference between SC-CO$_2$- and hexane-extracted oils.

13.4 Conclusions

In the extraction of evening primrose oil with compressed CO$_2$, the solubility and extraction rate of the oil were affected by the pressure and temperature of CO$_2$. Under the range covered in this study, the highest yield of extraction was about 21 wt% at 50°C and 300 bar. The mass transfer parameter of the oil from the seeds to the CO$_2$ phase was found to increase linearly with the superficial velocity of CO$_2$. The oil extracted with the supercritical CO$_2$ was found to have comparable extraction yield and significantly less phosphorus content, compared with that obtained by *n*-hexane.

Nomenclature

a surface area per unit volume of seed (m^2/m^3)

C concentration of oil in CO_2 phase (kg of oil/m^3 of CO_2)

C^* equilibrium value of C (kg of oil/m^3 of CO_2)

C_o loading, i.e. concentration of oil in CO_2 stream leaving the extractor (kg of oil/m^3 of CO_2)

k_c overall mass transfer coefficient

t_R residence time of CO_2 in the extractor (min)

u_S superficial velocity (cm/min)

References

de Filippi, R. P. (1982) CO_2 as a solvent: application to fats, oils and other materials. *Chem. Ind.* **19**, 390.

Eggers, R. Sievers, U. and Stein, W. (1985) High pressure extraction of oil seed. *J. Am. Oil Chem. Soc.* **62**, 1222.

Friedrich, J. P. and List, G. R. (1982) Characterization of soybean oil extracted by supercritical carbon dioxide and hexane. *J. Agric. Food Chem.* **30**, 192.

Friedrich, J. P. and Pryde, E. H. (1984) Supercritical CO_2 extraction of lipid-bearing materials and characterization of the products. *J. Am. Oil Chem. Soc.* **61**, 223.

Friedrich, J. P., List, G. R. and Heakin, A. J. (1982) Petroleum-free extraction of oil from soybeans with supercritical CO_2. *J. Am. Oil Chem. Soc.* **59**, 288.

King, M. B., Bott, T. R., Barr, M. T. and Mahmud, R. S. (1987) Equilibrium and rate data for the extraction of lipids using compressed carbon dioxide. *Sep. Sci. Tech.* **22**, 1103.

McHugh, M. A. and Krukonis, V. J. (1986) *Supercritical Fluid Extraction, Principles and Practice.* Butterworths, Stoneham, MA.

Mehlenbacher, V. C., Hopper, T. H., Sallee, E. M. and Link, W. E. (1973) *Official and Tentative Methods of the American Oil Chemists' Society*, American Oil Chemists' Society, Champaign, IL.

Stahl, E., Schuetz, E. and Mangold, H. (1980), Extraction of seed oils with liquid and supercritical carbon dioxide. *J. Agric. Food Chem.* **28**, 1153.

14 High pressure extraction of organics from water

Ž. KNEZ, F. POSEL and I. KRMELJ

Abstract

Possible applications of supercritical fluid extraction (SFE) for the separation of organics from water, properties of supercritical fluid solvents, and experiments on separation bioactive substances and fermentation products from water are presented. Supercritical extraction has been applied to remove toxic organics from pesticide polluted water and some fermentation products such as ethanol, 2-propanol and some antibiotics from water solutions. Small-scale semi-batch experiments were performed at various temperatures and pressures, the contents of active substances in raffinate were determined by GLC or HPLC, and total organic contamination by chemical oxygen demand. The laboratory scale results show that this separation process is an effective method for the separation of organics from water solutions.

14.1 Introduction

Separation of organics from water with liquefied (subcritical) or supercritical gases is currently an interesting area of research. There are two main directions in this research: (a) separation of chemicals produced by fermentation, and (b) purification of polluted effluent waters (Knez *et al.*, 1988, 1990).

Various chemicals, such as alcohols or organic acids, are produced by fermentation. Most of the bio-routes suffer from the limitation that the chemicals are normally formed in the fermentation broth at a relatively low concentration. The separation of these chemicals from water by distillation or by evaporation requires a large input of energy per unit of product. Extraction with supercritical fluid (SCF) solvent offers the potential for low energy separation, especially if the distribution coefficient for the organic in the SCF/water system is favorable.

Although many polar compounds are very soluble in carbon dioxide, the high degree of water affinity, characterized empirically by a low value of the distribution coefficient, can make extraction by carbon dioxide unfavorable. On the other hand, distribution coefficients for such organic compounds as the higher alcohols, esters, and various aromatics, are much higher; even values of 10 or more have been reported.

The use of supercritical CO_2 (SC-CO_2) as a solvent for extraction of natural products (especially solids) in the food industry has advantages over other solvents because CO_2 is non-toxic, non-flammable, non-corrosive, cheap and readily available in large quantities and in high purity. Since CO_2 also has a relatively low critical pressure and critical temperature, it is easy to handle and to remove from any solute.

The knowledge of phase equilibria and mass transfer rates is essential for the design of process equipment (Brunner, 1984; Lack and Marr, 1986). The objective of mass transfer investigations is to find the optimum specific solvent flow rate of the process. The specific solvent flow rate is defined as mass flow of the solvent per amount of feed and extraction time for a desired extraction efficiency. The energy consumption is linearly dependent on the mass flow rate of the solvent.

14.2 Apparatus and methods

Experiments were performed on a semi-batch operated apparatus produced by UHDE-GmbH-Hagen. The technical data of the plant are: volume of the extractor $V_1 = 4$ l and $V_2 = 0.5$ l; the maximal working pressure 500 bar; the maximal working temperature 200°C. The flow rate can be varied between 5 and 33 kg/h. CO_2 was supplied by Rogaška Slatina-Slo and was 99.94 vol% pure.

An experimental study was performed on a 5-cm diameter and 0.5-l volume extractor device operated in the semi-batch way. The SFE systems studied were ethanol/water/carbon dioxide and 2-propanol/water/carbon dioxide at 70–400 bar and 25–95°C. In order to observe the extraction efficiency, the supercritical solvent flow rate was varied.

In order to obtain information on technical and economic availability of SFE, as applied to separation of active substances and organic solvents from pesticide polluted effluent waters, small scale batch experiments with a model solution of commercially available pesticides were performed at 40°C and 150 bar. Extraction time was 50 min.

14.3 High pressure extraction of water solutions

The method of mass transfer calculations used in this work is identical to that of a spray column contactor. The equations generally used for calculations of overall mass transfer coefficients are (see e.g. King, 1980).

$$\frac{1}{K_{of} a} = \frac{1}{k_f a} + \frac{m_{fl}}{k_l a} \qquad (14.1)$$

for the fluid phase and

$$\frac{1}{K_{ol} a} = \frac{1}{k_l a} + \frac{1}{m_{fl} k_l a} \qquad (14.2)$$

for the liquid phase, where k_{fa} and k_{la} are the volumetric mass transfer coefficients in the fluid and the liquid phase, respectively. The total solute mass flow rate is defined as

$$q_m = (K_{of} a)((C_f^* - C_f)V) = (K_{ol} a)((C_l - C_l^*)V) \qquad (14.3)$$

The overall volumetric mass transfer coefficient was determined as

$$K_{of} a = q_m/(C_f^* - C) V \qquad (14.4$$

where q_m is the average solute mass flow rate and $(C_f^* - C_f)$ is the average concentration potential, determined by the measurement of the aqueous feed concentration of the solute at the beginning and the end of the time interval. V is the volume of liquid in the extractor.

14.4   Results and discussion

14.4.1   Supercritical extraction of pesticides from water

Analysis of raffinates gave the following results:

1. Aniten DS: the decline in fluorenol content was 11.2%;
2. Betanal Am-11: the decline in chemical oxygen demand (COD) was 22.0%;
3. Lasso EC: the decline in COD was 21.0% and the decline in content of alachlor in water was 30.0%;
4. Volaton 500 EC: the decline in COD was 9.5%.

Our knowledge about the economics of the new SCF process is incomplete because several data are missing. Nevertheless, the first economic evaluation shows that the SFE process has an economic advantage over distillation processes, incineration or adsorption on active carbon (Knez *et al.*, 1988). The comparison of operating costs shows a marked advantage of the SFE process over quoted classical processes; the comparison of capital costs shows that only adsorption on active carbon has an economic advantage over the new process.

A feasibility study for the possible purification of pesticide contaminated effluent water with SFE, which is currently purified by adsorption on active carbon, was carried out. It was found that costs of the extraction process with

Table 14.1 Calculations of volumetric mass transfer coefficients for EtOH/H_2O/CO_2 systems

t (°C)	p (bar)	ρ (kg/m^3)	$q_{v(CO_2)}$ (l/h)	v (cm/s)	m	K_{ofa} (h^{-1})
50	150	698	15.7	0.3	0.09	4
50	150	698	26.2	0.5	0.09	12
50	150	698	36.6	0.7	0.09	30
20	100	857	32	0.5	0.06	8
50	280	857	32	0.5	0.09	27
70	400	857	32	0.5	0.12	27
50	100	390	15	0.5	0.09	6
50	100	390	30	1.0	0.09	17
50	80	220	16	1.0	0.09	13
50	80	220	25	1.5	0.09	17
50	80	220	33	2.0	0.09	21
50	70	172	17	1.3	0.09	18
50	75	194	25	1.7	0.09	20

CO_2 are 2.5–3 times lower than those of active carbon adsorption. This promises a saving of at least US$33.7 per 1 m³ waste water.

14.4.2 Ethanol/water/CO_2

The content of ethanol in raffinate as a function of time and pressure shows that with increasing pressure at constant temperature (up to 200 bar), the extraction rate is increased. With further increase in pressure, the extraction rate decreased, probably due to the influence of density and viscosity of supercritical CO_2. The optimum pressure for extraction of ethanol from 10% (by wt) aqueous solution at 50°C was found to be 200 bar. The influence of temperature was also observed and it was found to increase with temperature. With increasing flow rate, the efficiency of extraction increased, but the optimum was found to be 28.5 kg CO_2/h (related to 0.43 cm/s velocity through the aqueous feed in the extractor).

Volumetric overall mass transfer coefficients were determined from equation (14.4) and are presented in Table 14.1. The distribution coefficients for the ethanol system were taken from the literature (Bunzenberger, 1988).

At 150 bar and 50°C, higher flow rates increase the overall volumetric mass transfer coefficients. At constant CO_2 density (857 kg/m³), the elevated temperature, which causes the increase in pressure, increases the overall volumetric mass transfer coefficient only when the temperature is over its critical value. The lowest value of the coefficient is at 20°C, whereas at 50 and 70°C the coefficients have the same value. At 50°C and at pressures of 70, 75, 80 and 100 bar, higher flow rates increase the overall volumetric mass transfer coefficient, but the influence is lower than that at 150 bar and at the same temperature. From Table 14.1, it is evident that at lower fluid (CO_2) densities, the influence of the solvent flow rate is lower because of the lower solvent

Table 14.2 Calculations of volumetric mass transfer coefficients for the 2-propanol/H_2O/C system

t (°C)	p (bar)	ρ (kg/m^3)	$q_{v(CO_2)}$ (l/h)	v(cm/s)	m	$K_{of}a$ (h^{-1})
50	150	698	16	0.3	0.2	2
50	150	698	26.2	0.5	0.2	19
50	150	698	36.6	0.7	0.2	36
20	100	857	32	0.5	0.2	14.3
50	280	857	32	0.5	0.2	21.4
70	400	857	32	0.5	0.2	28
50	100	390	15	0.5	0.2	2.5
50	100	390	30	1.0	0.2	14
50	80	220	16	1.0	0.2	4
50	80	220	25	1.5	0.2	15
50	80	220	33	2.0	0.2	50
50	70	172	17	1.3	0.2	11
50	75	194	25	1.7	0.2	19

capacity. At a constant temperature of about 50°C and at constant fluid velocities (0.5 in 1.0 cm/s), the increase in pressure increases the overall volumetric mass transfer coefficients.

14.4.3 Propanol/water/CO$_2$

The volumetric overall mass transfer coefficient was determined by the same method as for the ethanol/water/CO_2 system. The results are presented in Table 14.2. The distribution coefficients were taken from the literature (Paulaitis *et al.*, 1985; Lahiere and Fair, 1989). At 150 bar and 50°C, it was found that the higher flow rate increased the efficiency of the extraction process. The values are comparable with the values for the ethanol/water/CO_2 system.

At a constant SCF density of 857 kg/m³, the increase in temperature (which is also reflected in higher pressure) increases the volumetric overall mass transfer coefficients, but this effect cannot be compared with the effect of a higher flow rate. At a temperature of 50°C and pressures of 70, 75, 80 and 100 bar, a greater influence of the flow rate was established. It was found that the values of overall volumetric mass transfer coefficients are not very sensitive to pressure changes. At a constant temperature of 50°C and at constant fluid velocities (0.5 and 1 cm/s), the increase in pressure also increases the overall volumetric mass transfer coefficients.

14.5 Conclusion

Experiments on the treatment of pesticide and phenol polluted/water and on the separation of fermentation products such as ethanol, 2-propanol and some antibiotics with SC-CO$_2$ show that this separation process is very effective. Further experiments on a counter current column are being carried out.

Nomenclature

C_f concentration of component in the fluid phase
C_f^* equilibrium concentration of component in the fluid phase
C_l concentration of component in the liquid phase
C_l^* equilibrium concentration of component in the liquid phase
k_{fa} volumetric fluid phase mass transfer coefficient
k_{la} volumetric liquid phase mass transfer coefficient
K_{ofa} overall volumetric mass transfer coefficient
m_{fl} distribution coefficient between fluid and liquid phase
q_m mass flow rate
V volume of liquid in the extractor

References

Brunner, G. (1984) Mass transfer from solid material in gas extraction, *Ber. Bunsenges. Phys. Chem.* **88**, 887.

Bunzenberger G. (1988) *Zur Stofftrennung mit Kohlendioxide unter Druck in Fluessig-Fluessig und Fluessig-Fest Systemen*, Der Fakultet für Maschinenbau der Technischer Universitaet Graz, Austria.

King, C. J. (1980) *Separation Processes*, Mc Graw-Hill, New York.

Knez, Ž., Posel, F. and Senčar, P. (1988) Treatment of pesticide polluted water with supercritical CO2, in *Proc. Int. Symp. on Supercritical Fluids*, Vol. 2, ed. M. Perrut, Institut National Polytechnique de Lorraine, Nice, pp. 777–782.

Knez, Ž., Golob, J., Posel, F. and Krmelj, I. (1990) Extraction of water solutions and solid substances with dense CO2, in *Abstract Handbook 22 Int. Symp. on High Pressure Chemical Engineering*, ed. G. Vetter, Dechema-VDI, Erlangen, pp. 243–249.

Lack, E. and Marr, R. (1986) Mass transfer problems in high pressure extraction, in *ISEC 86 Dechema*, Vol. 3, Munich, pp. 645–651.

Lahiere R. J. and Fair, J. R. (1989) Novel technique to measure equilibria of supercritical solvents and liquid mixtures. *J. Chem. Eng. Data* **34**, 275.

Paulaitis, J. M., J. R. Diandreth, J. R. and Kander, R. G. (1985) An experimental study of phase equilibria for isopropanol–water–CO2 mixtures related to supercritical fluid extraction of organic compounds from aqueous solutions, in *Supercritical Fluid Technology*, eds. J. M. L. Penninger *et al.*, Elsevier, Amsterdam, pp. 149–160.

15 Production of low-fat and low-cholesterol foodstuffs or biological products by supercritical CO₂ extraction: processes and applications

A. CASTERA

Abstract

Food, cosmetic and pharmaceutical industries have taken increasing interest in obtaining high-value products, free of specific undesirable components with unchanged physical properties. Under special conditions of pressure and temperature, processes such as delipidation or decholesterification, are possible by supercritical fluid extraction (SFE). Supercritical pilot-extractor, feasibility of the process and some examples of application are described: low-fat potato chips, defatting and deodorizing corn germs, purified lecithins, low-fat and low-cholesterol yolk powder, dry meat, fish or milk powder, low-cholesterol and fractionated butter. The processing conditions and experimental results are given.

15.1 Introduction

In the last decade, a range of new 'light' food products, especially 'low-fat' and 'low-cholesterol' products, have been available to consumers, who have become more and more concerned about health and nutrition. An extraction process using supercritical fluids (SCF) has been newly developed for the purification or fractionation of biological substrates without changing their physical or chemical properties.

Supercritical carbon dioxide ($SC\text{-}CO_2$) is the most useful SCF because it is non-toxic, inexpensive, non-explosive, and its critical point is easy to reach (about 31°C at 73 bar). $SC\text{-}CO_2$ can extract apolar compounds such as fat and oil, and more specifically triglycerides. However, proteins, carbohydrates, mineral salts or polar lipids (phospholipids or glycolipids) are not or are only slightly soluble in $SC\text{-}CO_2$. $SC\text{-}CO_2$ is also a more versatile solvent than liquid organic solvents because extraction conditions and selectivity can be varied over a wide range of temperature and pressure. This is particularly relevant in the near critical area where small changes in pressure and temperature correspond to large density variations. Another advantage is the easy removal of the solvent from extracted products, by isothermal pressure reduction or by isobaric heating. Accordingly, the food industry has taken more and more interest in SC-CO_2 extraction, especially for delipidation or, in the case of animal products, decholesterification. The first application in the oil and fat industry was the production of partially refined oils, but in this case, the high-value products were not the extracts but the defatted residues.

The production of low-fat and low-cholesterol feedstuffs or biological products by SC-CO_2 has been reported in many patents and papers, especially from Japan and the United States. Generally, the delipidation is partial because fat is very important for the flavor, the texture and the palatability of the final product. The temperature of the process is moderate in order to preserve the thermosensible biochemical compounds, especially to avoid protein denaturation.

The yield by SC-CO_2 extraction depends on the constitution of the product, particularly its water content, the surface area (it is easier to remove fat from a powder than from a paste) and the experimental conditions (temperature, pressure, CO_2 flow rate, extraction time, addition of co-solvent or adsorbant, etc.). The processes generally consist of solid/SC-CO_2 batch extraction, and, in the case of pure fats, SC-CO_2 selectivity also gives a fractionation effect.

15.2 General process

While some differences in laboratory scale or industrial equipment can be found, the general flowchart is basically the same. Therefore, we can describe our SFE plant as an example. The capacity of the extraction and separation vessels, the specification of the compressors, the pressure and temperature control systems and/or the number of separators can change without modifying the fundamental process.

As shown in Figure 15.1, the CO_2 is pumped from a bottle of gaseous CO_2 at about 45–55 bar and this can be purified through an active charcoal bed to filter hydrocarbon residues or water traces. It is then condensed through a heat exchanger and stored in the liquid state. A liquid polar co-solvent such as an alcohol can be introduced to the liquid CO_2 tank with a low-pressure pump (80 bar, maximum flow rate of 400 ml/h). The level of co-solvent in SC-CO_2 is generally in the range of 1–10% (w/w). Liquid CO_2 with or without co-solvent is then pumped with a high pressure compressor that is equipped with a metallic membrane able to deliver CO_2 at a pressure up to 450 bar and a maximum flow rate of 19 l/h. The compressed liquid CO_2 is then heated up to the extraction temperature. Here, the fluid has reached a supercritical state.

The sample charge, generally in a dehydrated and ground form, is introduced into the extractor vessel, a thermostated cylinder with a capacity of 400

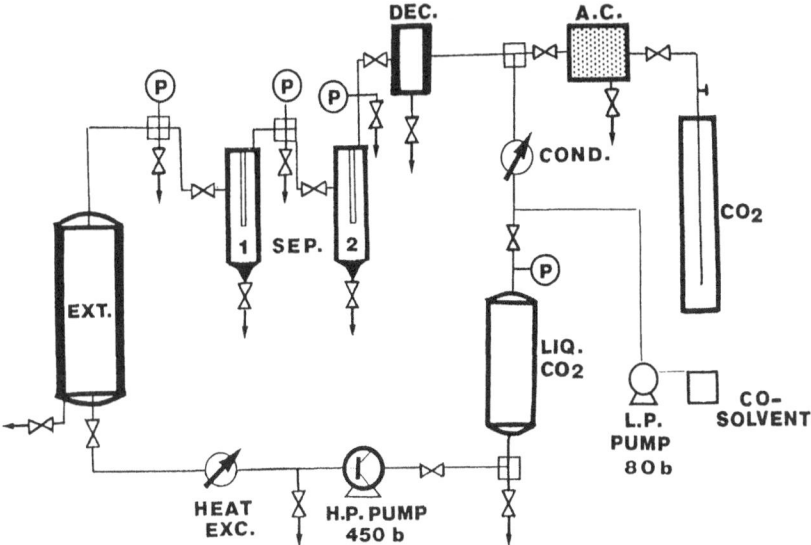

Figure 15.1   Apparatus for SC-CO_2 extraction. A.C., active charcoal tank; COND., condensor; LIQ.CO_2, liquefied CO_2 storage tank; L.P. PUMP, low-pressure co-solvent pump; H.P. PUMP, high-pressure pump; HEAT EXC., heat exchanger, EXT., extraction vessel; SEP. separators 1 and 2; DEC., decantation tank.

ml. Filters or glass wools can be packed into the bottom and the top of the extractor to retain small particles. To increase the contact surface, glass rings or a dispersant such as anhydrous sodium sulfate can be added to the charge. A thermocouple can be inserted into the extraction vessel to monitor its internal temperature. The SC-CO_2 passes through the charge from the bottom to the top of extractor and dissolves lipidic components. Then the discharged CO_2 containing the extracted lipids is pumped into a separator where the CO_2 is depressurized with slight heating to prevent rapid cooling of the solute and CO_2 stream. The cyclone design and the size of the separators help eliminate the risks of dispersion or freezing and therefore minimize the loss in volatile components during the CO_2 vaporization. The pressure in the separators varies from 45–50 bar to 250 bar. Two serial separators allow the extract to fractionate as a function of pressure. Generally the last separator or a decantation vessel is regulated at the CO_2 bottle pressure to condensate the last fractions of extract, co-solvent and water. Then the CO_2 is recycled in the circuit. The extract can be drawn off the bottom of the separators during extraction or recovered in a metallic cylinder flask put inside the separators. The pressure in both the extraction and separation vessels can be controlled by means of manually adjusted back-pressure regulators. Rupture disks and atmospheric pressure valves are installed to provide overpressure protection for the vessels and pressure controllers. Between two experimental batches, the whole extraction system has to be cleaned with compressed air and solvent.

15.3 Food products: defatting applications

One of the first reported applications of SC-CO$_2$ delipidation is the production of low-fat potato chips (Hannigan, 1981). Some previous results have demonstrated a reduction of total fat from about 40% fat by weight to about 20%, without affecting the structure and flavor characteristics of the original product. The total calorie count is cut by about 20% and the protein content is slighty increased from about 1.5 to 2%. The SC-CO$_2$ process involves simply conveying chips from the frier to an extraction vessel through which a stream of CO$_2$ is passed, then oil dissolved in CO$_2$ is routed to a separator before fluid recycling. Although the feasibility of the extraction is evident because fried chips are thin with a large surface area, until now, there has been no industrial production in France. Probably one of the reasons is that the traditional defatting process by drying just-fried chips is less expensive and does not require high pressure equipment.

A process of defatting and deodorizing corn germs by extracting the oleaginous material with SC-CO$_2$ and ethyl alcohol as an extraction aid, was recently described by Yamaguchi (1989). These highly defatted and deodorized corn germs are useful as a raw material for dietary fibers, used as an important constituent of healthy food or as an additive in various kinds of food preparations such as cookies, ice creams, soups, drinks, etc. With only SC-CO$_2$, the defatting of the germs is insufficient and the product retains a relatively strong unpleasant odor. Among several organic solvents such as n-hexane, acetone, methyl alcohol and other alcohols, it has been found that ethyl alcohol has the highest efficiency with respect to the defatting and deodorization effect and is preferable with regard to toxicity. Prior to SC-CO$_2$ treatment, it is preferable to extract dried corn germs by compression or extraction with hexane. Then the supercritical extraction is performed under a pressure of 250–400 bar, at a temperature of 35–50°C and with an amount of ethyl alcohol ranging from 3 to 15% by weight in CO$_2$. The defatted corn germs obtained contain only 0.3% oil by weight or less, from 20 to 25% of protein and 60% of fibrous material (cellulose, lignin, etc.). More advantageously, the defatted corn germs can absorb twice as much water compared to the water absorption in other types of dietary fiber. The processing conditions are given in Table 15.1. No kinetic indication about extraction times was given in the patent.

In a prior patent, Christianson and Friedrich (1985) indicated that the residual lipid of the SC-CO$_2$ defatted corn germ and the peroxidase activity responsible for development of off-flavors during storage were reduced to a fraction of the levels obtainable by conventional hexane extraction methods. Thus, the corn germ flour obtained has an extended shelf-life with acceptable flavor.

In a similar manner, defatting by SC-CO$_2$ extraction can be applied to other starch-containing vegetable matter such as cereal grains or to food oilseeds

Table 15.1 SC-CO_2 extraction of corn germs: experimental conditions and defatting results (Yamaguchi, 1989)

Oil in initial germs (%)	Extraction pressure (bar)	Extraction temperature (°C)	Co-solvent	Oil in defatted germs (%)	Defatted germs aspect	Water absorption (ml/g)
10	250	40	Ethanol 3%	0.29	No odor pale color	11.0
25	350	38	Hexane 3%	0.67	Slight odor pale color	5.5
50	250	40	None	1.51	Strong odor dark color	4.0

such as nuts, almonds, peanuts, etc. (Schwengers, 1976). To maintain organoleptic quality and structure of the product, it is preferable to remove the fat at low temperature, even below the critical point. Thus, the fluid is not in supercritical state but serves as a high pressure liquid. Consequently, the yield depends on the extraction pressure and time, but a partial delipidation is more advantageous because a total defatted grain is crumbly and tasteless. These light-fat products can be used as raw material for cookies, breakfast cereals, sweets, chocolates, etc.

Eldridge *et al.* (1986) and Friedrich and Eldridge (1985) described the preparation and quality evaluation of SC-CO_2 defatted soybean flakes. The conventional hexane method leaves undesirable constituents and traces of residual hexane in the extracted soybean meal giving raw, grassy, beany and bitter flavors at unacceptable levels for human consumption without significant degradation of the nutritional properties. If the moisture content of flakes is first adjusted to about 6.5–15% and then extracted with SC-CO_2, a meal with high organoleptic and protein solubility is produced. The optimal extraction parameters to obtain defatted soybean meal with high nitrogen solubility (greater than 70%) and flavor score greater than 6.5 on a scale of 1 (poor quality) to 10 (excellent quality) are about 830 bar, 85°C and moisture levels of 10.5–11.5%. The presence of water in the flakes causes denaturation of lipoxygenase with a concurrent increase in flavor score, but also a slight decrease in protein solubility. Isolates and alcohol-washed concentrates prepared from SC-CO_2 defatted flour have good cereal-like flavor even after storage at 37°C for 2 months. These high organoleptic and high protein quality soybean products can be used in beverages, soups and baked goods.

Another example of delipidation in the food industry is the purification of crude lecithins. Commercial lecithins or phosphatides are a mixture of phospholipids (about 60–70% by weight), triglycerides (about 30–35% by weight) and some other slightly or very polar lipids such as free fatty acids of glycolipids. Lecithins are largely used in many food, dietary, pharmaceutical and cosmetic products. They are produced in soybean crude oil degumming

Table 15.2 SC-CO_2 extraction of crude soya lecithin: experimental conditions and acetone insoluble content (Castera and Baa-Puyoulet, 1991)

Ratio free volume/sample volume	Extraction pressure (bar)	Extraction temperature (°C)	CO_2 flow-rate (l/h)	Time of extraction (h)	Acetone insoluble content of residue (%)
3/1	300	32	9.5	2.5	94
6/1	400	60	14	3.5	97

operations where phosphatides are hydrated and separated from oil by centrifugation. Because of their polarity, phospholipids are insoluble in acetone and also in SC-CO_2 when triglycerides dissolve in carbon dioxide. Generally de-oiled lecithin is obtained by treating the crude lecithin with acetone and 90–95% phospholipid concentration can be achieved quite readily. The advantage of CO_2 treatment is to obtain pure fractions without residual solvent, so lecithins are directly usable for physiological purposes, and with the physically attractive aspect of a powder. The atomization effect of the SC-CO_2 extraction has been reported in other applications and in this case, it permits the use of de-oiled lecithins in solid foods such as soups or milk powder. The experimental optimum conditions are relatively high temperature and pressure, about 50–60°C and 350–400 bar, to increase triglyceride solubility in CO_2 (Heigel and Hueschens, 1983).

Some results obtained in our laboratory are reported in Table 15.2 (Castera and Baa-Puyoulet, 1991). Commercial liquid lecithin Lucas Meyer with an acetone insoluble content of 62.6%, including 86% of phospholipids and 14% of glycolipids, a triglycerides content of 36.7% and a water content of 0.7% was used. To obtain the atomization effect during the extraction process, the free volume ratio in the extraction vessel/sample volume should be as high as possible. This ratio limits the foam formed, particularly in the first step of the process, because of the emulsifying properties of lecithins. The extraction cylinder was also equipped with several cellulose filters on the top and the bottom and filled with glass rings. The extraction was accomplished in 2–4 h, whereas the kinetic study shows a total removal of the oil after 60 min. The dissolved oil is separated from the solvent at 50 bar and 50°C. The de-oiled lecithin obtained has an acetone insoluble content of about 95–97%. The de-oiled lecithin obtained by SC-CO_2 extraction is more colored but less soya-smelling than those obtained by acetone insolubilization. The residue has the aspects of a friable crystallized foam. The de-oiled lecithin has the same phospholipidic composition as the initial crude lecithin. Glycolipid content in the total polar lipids increases because phospholipids pass with triglycerides in SC-CO_2 ('third solvent' effect) while polar glycolipids are absolutely insoluble in SC-CO_2.

15.4 Food products: defatting and decholesterol applications

In the case of some animal products, i.e. egg, milk powder, fish or meat muscle, and butterfat, the SC-CO_2 extraction enables us to decrease the fat content and reduce the cholesterol content. In recent years, doctors agree that both food cholesterol and saturated fat have a direct effect on blood cholesterol and therefore on the risk of atherosclerosis and cardiovascular diseases. This finding was proved in several epidemiological surveys and animal experimentations. However, at present, this 'lipidic theory' is largely contested because for average people, blood cholesterol does not increase with alimentary cholesterol (CIDIL, 1990), but for the media and consumers, 'cholesterol' in food components is still a 'bad word'. In 1988, a *Wall Street Journal* survey indicated that 60% of American people thought that fat and cholesterol were dangerous for health. In 1987, we read in the *Journal of the American Medical Association* that 60% of people surveyed believed that decreasing food cholesterol could decrease blood cholesterol.

The FAO/OMS committee recommends a cholesterol consumption of 300 mg per day. Lamb's brain and egg yolk have the highest cholesterol content with about 1500–2000 mg/100 g, fish eggs, liver and poultry eggs have a content ranging from 400 to 700 mg/100 g, butter contains about 250 mg/100 g and tallows, lards, fish, shellfish, cheeses, pork products and milk contain less than 100 mg/100 g.

Food research laboratories have developed some processes to reduce the cholesterol content without changing the physical and chemical properties of the products. Because of the relative polarity of the cholesterol molecule due to the free hydroxy group and because of the molecular weight which is more than twice that of triglycerides, the processes to selectively eliminate cholesterol are molecular distillation, precipitation, liquid extraction, adsorption or cyclodextrin complexation. Since 1984, several patents have described removal of cholesterol from foods by SC-CO_2 extraction (MacLachlan *et al.*, 1989; MacLachlan and Catchpole, 1990). SC-CO_2 extraction offers the advantage of removing total lipids and cholesterol in one step, with in some cases a fractionation of fat as a function of unsaturation of the fatty acids and molecular mass of the triglycerides.

As shown in Figure 15.2, the solubility of pure triglycerides increases linearly with increase in pressure. For instance, the solubility of dehydrated butter is 0.41% (w/w) at 150 bar and 40°C while that at 300 bar reaches 1.88% (w/w) (Shishikura *et al.*, 1986). In the presence of water in oil, the solubility increases until 200 bar and then decreases with increasing pressure. The optimum condition of SC-CO_2 extraction at 40°C for the highest selectivity of cholesterol over triglycerides is found to be 150 bar.

In 1987, a Japanese patent (Q.P. Corp, 1987) describes a method that involves contacting dry cholesterol-containing food with SC-CO_2. Under previous extraction, the food was dried by heat-, spray-, air- or freeze-treatment.

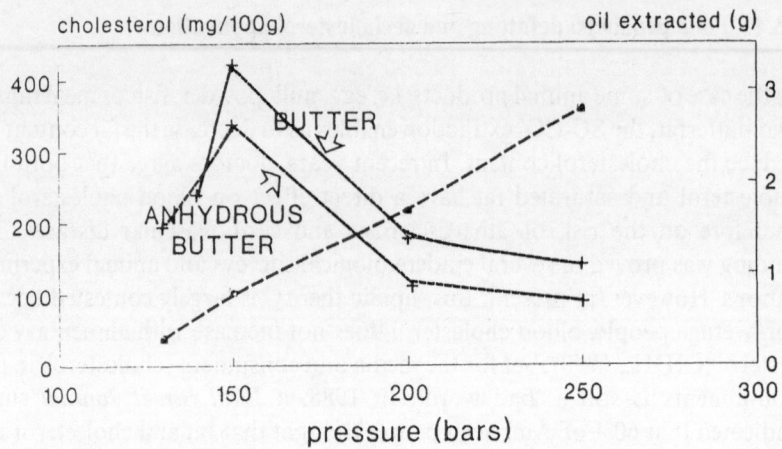

Figure 15.2 Effect of SC-CO2 pressure at 40°C on the extraction efficiency of the cholesterol and the oil from butter.+, cholesterol; – –, butter oil.

The moisture content was reduced below 15%, preferentially to 2–8% because it seems difficult to extract cholesterol from food with a high moisture content. The experimental temperature was 30–45°C and pressure ranged between 130 and 250 bar. In these conditions, cholesterol can be selectively removed without heat-denaturation of proteins. This process can be applied to yolk powder, egg powder, dry meat or milk powder. Another patent (Q.P. Corp, 1984) indicates the possibility of using a vegetable oil as 'third solvent' or 'co-solvent' for facilitating the cholesterol extraction. The decholesterified and partially defatted products obtained can be used for atherosclerosis-risk patients.

In 1990, Froning tested different combinations of pressure and temperature on dried egg yolk as shown in Table 15.3 (Froning *et al.*, 1990). Egg yolk contains about 35% lipids consisting of triglycerides (65%), phospholipids (30%), cholesterol (4%) and carotenoids (traces). When the temperature increases from 40°C to 55°C and the pressure increases from 163 bar to 374 bar more lipids and cholesterol are removed. These results can be explained by increasing fluid density and cholesterol volatility under high temperature/pressure values. Extraction at about 300 bar and 45°C removes approximately two-thirds of the cholesterol and one-third of the total neutral lipids.

A great advantage of SC-CO$_2$ as an extraction solvent is the absence of change in functional and sensory properties because phospholipids and native proteins are not denatured. For instance, the stability of mayonnaise prepared with the lower fat and lower cholesterol egg yolk is better than emulsion prepared with raw egg yolk except for the 374 bar/55°C conditions. Sponge cake volume increases after SC-CO$_2$ extraction except for the 374 bar/55°C conditions, which involve partial denaturation of the proteins. On the other

Table 15.3 SC-CO_2 extraction of spray-dried egg yolk: experimental conditions and composition results (Froning *et al.*, 1990) (CO_2 flow-rate is 45 g/g powder)

Extraction pressure (bar)	Extraction temperature (°C)	Total lipids decrease (%)[a]	Cholesterol decrease (%)[a]	Protein increase (%)[a]	Lecithin increase (%)[a]
163	40	2	16	7	12
238	45	12	28	25	31
306	45	34	66	60	79
374	55	36	66	64	80

[a] Results are expressed with regard to the initial contents of a control product.

hand, SC-CO_2 extraction produces a lighter color dried yolk. Froning's study reveals that the best combination of extraction efficiency and functional quality can be achieved with conditions of 45°C and 300 bar.

In a recent research paper, Rossi *et al.* (1990) describe the SC-CO_2 extraction of egg yolk powder in the presence of ethanol as polar entrainer. Most of the native lipids are bound to the proteins in the form of lipoprotein complexes which represent about 60% of total protein. Increasing the polarity of CO_2 by adding alcohol has a positive effect on the extraction yield of lipid-bearing materials but reduces drastically the selectivity for cholesterol and lecithin. The experimental conditions are 150 bar, 50°C, an extraction time of 1 h and a CO_2 flow rate of about 2 kg/h. Extraction of total lipids increases from about 3% when pure CO_2 is used to about 30% when 7% ethanol is added to SC-CO_2, while protein concentration increases from 40% to 53% and cholesterol content decreases from 2.2% to 1.4% in the residue. Part of the proteins and phospholipids are removed when ethanol is added; then the functional properties of the final product can be changed.

For the same nutritional and marketing purposes as described above, there is great interest in cholesterol reduction and protein concentration of fish or meat muscle to obtain functional ingredients for food manufacturing. On the other hand, some meats and fish with a high proportion of unsaturated fatty acids have relatively short oxidation stability. The rancidity reactions can give unpleasant flavors. In some other cases, such as lamb meat, a strong natural unacceptable smell is associated with the meat fat. The removal of lipids together with the removal of cholesterol has great potential for lipid-containing foods such as meats, fish or dairy products.

SC-CO_2 extraction was studied by Hardardottir and Kinsella (1988) as a method removing lipids and cholesterol from fish muscle. Prior to extraction it is necessary to increase surface area by freezing, then grinding and drying the fish muscle. In this case, results show that the amount of lipids removed from the fish is almost constant regardless of the pressure levels from 135 bar to 324 bar and temperature levels from 40°C to 50°C. Increasing the extraction time from 3 h to 9 h and addition of 10% ethanol in CO_2 increase lipid removal. With an alcoholic entrainer, variations of pressure,

Table 15.4 SC-CO_2 extraction of fish muscle at 275 bar and 40°C: experimental conditions and extraction results (Hardardottir and Kinsella, 1988)

Solvent	Time extraction (h)	Total lipids decrease (%)[a]	Cholesterol decrease (%)[a]
Pure CO_2 780 g/g	9	78 (triglycerides)	97
CO_2 260 g/g + 10% ethanol (w/w)	6	97 (triglycerides + phospholipids)	99.7

[a] Results are expressed with regard to the initial contents of a control product.

Table 15.5 SC-CO_2 Extraction of meat samples at 220 bar and 45°C: composition results (MacLachlan and Catchpole, 1990)

Sample	Initial moisture content (%)	Total lipids decrease (%)[a]	Cholesterol decrease (%)[a]
Chicken	76.2	38.0	3.5
Cooked beef	58.1	47.6	21.5
Dried lean meat	54.0	67.9	38.5
Dried lean meat	45.3	93.7	63.9

[a] Results are expressed with regard to the non-extracted product.

temperature, time or sample size do not significantly affect lipid removal. Ethanol allows the dissociation of phospholipid–protein complexes and thereby improves polar lipids extraction.

As shown in Table 15.4, the yields of defatting and decholesterification are very high. The SC-CO_2 and SC-CO_2/ethanol extracted fish contain 97% and 99% protein, respectively. The increase in protein results from the removal of both fat and moisture. The solubility of the proteins decreases during extraction apparently because of myosin aggregation. Formation of intermolecular hydrogen bonds and hydrophobic interactions are responsible for the insolubilization of some protein components. Of course, this finding limits further food applications for these fish products.

A recent New Zealand patent describes SC-CO_2 defatting of meat samples including the following three steps: reduction in particle size by slicing or flaking frozen meat; moisture removal; and SC-CO_2 extraction (MacLachlan and Catchpole, 1990). The meat flakes are dried to a moisture content in the range 30–55 % w/w by removing all the 'free water'. This is possible by freezing in an inert atmosphere, using a supercritical fluid or cooking. The extraction conditions are pressure in the range 200–300 bar and temperature in the range 30–50°C. As shown in Table 15.5, the moisture content has a great influence on fat and cholesterol extraction efficiency. Drying at lower moisture levels than in the previous QP Corp patent can produce changes in the protein structure and some oxidation and non-enzymatic browning can occur to fat, protein and pigments during storage. The invention also provides a

Table 15.6 SC-CO_2 extraction of ground beef: experimental conditions and defatting results (Chao *et al.*, 1991)

Extraction pressure (bar)	Extraction temperature (°C)	Total lipids decrease (%)[a]	Cholesterol decrease (%)[a]
172	40	17	21
172	50	38	40
310	35	57	26
310	50	73	37

[a] Results are expressed with regard to the initial contents of a control product.

reconstituted meat product, suitable for hamburgers or sausages with low-fat and low-cholesterol content. It has been found that the reconstituted product has an acceptable texture and an improved flavour. After separating cholesterol from the extracted fat by specific adsorption for example on a calcium carbonate bed, some of the cholesterol-free fat can be added back to the protein product prior to or during reconstitution of food product. It has been found that SC-CO_2 also has bacteriostatic properties.

In a very recent paper, Chao *et al.* (1991) discuss the possibility of obtaining beef with the minimum fat necessary for palatability and no cholesterol. The initial moisture content is not discussed. Frozen ground beef containing 19% of fat is extracted by SC-CO_2 with a pressure in the range of 150–300 bar and a temperature in the range of 30–50°C. As shown in Table 15.6, the predominant factor in increasing fat extraction is higher pressure rather than temperature. The lower pressure could be used to preferentially concentrate cholesterol. The distribution of cholesterol in the extracted lipids generally decreases in the initial stage of the extraction and then increases with CO_2 used. In a previous work, King *et al.* (1989) indicated some results obtained on dehydrated and low-fat containing meat samples (with a moisture content of 2% and a fat content of 2–20% by wet weight), such as lard, ham or sausage. Experiments conducted at 345 bar or 690 bar and at 80°C gave a yield of over 96% of the theorical fat content of the meat samples within less than 1 h. The higher the initial water content, the less the total extracted fat. These results are contradictory with some of the previous works. We can conclude that SC-CO_2 extraction efficiency depends largely on the quantity of lipids and water in the sample and also on the presence of lipoprotein and proteo–phospholipid complexes. The level of intramuscular bound lipids is variable with the type of matrix. Therefore, water has two opposite effects: increasing fat removal by cutting polar lipids bound and decreasing fat removal by inhibiting SC-CO_2 contact with the lipid phase. Thus, it is very difficult to estimate the qualitative and quantitative results of a SC-CO_2 extraction before testing it. It is possible to apply the thermodynamic properties of the SCF on pure components and pure fluids such as a study on molecule solubility in SC-CO_2 as a function of the pressure or the CO_2 flow rate. However, in the case of natural products such as foods, the system is not a binary solute/SC-CO_2 system but a more complicated solute/water/SC-CO_2 system with interferences among the different food components.

The last food application discussed here is the removal of cholesterol from butterfat. Butterfat can be fractionated as a function of the molecular weight and the unsaturation of the triglycerides and a more spreadable butter can be obtained. Many papers describe the fractionation of anhydrous butterfat by SC-CO_2 (Kaufmann et al., 1982; Shishikura et al., 1986; Arul et al., 1987; Kankare and Antila, 1989). The short chain triglycerides (C24–C34) are preferentially extracted at pressures in the range of 100–150 bar and temperatures in the range of 40–50°C. The medium chain triglycerides (C36–C42) are more concentrated in oil fractions obtained at 200–250 bar, while triglycerides with a carbon number more than C46 require pressures of about 350 bar and 70°C to be soluble in SC-CO_2. Shishikura et al. (1986) have proposed a SC-CO_2 extraction at 150 bar and 40°C for 4 h. The extract contains 92% of the initial cholesterol and appears as a liquid oil; 75% of the triglycerides have a carbon number less than C40. The residue contains only 8% of the initial cholesterol but appears as a very hard and flavorless butter; 25% of the triglycerides have a carbon number more than C46. From these experiments, it is concluded that the preparation of a low-cholesterol butter oil by simple extraction with SC-CO_2 is impractical. Bradley (1989) proposes two-stage processing at a higher temperature (80°C). The lower pressure first stage (less than 160 bar) allows the extraction of flavors and low melting triglycerides. The residue with cholesterol is conveyed to the second stage higher pressure unit (greater than 170 bar) where cholesterol is removed while high melting triglycerides are not extracted. The light extract and the heavy raffinate both with low-cholesterol content can be mixed to obtain a melting-point range fat equivalent to butter.

The use of a column of an adsorbant can selectively separate cholesterol from neutral oil. The use of silica gel can adsorb 94% and 75% of the initial cholesterol of anhydrous butter oil extracted by SC-CO_2 at 300 bar and 40°C when a silica gel/butter ratio of 3:1 (w/w) and 1:1 (w/w), respectively, is used. The extract is a fluid 'butter-oil' with good taste. On the other hand, the total yield of extracted oil is limited to only 50% when a ratio of 3:1 is used and reaches 80% of the applied butterfat when a ratio of 1:1 is used (Shishikura et al., 1986). A world patent describes the possibility of using a basic adsorbant such as calcium hydroxide, calcium oxide, calcium or magnesium carbonate or the like (MacLachlan and Catchpole, 1990). A basic adsorbant is more suitable than an acid adsorbant because triglycerides, flavor components and pigments are not adsorbed. The experimental conditions are an extraction pressure in the range of 200–300 bar, an extraction temperature in the range of 30–50°C and a fat/adsorbant ratio of 0.1–10 g/100 g. The moisture content of the adsorbant has to be as low as possible. As shown in Table 15.7, by using a calcium hydroxide adsorbant, 100% of the cholesterol is retained in the bed

Table 15.7 SC-CO$_2$ extraction of butterfat and lard: experimental conditions and adsorption results (MacLachlan and Catchpole, 1990)

Sample	Initial cholesterol content (mg/100 g)	Adsorbant and fat/adsorbant ratio (g/g)	Pressure (bar)	Temperature (°C)	Triglycerides recovered (%)	Cholesterol adsorbed (%)
Butter	242	Ca(OH)$_2$ 0.5/45	220	45	80	100
Lard	991	Ca(OH)$_2$ 0.5/40	220	35	97.7	100
Lard	991	CaCO$_3$ 0.5/40	220	45	100	99.7
Butter	242	MgO 2/45	220	35	59.7	100

of adsorbant and 80% of the triglycerides are recovered. The extracted butter oil has a lower melting point than initial butterfat because of simultaneous fractionation. At present, in France, two low-cholesterol content butters are proposed for consumers; neither is produced by SC-CO$_2$ extraction but they are both obtained by cyclodextrin complexation. Several industrial laboratories have tried to optimize SC-CO$_2$ decholesterification of butterfat but the results have not yet been published.

15.5 Biological products: decholesterol applications

SC-CO$_2$ extraction provides a method of obtaining the desired components of animal tissues, cells and organs (Kamarei, 1987). Lipid rich organs such as liver, brain, kidney and epithelial tissues are rich sources of necessary biological products such as sphingolipids, steroids, sterols, vitamins, pigments, terpenes, prostaglandins, waxes, etc. According to the choice of SCF, experimental conditions and animal source material, peptides, hormones, enzymes, saccharides, amino acids can also be extracted. Different fluids affirmed as safe by the FDA, such as CO$_2$, N$_2$, He, C$_3$H$_8$, can be used according to each tissue and each desired extract. Undesirable substances may be removed from a sample following the same process. In some high-value natural lipidic extracts rich in polar lipids such as glycolipids, sphingomyelin, cerebrosides, the cholesterol is particularly undesirable for further pharmaceutical or cosmetic applications. The same SC-CO$_2$ processes as described before can be applied.

In our laboratory, SC-CO$_2$ decholesterification was applied to solvent extracted brain polar lipid mixtures. The extracts contained cholesterol in the range from 15% to 40% (w/w). The initial moisture content was below 5% (w/w). After freezing and grinding, the product was extracted under different experimental conditions, i.e. extraction pressure ranging from 150 bar to 300 bar and extraction temperature ranging from 35°C to 55°C. Time of extraction was varied from 3 h to 8 h. As shown in Fig. 15.3, the cholesterol extraction efficiency depends greatly on the time of extraction and also on the initial

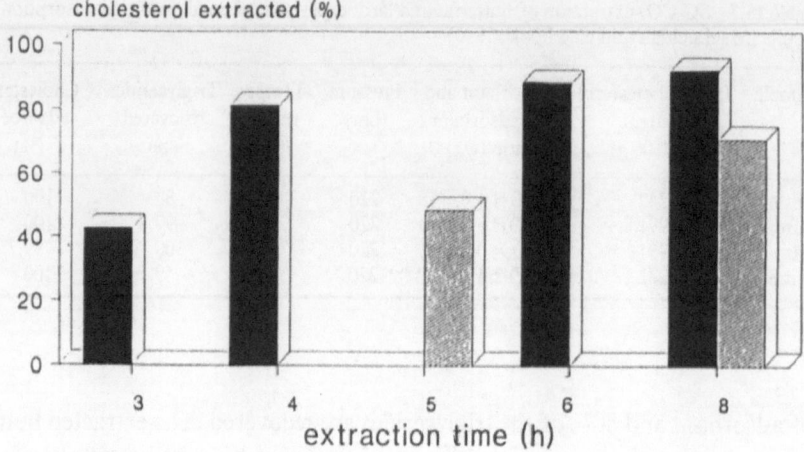

Figure 15.3 Effect of SC-CO$_2$ extraction time and initial cholesterol content at 150 bar and 50°C on the extraction efficiency of the cholesterol from brain lipidic extracts. Solid blocks, initial chol. 20–25%, shaded blocks, initial chol. 40%.

cholesterol content. For a cholesterol content of about 20–25% (w/w), 88% and 92% of the cholesterol was extracted in 6 h and 8 h at 150 bar and 35°C, respectively. Under the same experimental conditions, 92% and 71% of the cholesterol was extracted from a 20% and 40% (w/w) initial cholesterol content, respectively. An increase in pressure reduced total extracted cholesterol as seen previously in other results. On the other hand, an increase in temperature increased the cholesterol extraction efficiency: 43% and 76% of the initial cholesterol were extracted at 40°C and 55°C, respectively. The decholesterified product is a fine colorless powder. The atomization effect was improved by the sample mass/extractor volume ratio. In our experiments, charges of about 10–25 g were introduced to a 400-ml extraction vessel. These results have proved the feasibility of cholesterol removal by SC-CO$_2$ extraction of complex natural mixtures. Because SC-CO$_2$ is non-toxic and is totally removed from the extract and the product, such a process can be applied to biological materials.

References

Arul, J., Boudreau, A., Makhlouf, J., Tardif, R. and Sahasrabudhe, M. R. (1987) Fractionation of anhydrous milk fat by supercritical carbon dioxide. *J. Food Sci.*, **52**, 1231.

Bradley, R. L., (1989) Removal of cholesterol from milk fat using supercritical carbon dioxide. *J. Dairy Sci.* **72**, 2834.

Castera, A. and Baa-Puyoulet, P. (1991) Possibilités de purification et de fractionnement des lécithines de soja à l'aide de CO$_2$ supercritique. Unpublished.

Chao, R. R., Mulvaney, S. J., Bailey, M. E. and Fernando, L. N. (1991) Supercritical CO$_2$ conditions affecting extraction of lipid and cholesterol from ground beef. *J. Food Sci.* **56**, 183.

Christianson, D. D. and Friedrich, J. P. (1985) Production of food-grade corn germ product by supercritical fluid extraction, US patent 4, 495, 207.

CIDIL, UNAFORMEC (1990) *Cholesterol et Prévention Primaire*. Colloque International, Paris, 160 pp.

Eldridge, A. C., Friedrich, J. P., Warner, K. and Kwolek, W. F. (1986) Preparation and evaluation of supercritical carbon dioxide defatted soybean flakes. *J. Food Sci.* **51,** 584.

Friedrich, J. P. and Eldridge, A. C. (1985) Production of defatted soybean products by supercritical fluid extraction, US patent 4,493,854.

Froning, G. W., Wehling, R. L., Cuppett, S. L., Pierce, M. M., Niemann, L. and Siekman, D. K. (1990) Extraction of cholesterol and other lipids from dried egg yolk using supercritical carbon dioxide. *J. Food Sci.* **55,** 95.

Hannigan, K. J. (1981) Extraction process creates low-fat potato chips. *Food Eng.* **7,** 77.

Hardardottir, I. and Kinsella, J. E. (1988) Extraction of lipid and cholesterol from fish muscle with supercritical fluids. *J. Food Sci.* **53,** 1656.

Heigel, W. and Hueschens, R. (1983) Process for the production of pure lecithin directly usable for physiological purposes, US patent 4, 367, 178.

Kamarei, A. (1987) Supercritical fluid extraction of animal derived materials, World patent 87/02697.

Kankare, V. V. and Antila, V. (1989) Extraktion von milchfett mit überkritischem kohlendioxid. *Fat Sci. Technol.* **91,** 485.

Kaufmann, W., Biernoth, G., Frede, G., Merk, W., Precht, D. and Timmen, H. (1982) Fraktionierung von butterfett durch extraktion mit überkritischem CO_2. *Milchwissenschaft* **37,** 92.

King, J. W., Johnson, J. H. and Friedrich, J. P. (1989) Extraction of fat tissue from meat products with supercritical carbon dioxide. *J. Agric. Food Chem.* **37,** 951.

MacLachlan, C. N. S. and Catchpole, O. J. (1990) Separation of sterols from lipids, World patent 90/02788.

MacLachlan, C. N. S., Catchpole, O. J. and Nicol, R. S. (1989) Removal of lipids from foodstuffs, European patent 0, 356, 165.

Q. P. Corp (1984) Production of oil or fat containing larger amount of cholesterol by extraction dried cholesterol source with supercritical dioxide, Japanese patent 59140299.

Q. P. Corp (1987) Low cholesterol food preparation by cholesterol extraction with supercritical carbon dioxide, Japanese patent 87051092.

Rossi, M., Spedicato, E. and Schiraldi, A. (1990) Improvement of supercritical CO_2 extraction of egg lipids by means of ethanolic entrainer. *Ital. J. Food Sci.* **4,** 249.

Schwengers, D. (1976) Extraction of fat from starch-containing vegetable matter, US patent 3 939 281.

Shishikura, A., Fujimoto, K., Kaneda, T., Arai, K. and Saito, S. (1986) Modification of butter oil by extraction with supercritical carbon dioxide. *Agric. Biol. Chem.* **50,** 1209.

Yamaguchi, M. (1989) Method for the preparation of defatted corn germs, European patent 0,367,128.

16 Fractionation of beef tallow with supercritical CO$_2$

R. R. CHAO, S. J. MULVANEY, M. E. BAILEY and
H. HUANG

Abstract

Fractionation of beef tallow (BT) with supercritical CO$_2$ (SC-CO$_2$) was studied using extraction pressures of 138, 241, and 345 bar at 40 and 50°C, and a separation pressure of 34.5 bar at 40°C. Lipid extractability increased with increased pressure. A retrograde phenomenon of lipid extractability was observed at 170–175 bar. Beef tallow fractionated at lower pressures, however, contained higher amounts of cholesterol concentration, [chol], than that extracted at higher pressures. Beef tallow was also extracted at 345 bar/40°C and fractionated using three separators connected in series with a decreasing pressure profile of 173 bar, 117 bar and 34.5 bar at 40°C. The results indicated that the lipid fractions with higher [chol] were obtained from the separator at 34.5 bar/40°C. The relationship between the affinity of cholesterol for different triglycerides is discussed with respect to the fatty acid composition and differential scanning calorimetry (DSC) thermograms for selected fractions.

16.1 Introduction

BT is commonly used as a fat medium in commercial frying operations not because it is cheaper than vegetable oils, but simply because of 'customer preference' for the unique and desirable flavors it contributes to the fried products, which partially hydrogenated vegetable fats are not able to provide (Ha and Lindsay, 1991). BT is also an excellent source of highly functional products for bakery applications because of its unique triglyceride composition. In terms of *trans* isomeric monoenes, which may have adverse nutritional and biochemical effects on humans, BT contains only 4% naturally occurring *trans* fatty acids calculated as elaidic acid compared with some hydrogenated vegetable oils which may contain up to 20–40% *trans* fatty acids (Luddy *et al.*, 1973). Over the past several years, however, usage of BT has declined. This is attributed to continued health concerns over meat fat consumption reactive to various questions relative to cholesterol risks with heart disease. Moreover, studies cited by Ryan *et al.* (1981) Ryan and Gray (1984) and Nawar *et al.* (1991) have shown that when cholesterol is heated in air it readily undergoes oxidation to form products that might be angiotoxic, cytotoxic, and/or carcinogenic. The reduced usage of BT and the keen competition from hydrogenated vegetable oils have prompted a renewed interest in developing fractionated BT, with altered cholesterol and fatty acid composition distribution, so that it might be used more satisfactorily as a fat frying medium.

Recently, research interest has focused on using supercritical CO_2 (SC-CO_2) to fractionate fats and oils in an attempt to stimulate new uses for these commodities and also to recycle by-products of processes utilizing these fats. Diverse applications of supercritical fluid extraction (SFE) to obtain edible oils and fats from natural products have been studied including milk fat (Biernoth and Merk, 1985; Shishikura *et al.*, 1986; Arul *et al.*, 1987; Chao *et al.*, 1988; Kankare *et al.*, 1989), oil seeds (Friedrich and Pryde, 1984; List *et al.*, 1984; Christianson and Friedrich, 1985; Eldridge *et al.*, 1986; Favati *et al.*, 1991), color seeds (Chao *et al.*, 1991b), meats (Hardardottir and Kinsella, 1988; Chao *et al.*, 1991a), marine oils (Eisenbach, 1984; Nilsson *et al.*, 1988, 1989; Rizvi *et al.*, 1988; Yeh *et al.*, 1991), and rice bran (Zhao *et al.*, 1987). It is evident from data presented in these studies that SFE represents a viable alternative process for the fractionation and separation of fats and oils. The aim of this study was to develop a suitable method using SC-CO_2 to fractionate BT into 'tailor-made' edible ingredients that may be used as ingredients in food manufacturing. This work was designed to determine the effects of different SFE operating conditions on the cholesterol, fatty acid and other fat components of BT extracts.

16.2 Materials and experimental methods

16.2.1 Supercritical fluid extraction (SFE)

A SFE unit manufactured by Newport Scientific Inc. (Jessup, MD) and modified by our laboratory to include two accessory separators (Figure 16.1) was used to extract the soluble solutes from BT. Since each separator was equipped with a back pressure regulator, the pressure of the separators could be individually adjusted from 0 to 414 bar. A detailed description of the operation of the system has been presented elsewhere (Chao *et al.*, 1991a), so only a brief summary of the operation is given here for its relevance to the discussion of the results. Commercial grade crude beef tallow was used for the studies and carefully melted, mixed, and subdivided into smaller plastic containers of 200 g BT each and stored at −27°C until used.

For each experimental trial, about 100 g or 200 g of BT was charged into a sealed extraction vessel. SC-CO_2 at the desired pressure and temperature was used for extraction at 138–345 bar and 40/50°C, respectively. The solute-laden

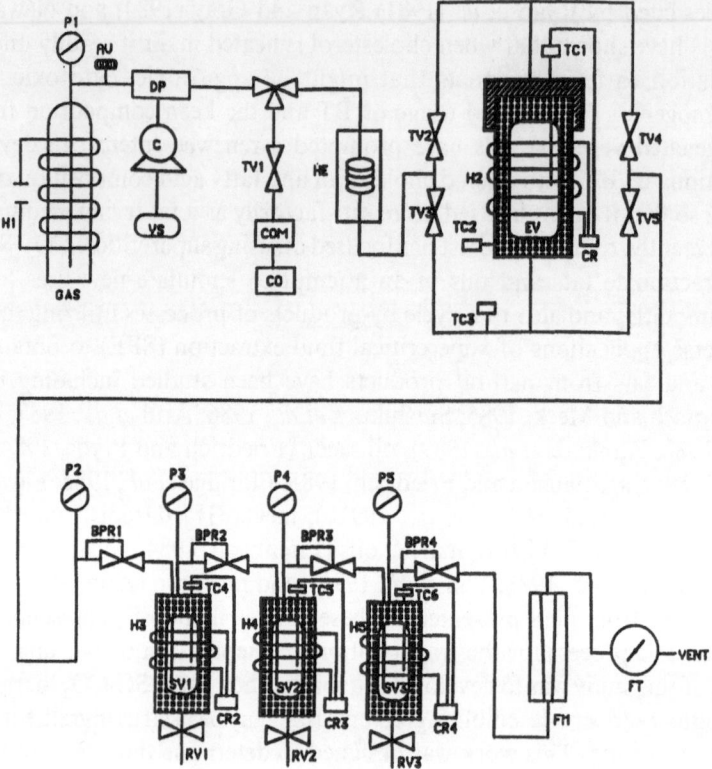

Figure 16.1 SFE unit with three separators. AV, adjusting valve; BPR1–BPR4, back pressure regulators; CO + COM co-solvent pump; EV, extraction vessel; FM, flow meter; FT, flow totalizer, H1–H5, heaters; HE, water bath; P1–P5, pressure gauges; RV1–RV3, pressure relief valves; TC1–TC6, thermocouples; TV1–TV5, two-way through valves; VS + C + DP, variable speed diaphragm compressor.

SC-CO_2 stream leaving the extractor was then reduced through a back pressure regulator to 34.5 bar/40°C, where the solutes were precipitated and collected in one or all three separators, depending on test requirements.

The flow rate and volume of CO_2 were measured with a flow meter and a totalizer, respectively. The mass flow rate of CO_2 measured at 20°C was maintained at 3.40 kg/h and the extraction time ranged from 4 to 6 h or until BT was totally extracted. Extracted samples were collected for each 1.57 kg of CO_2 used and stored in a dark room at −27°C until used for chemical analysis.

16.2.2 *Determination of cholesterol content*

Preparation of samples for cholesterol content was based on a modified AOAC method (AOAC, 1984), using trimethylsilyl (TMS) ethers and derivatives. To make the TMS derivatives, 60 μl TMS imidazole and 40 μl *N*, *O*-bis-trimethylsilyltrifluoroacetamide (containing 1% TMS chloride) (Pierce Chemical Co.) was added to the dried cholesterol sample in a 1.0 ml vial. The sample was injected into a Supelco SPB-1, fused silica capillary column (30 m × 0.32 mm i.d.) in a Varian Model 3700 Gas Chromatograph equipped with dual flame ionization detectors. The initial holdup time was 4 min at 270°C followed by a programmed temperature ramp to 300°C at a ramp rate of 10°C/min. Helium flow rate and split ratio were set at 1.5 ml/min and 50:1, respectively, while the injector/detector temperature was set at 310°C.

16.2.3   Determination of fatty acid content

The preparation of samples for total fatty acid analysis was modified from the AOAC method 28.056 (AOAC, 1984) by changing heptane to hexane and employing a Supelco SP-2330 capillary column, 30 m × 0.32 mm i.d. The preparation of the sample for cholesterol determination was modified from JAOAC method 28.110 (AOAC, 1984). Using the same Varian Model 3700 Gas Chromatograph discussed above, samples were injected onto a Supelco SPB-1, fused silica capillary column (30 m × 0.32 mm i.d.) at 27°C with an initial holdup time of 4 mins, then programmed to 300°C at a ramp rate of 10°C/min. The helium flow rate was 1.5 ml/min and the split ratio was 50:1. The temperature of injector and detector were both set at 310°C. Regular calibration was performed for both the cholesterol and fatty acids methods to maintain the standard deviation of accuracy of GC results at 1%.

16.2.4   Differential scanning calorimetry of lipid samples

DSC thermograms were measured on 7–9 mg lipid samples with a Perkin-Elmer (Norwalk, CT) DSC-7 instrument. The instrument was calibrated regularly against standards of pure indium. The method developed by Bentz and Breidenbach (1969) was followed with minor modification for sample conditions. Each sample was cooled using the built-in refrigeration system to 0°C for 3 min before increasing the temperature from 0 to 50°C with a scanning rate of 5°C/min. Purge gas was 99.99% nitrogen and the gas flow rate was 25–30 ml/min.

16.3   Results and discussion

16.3.1   Single-pass SFE

16.3.1.1   Extraction yield of lipid. The cumulative yields of the lipid fractions collected during extraction of 100 g charged weight of BT for various operating conditions are shown in Figure 16.2. The SC-CO$_2$ needed to extract

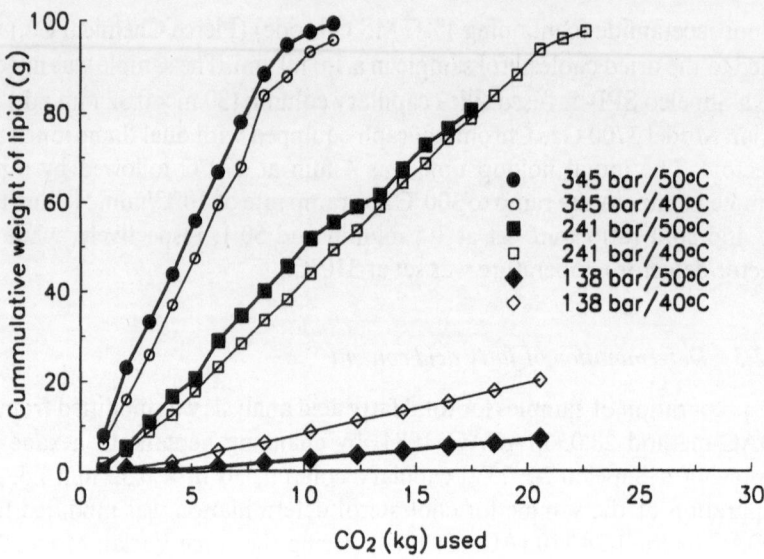

Figure 16.2 Cumulative weight of beef tallow extracted by CO_2 with extraction pressure and temperature varied from 138 to 345 bar and 40–50°C, respectively. ● (50°C), O (40°C), 345 bar; ■, □, 241 bar; ♦, ◊, 138 bar.

all BT charged in the extractor at 345 bar and 241 bar was approximately 10 kg and 22 kg, respectively. However, at the lower pressure (138 bar), only 22% of total BT was extracted after 20 kg CO_2 was used. The estimated constant extraction rate of BT for the run at 345 bar was 11.5 g/kg CO_2, for up to 8 kg CO_2 passed through the extractor. The constant extraction rate for the run at 241 bar was 5.0 g/kg CO_2 for up to 20 kg CO_2 used and 0.4–0.8 g/kg CO_2 for the 138 bar runs, depending upon the temperature. The results indicated that the lipid solubility of SC-CO_2 is heavily dependent upon the applied pressure/temperature conditions corresponding to different densities of SC-CO_2. Examining the effect of temperature on the extractability of lipid yield revealed that increasing the extraction temperature from 40 to 50°C at 138 bar decreased the lipid yield from 22 to 5%. For extractions carried out at 241 bar or 345 bar, the lipid yield extracted at 50°C was generally higher than that extracted at 40°C. This is evidence for the existence of a retrograde extraction pressure around 170–175 bar with respect to lipid solubility as reported by Stahl *et al.* (1988) for soybean and jojoba oils.

16.3.1.2 Extraction yield of cholesterol. The results of cholesterol concentration, [chol], expressed as mg cholesterol/100 g lipid for selected fractions from various experimental trials are given in Figure 16.3. As expected, the [chol] in fractions extracted at 345 bar was the lowest, followed by that extracted at 241 bar, and the highest concentration was at 138 bar. Therefore, extracting BT with the lowest pressure led to the highest selectivity for cholesterol.

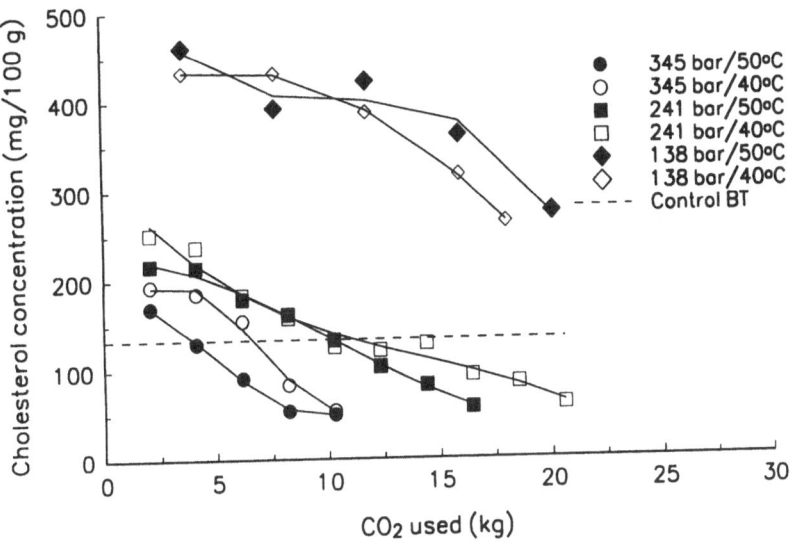

Figure 16.3 Cholesterol concentration of selected fractions of beef tallow extracted by CO_2 with extraction pressure and temperature varied from 138 to 345 bar and from 40 to 50°C, respectively. ● (50°C), ○ (40°C), 345 bar; ■, □, 241 bar; ◆, ◊, 138 bar; – –, Control BT.

Comparing the results of the yields of extracted lipid and cholesterol in Figures 16.2 and 16.3 it was found that within the range of constant extraction rates at 241 and 345 bar at 40°C, i.e. 5.0 and 11.5 g/kg CO_2, the average cholesterol yield was 7–9 mg and 18–20 mg, respectively. These results indicated that the yield of extracted cholesterol increased with increased lipid yield. In addition, while the lipid yield was essentially constant for fractions collected after 7 kg CO_2 for the 345 bar/40°C run and 15 kg CO_2 for the 241 bar/40°C run, their corresponding [chol] were notably decreased. A material balance revealed that this decrease in [chol] for both runs did not occur until the [chol] of the BT remaining in the extractor was reduced from the original 130–160 mg/100 g to 70–80 mg/100 g lipid.

16.3.1.3 Compositional variation of fatty acids. The effects of extraction conditions on the variation of lipid composition, especially the fatty acids of nutritional concern, need to be evaluated. Table 16.1 gives a list of the initial and latter fractions of the runs to reveal the general trend in the variation in fatty acid content. In the case of the extractions at 241 bar and 345 bar, the concentration of myristic acid, [C14:0], from those fractions collected at 2 or 4 kg CO_2 used were generally lower than those collected at 10 or 16 kg CO_2 used, despite the difference in temperatures. Similarly, the concentrations of palmitic acid, [C16:0], and palmitoleic acid, [C16:1], were also slightly decreased. In contrast to the trend that occurred in [C14:0], the concentrations

Table 16.1 Total fatty acid distribution of fractions of beef tallow fractionated by supercritical CO_2 at different conditions (charged weight 100 g)

Pressure (bar)	Temperature (°C)	CO_2 (kg)	Fatty acid composition (%)						
			C14:0	C16:0	C16:1	C18:0	C18:1	C18:2	C18:3
138	40	4	4.0	22.4	3.7	11.2	38.9	3.5	0.6
		16	4.0	23.5	3.8	12.1	40.9	3.7	0.9
	50	4	3.9	21.0	3.3	10.5	45.7	3.1	0.6
		16	4.4	23.5	3.7	11.5	36.7	3.3	0.6
241	40	4	3.8	22.9	3.5	12.9	38.9	3.2	0.7
		16	2.3	22.4	2.9	17.7	44.9	3.9	1.2
	50	4	3.7	22.6	3.3	12.5	37.6	3.1	0.7
		16	2.7	22.9	3.1	16.5	43.2	3.4	0.5
345	40	2	3.5	22.8	3.2	13.3	39.2	3.7	0.7
		10	2.0	20.9	2.5	19.2	44.3	3.6	1.0
	50	2	3.6	23.4	3.3	14.1	38.4	3.2	0.5
		10	1.9	21.4	2.4	20.6	44.4	3.0	0.6
Unextracted BT			3.3	22.5	3.1	14.3	40.4	3.5	0.5

of both stearic acid, [C18:0], and oleic acid, [C18:1], were generally increased with increased extraction pressures, especially in the latter fraction collected at 345 bar. However, no similar changes were observed for the essential fatty acids linoleic, C18:2, and linolenic, C18:3, although they are homologous to C18:0 and C18:1. More research is needed to establish the relationship of the degree of unsaturation among C18 homologs and the selectivity of extraction by SC-CO_2 under different extraction conditions.

The compositional variation of the fatty acids described above were generally small for the 138 bar runs despite the long extraction time and different temperatures used. The reason for the erratic change in [C18:1] was not known and further study is needed to determine why the concentration of [C18:1] is erratic during extraction. Compared with the unextracted BT, the extractions obtained from the 138 bar runs generally contained higher [C14:0] and lower [C18:0], but the same level of other fatty acids.

16.3.2 Use of the multiple separators

The fractions in the above SFE runs were collected in only one separator at 34.5 bar/40°C. It is well known that the lipid solubility of CO_2 generally depends upon the density of CO_2 which can be manipulated through the variation of pressure. Hence, with the use of multiple separators, the pressure drop from the extractor to the multiple separators can be reduced gradually by adjusting the individual back pressure regulator of each separator. Using this procedure, the soluble lipids in SC-CO_2 leaving the extractor can be further separated into the multiple separators. In this study, the separation pressures

Table 16.2 Weights of lipid and cholesterol of selected beef tallow fractions extracted at 345 bar/40°C and fractionated by the order of 172 bar (S1), 103 bar (S2) and 34.5 bar (S3) at 40°C (charged weight of BT 200 g)

CO$_2$ (kg)	Separator	Weight of fraction (g)	[chol] (mg/100 g)	Weight of cholesterol (mg)
6	S1	10.2	128	13.0
	S2	4.1	171	6.9
	S3	7.1	433	30.7
12	S1	11.2	92	10.3
	S2	5.1	126	6.4
	S3	10.3	376	38.8
18	S1	11.9	49	5.8
	S2	4.4	75	3.3
	S3	10.7	272	29.2

of the three separators were adjusted at 170 bar/40°C for the first separator (S1), 102 bar/40°C for the second separator (S2), and 34 bar/40°C for the third separator (S3). The yields of lipid and cholesterol from the selected fractions collected from the three separators are given in Table 16.2. In order to accumulate enough sample, the fractions from S3 were collected at every 6 kg CO$_2$ used, while the fractions from S1 and S2 were collected at every 2 kg CO$_2$ used. The data revealed that the fractions from S3 contained significantly higher [chol] than the fractions from S2 and S1. Furthermore, for both fractions S1 and S2 the extracted lipid weight of fractions was maintained fairly constant as extraction ensued, but the [chol] actually decreased. A material balance for this experiment showed that the fractions obtained from S3 contained 17% of the extracted components, but 31–38% of the cholesterol based on the total accountable yield of lipid and cholesterol, respectively. Apparently, more cholesterol was carried over by SC-CO$_2$ to S3 than that separated in S1 and S2. The application of multiple separators, therefore, allows one to obtain relatively small quantities of lipid with markedly higher amounts of cholesterol. Therefore, a more selective separation of cholesterol from BT with a higher yield of cholesterol-reduced lipid is possible.

Since the solute-laden SC-CO$_2$ leaving the extractor was further fractionated with multiple separators, the effect of separation pressure on the compositional variation of fatty acids of the fractions from each separator was determined. Table 16.3 shows the composition of fatty acids for three fractions from three separators collected at 6 (F6), 12 (F12), and 18 (F18) kg CO$_2$ used. For each fraction, [C14:0] and [C16:1] consistently increased, whereas [C18:0] and [C18:1] decreased in order from S1 to S3. At the same time, [C16:0], [C18:2], and [C18:3] were unchanged. When the fractions with respect to each separator were compared, [C14:0] and [C16:1] decreased whereas [C18:0] and [C18:1] increased. Therefore, as extraction progressed, fractions consisting of

Table 16.3 Total fatty acid distribution of selected fractions of beef tallow extracted by CO_2 at 345 bar and 40°C using three separators (charged weight 200 g)

Fraction number	Separator	Fatty acid composition (%)						
		C14:0	C16:0	C16:1	C18:0	C18:1	C18:2	C18:3
F6	S1	3.2	23.0	3.1	16.2	36.4	2.5	0.3
	S2	4.1	24.4	3.6	14.2	35.1	2.5	0.3
	S3	4.8	22.1	3.8	13.8	33.4	2.5	0.3
F12	S1	2.7	22.3	2.8	17.6	36.7	2.5	0.3
	S2	3.5	23.6	3.3	15.2	35.4	2.6	0.3
	S3	4.8	23.4	3.7	11.4	31.0	2.4	0.3
F18	S1	2.1	21.3	2.5	21.1	38.8	2.6	0.3
	S2	3.0	24.1	3.1	18.4	39.2	2.7	0.4
	S3	4.2	23.3	3.4	12.9	32.4	2.6	0.3
Control		3.3	22.5	3.1	14.3	40.4	3.5	0.5

triglycerides with more fatty acid moieties of C18:0 and C18:1 and less C14:0 and C16:1 were concentrated in S1. At the same time, triglycerides containing higher C14:1 and C16:1 but less C18:0 and C18:1 were carried through S1 and S2 by SC-CO_2 and eventually precipitated in S3.

The results presented here for the variation of the composition of fatty acids and the extractability of cholesterol in BT during extraction agrees with results reported by Arul *et al.* (1987) for milk fat. While fractionating milk fat into liquid (12%), intermediate (38%), and solid (50%) portions with SC-CO_2 at 200 bar/80°C, Arul *et al.* reported that about 36% of the total cholesterol was found in the liquid portion, which was rich in low- and medium-chain triglycerides and fatty acids. They concluded that cholesterol had high affinity for the low- and medium-chain triglycerides. They also explained that because of the low density of SC-CO_2 used at 200 bar/80°C, long-chain triglycerides with unsaturated fatty acids, which also had high affinity for cholesterol were not easily extracted. The pressure used in our study was 345 bar, which was higher than that used by Arul *et al.* (1987). Our results revealed a high [chol] in fractions from S3 compared to S1 and S2. It is possible that the large portion of C18 homologs contained in BT relative to milk fat, may enhance the aggregation or affinity of cholesterol for long-chain triglycerides which makes it difficult for SC-CO_2 to extract cholesterol.

The same lipid fractions mentioned in Table 16.2 were also analyzed by DSC to determine the effect of composition on their melting behavior from 0 to 50°C. To simplify the discussion, reported results have been limited to the variation of the lipid components between the fractions from S1 and S3. The DSC thermograms presented in Figure 16.4 reveal that heat flow around the temperature zone of 35–45°C were notably increased as one compared F18S1 (the fraction collected from S1 after 18 kg CO_2 was used) with F6S1 (the

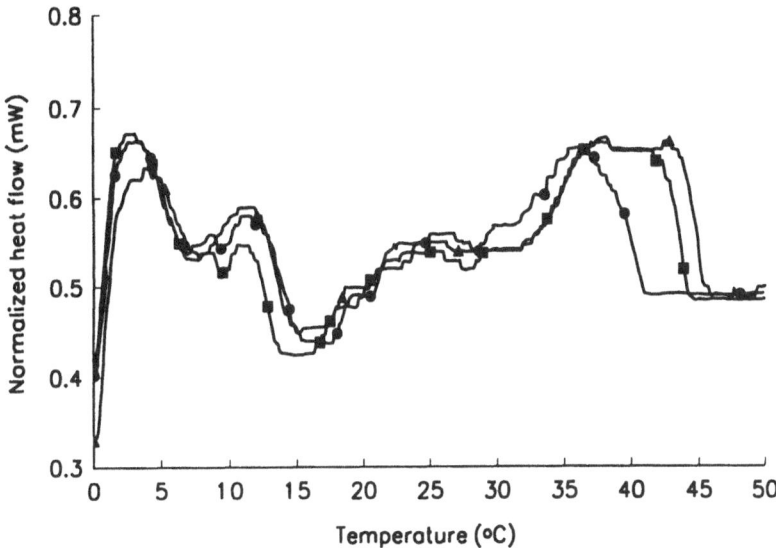

Figure 16.4 DSC thermograms of three beef tallow fractions, F6S1 (●), F12S1 (■), and F18S1 (▲) extracted by CO₂ at 345 bar/40°C and collected from the first separator at 117 bar/40°C after the passage of 6, 12, and 18 kg CO₂, respectively.

fraction obtained from S1 after 6 kg CO₂ was used). Apparently, extraction with SC-CO₂ caused a gradual shift with time towards extracting lipid components melting at temperatures higher than 35°C. Comparing the DSC thermograms (Figure 16.5) between the fractions from S1 and S3, a broader peak at 10°C and a more distinct high peak at 35–40°C occurred. The results indicated that despite the use of multiseparators, the lipid components melting at 35–45°C were sharply reduced as the extraction progressed.

16.4 Conclusions

Manipulation of extraction and separation conditions resulted in lipid fractions from BT with markedly high cholesterol content and high yield of cholesterol-reduced BT. Essentially, high extraction pressure coupled with gradual reduction of the density of SC-CO₂ during separation is needed to fractionate cholesterol efficiently with short extraction time and with minimal use of CO₂. Those fractions containing high [chol] generally consisted of high [C14:1] and low [C18:0] and [C18:1]. The selectivity of extractable lipid components by SC-CO₂ can be fine tuned by the effect of retrograde behavior on the lipid solubility. Research is now being conducted to determine the effect of various adjustments of separation pressures on the separation of cholesterol from those lipid components responsible for the unique flavors of BT.

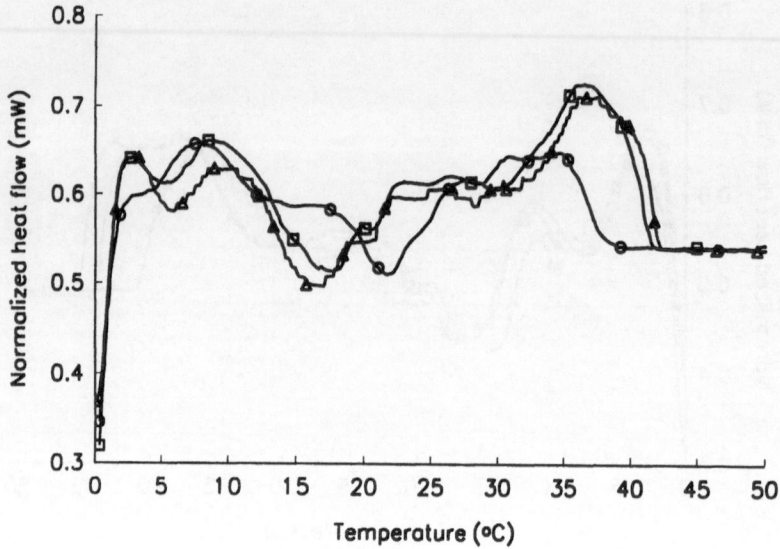

Figure 16.5   DSC thermograms of three beef tallow fractions, F6S3 (O), F12S3 (□), and F18S3 (△) extracted by CO_2 at 345 bar/40°C and collected from the third separator at 34.5 bar/40°C after the passage of 6, 12, and 18 kg CO_2, respectively.

Acknowledgements

Contribution from the Missouri Agricultural Experiment Station Journal Series No. 11616 is acknowledged. The authors are grateful to the Missouri Beef Industry Council for financial assistance (Grant no. C-5-30995). Thanks are also extended to technical persons of the Agricultural Experimental Station for their assistance in determining the cholesterol and fatty acid contents for this study.

References

AOAC (1984) Gas chromatography of sterols, in *Official Methods of Analysis*, 14th edition, Association of Official Analytical Chemists, Washington, DC.

Arul, J., Boudreau, A., Makhlouf, J., Tardif, R. and Sahasrabudhe, M. R. (1987) Fractionation of anhydrous milk fat by supercritical carbon dioxide. *J. Food Sci.* **52**, 1231.

Bentz, A. P. and Breidenbach, B. G. (1969) Evaluation of the differential scanning calorimetric method for fat solids. *J. Am Oil Chem. Soc.* **46**, 60.

Biernoth, G. and Merk, W. (1985) Fractionation of butterfat using a liquid gas in the supercritical state, U.S. Patent 4,504,503.

Chao, R. R., Sherbon, J. W., Tse, B. L. and Rizvi, S. S. H. (1988) Effect of supercritical CO_2 extraction with reflux on triglyceride distracts of anhydrous milk fat extract, presented at annual IFT meeting, New Orleans, Louisiana, paper no. 281.

Chao, R. R., Mulvaney, S. J., Bailey, M. E. and Fernando, L. N. (1991a) Supercritical CO_2 extraction of lipids and cholesterol from ground beef. *J. Food Sci.* **56**, 183.

Chao, R. R., Mulvaney, S. J., Tempesta, M. S., Sanson, D. R. and Hsieh, F.-H. (1991b) Supercritical CO_2 extraction and some of the characteristics of the color extracts. *J. Food Sci.* **56**, 80.

Christianson, D. D. and Friedrich, J. P. (1985) Production of food-grade corn germ product by supercritical fluid extraction, U.S. Patent 4, 495,207.

Eisenbach, W. (1984) Supercritical fluid extraction: a film demonstration. *Ber. Bunsenges. Phys. Chem.* **88**, 882.

Eldridge, A. C., Friedrich, J. P., Warner, K. and Kwolek, W. F. (1986) Preparation and evaluation of supercritical carbon dioxide defatted soybean flakes. *J. Food Sci.* **51**, 584.

Favati, F., King, J. W. and Mazzanti, M. (1991) Supercritical carbon dioxide extraction of evening primrose oil. *J. Am. Oil Chem. Soc.* **68**, 442.

Friedrich, J. P. and Pryde, E. H. (1984) Supercritical CO_2 extraction of lipid-bearing materials and characterization of the products. *J. Am. Oil Chem. Soc.* **61**, 223.

Glassner, D. A. and Grulke, E. A. (1986) The effects of surfactant concentration and crystal size on the olen yield from the detergent fractionation of tallow. *J. Am. Oil Chem. Soc.* **63**, 1066.

Ha, J. K. and Lindsay R. C. (1991). Volatile fatty acids in flavors of potatoes deep-fried in a beef blend. *J. Am. Oil Chem. Soc.* **68**, 294.

Hardardottir, I. and Kinsella, J. E. (1988) Extraction of lipid and cholesterol from fish muscle with supercritical fluids. *J. Food Sci.* **53**, 1656.

Kankare, V., Antila, V., Harvala, T. and Komppa, V. (1989) Extraction of milk fat with supercritical carbon dioxide. *Milchwissenschaft* **44**, 407.

List, G. R., Friedrich, J. P. and Pominski J. (1984) Characterization and processing of cottonseed oil obtained by extraction with supercritical carbon dioxide. *J. Am. Oil Chem. Soc.* **61**, 1847.

Luddy, F. E., Hampson, J. W., Herb, S. F. and Rothoritz, H. L. (1973) Development of edible tallow fractions for specialty fat uses. *J. Am. Oil Chem. Soc.* **50**, 250.

Nawar, W. W., Kim, S. K., Li, Y. K. and Vajadi, M. (1991) Measurement of oxidative interactions of cholesterol. *J. Am. Oil Chem. Soc.* **68**, 496.

Nilsson, W. B., Gauglitz, E. J. and Hudson, J. K. (1989) Supercritical fluid fractionation of fish oil esters using incremental pressure programming and a temperature gradient. *J. Am. Oil Chem. Soc.* **66**, 1596.

Nilsson, W. B., Gauglitz, E. J., Hudson, J. K., Stout, V. F. and Spinelli, J. (1988) Fractionation of Menhaden oil ethyl esters using supercritical fluid CO_2 *J. Am. Oil Chem. Soc.* **65**, 109.

Nilsson, W. B., Gauglitz, E. J., Hudson, J. K., Stout, V. F. and Spinelli, J. (1989) Fractionation of Menhaden oil ethyl esters using supercritical fluid CO_2 *J. Am. Oil Chem. Soc.* **66**, 1596.

Rizvi, S. S. H., Chao, R. R. and Liaw, Y. J. (1988) Concentration of omega-3 fatty acids from fish oil using supercritical carbon dioxide, in *Supercritical Fluid Extraction and Chromatography: Techniques and Applications*, eds B. A. Charpentier and M. R. Sevenants, ACS Symp. Ser. No. 366, American Chemical Society, Washington, DC, p. 89.

Ryan, C. C., Gray, J. I. and Morton, I. D. (1981) Oxidation of cholesterol in heated tallow. *J. Food Agric.* **32**, 305.

Ryan, T. C. and Gray, J. I (1984) Distribution of cholesterol in fractionated beef tallow. *J. Food Sci.* **49**, 1390.

Shishikura, A., Fujimoto, K., Kaneda, T., Arai, K. and Saito, S. (1986) Modification of butter oil by extraction with supercritical carbon dioxide. *Agric. Biol. Chem.* **50**, 1209.

Stahl, E., Quirin, K.-W. and Gerard, D. (1988) *Dense Gases for Extraction and Refining*, Springer-Verlag, Berlin, pp. 1–29.

Yeh, A., Liang, J. H. and Hwang, L. S. (1991) Separation of fatty acid esters from cholesterol in esterified natural and synthetic mixtures by supercritical carbon dioxide. *J. Am. Oil Chem. Soc.* **68**, 224.

Zhao, W., Shishikura, A., Fujimoto, K., Arai, K. and Saito, S. (1987) Fractional extraction of rice bran oil with supercritical carbon dioxide. *Agric. Biol. Chem.* **51**, 1773.

17 Supercritical CO_2 extraction of oil from a seaweed, *Palmaria palmata*

V. K. MISHRA, F. TEMELLI and B. OORAIKUL

Abstract

The extraction of oil from *Palmaria palmata*, with 45% eicosapentaenoic acid (EPA), was studied using supercritical CO_2 (SC-CO_2) in a once-through dynamic flow system at 20.78, 41.47 and 62.15 MPa and 35, 45 and 55°C. Over the range of pressures and temperatures studied, there was no significant change in the solubility. The observed solubility range was 2.88–3.63 mg/l of CO_2 at STP. The maximum concentration of EPA was found in the extracts obtained at 20.78 MPa. Addition of ethanol to the seaweed (10% w/w) led to a threefold increase in solubility of lipids and a modest increase in the yield of EPA. Fractional recoveries of major fatty acids, i.e. 14:0, 16:0 and 20:5 (ω-3) showed a decrease with time of extraction after reaching a maximum at the second hour of extraction for the saturated fatty acids, and the fourth hour for EPA.

17.1 Introduction

The omega-3 (ω-3) fatty acids have been found to have potential health benefits in various inflammatory diseases e.g. atherosclerosis, asthma, arthritis and cancer (Weaver and Holub, 1988). Among these polyunsaturated fatty acids (PUFA), eicosapentaenoic acid (EPA, 20:5 ω-3) and docosahexaenoic acid (DHA, 22:6 ω-3) are of major importance due to their immediate effects on plasma lipids (Kinsella, 1990). Currently, the major source of ω-3 fatty acids is fish oils. The ω-3 fatty acid content of fish oils can be as high as 25%. However, there is a considerable variation in the composition of fish oils depending on the species, parts of the fish, and location of the catch, etc. Since the demand for ω-3 fatty acids is expected to increase, and could not be met by fish oil alone, alternative sources of these fatty acids are currently being sought to fulfil the demand (Yongmanitchai and Ward, 1989).

Supercritical fluid extraction (SFE) is emerging as an important separation method in the field of food and pharmaceutical applications. With this technology, it is possible to separate heat sensitive compounds like ω-3 fatty acids and to avoid any toxic solvent residues in the product. The isolation and fractionation of ω-3 PUFA, using SC-CO$_2$, from fish, fish oil and esters have been studied by several researchers (Eisenbach, 1984; Yamaguchi et al., 1986; Nilsson et al., 1988, 1989; Rizvi et al., 1988). Eisenbach (1984) used SC-CO$_2$ at 15.2 MPa/50°C to fractionate ethyl esters of cod fish oil. A fraction containing C20 esters in 90% purity was achieved with a hot finger. Rizvi et al. (1988) compared the yield of EPA from various feedstocks, i.e. free fatty acids, esters of fatty acids and fish oil and showed that fatty acid esters had the highest solubility in SC-CO$_2$. The concentration of ω-3 fatty acids was increased from 40% to about 88% using a refluxing system when free fatty acids were used as feed material. Nilsson et al. (1988) reported 96% concentration of ω-3 fatty acids from urea pretreated menhaden oil fatty acid ethyl esters by a temperature gradient in a column at 15.16 MPa. Incremental pressure programming (13.1–15.1 MPa) at a maximum temperature of 80°C was also used to fractionate the same feed material by Nilsson et al. (1989).

Yamaguchi et al. (1986) were the first to extract fish oil enriched in EPA directly from krill meal and freeze dried krill. At 25.4 MPa and 80°C, the recovery of oil from freeze dried krill was higher than that from krill meal. Hardardottir and Kinsella (1988) used SC-CO$_2$ at 13.78–34.46 MPa and 40–50°C to extract lipids and cholesterol from trout muscle. The use of algae for the extraction of ω-3 oils is relatively new. Polak et al. (1989) studied SC-CO$_2$ extraction of lipids from freeze dried microalgae *Skeletonema costatum* (25% EPA) and *Ochromonas danica* (11% EPA) at 17–31 MPa and 40°C.

Palmaria palmata of the class Rhodophyta was selected as a raw material for extraction of oil because of its very high EPA content (45%) (Mishra et al., 1993). The objective of this work was to test the feasibility of extraction of lipids by SC-CO$_2$ from *P. palmata* at different extraction temperatures and pressures under equilibrium conditions.

17.2   Materials and methods

17.2.1   Material

Palmaria palmata used in this study was cultured under controlled laboratory conditions at the National Research Council, Halifax, NS. The conditions used for the cultivation have been described by Morgan et al. (1980). The harvested alga was air-shipped and freeze-dried immediately upon receipt in a RePP freeze dryer. The dried sample was ground under a spray of liquid nitrogen, vacuum packaged and stored at − 75°C.

17.2.2   Extraction system

The supercritical extraction unit was manufactured by Newport Scientific Inc.

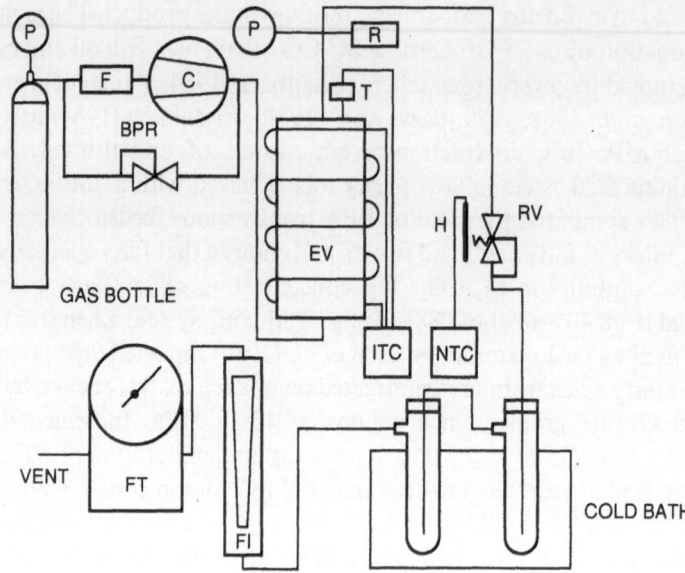

Figure 17.1   Schematic diagram of SC-CO$_2$ extraction apparatus. T, Thermocouple; F, filter; H, heater; R, rupture disc; RV, relief valve; NTC, non-indicating temperature controller; ITC, indicating temperature controller; FT, flow totalizer; C, diaphragm compressor; EV, extraction vessel; FI, flow indicator; P, pressure gauge; BPR, back pressure regulator.

(Jessup, MD). Carbon dioxide (99.5% purity) was compressed to the desired pressure by a diaphragm compressor with a maximum rating of 69 MPa. The extraction vessel (300 ml) was heated by a heating tape and temperature was controlled by a thermostat ($\pm 1°C$). Pressure was controlled by a back pressure regulator. A basket that fits into the extraction vessel was fabricated to facilitate loading and unloading of the sample. The salient features of the system are depicted in Figure 17.1.

17.2.3   Extraction procedure

Forty grams of ground sample was contacted with the upflowing SC-CO$_2$ at the set temperature and pressure in the extraction cell. The sample was held in place within the extraction basket by two metallic filters (10 µm) mounted on both ends of the basket to prevent any carry over of particles in the system. The oil-laden SC-CO$_2$ was depressurized to atmospheric pressure through a needle valve which was heated to the extraction temperature to prevent freezing. The extract was collected in glass tubes maintained at $-20°C$ in a cold bath using ethylene glycol and the CO$_2$ was vented. The extracts were stored at $-75°C$.

The pressure and temperature ranges studied were 20.68, 41.36, 62.15 MPa and 35, 45, 55°C, respectively. Solvent flow rate was maintained at 300 ml/min in order to approach equilibrium conditions (McHugh and Krukonis, 1986). The extracts were weighed periodically and the corresponding volume of CO_2 passing through the cell was recorded by a dry gas meter at ambient conditions. The solubilities were calculated from a plot of the amount of total lipid recovered versus the corresponding volume of CO_2 used. In a separate experiment, lipids from dulse were extracted with chloroform/methanol according to the method of Bligh and Dyer (1959) and infused in alumina beads. The beads were then extracted with SC-CO₂ at 62.15 MPa and 45°C. Lipids recovered from alumina beads were compared to those extracted from seaweeds under similar conditions. The fatty acid composition of the five sequential lipid fractions, collected every hour through a 5-h period from the alumina beads, was determined.

The system lacked the provision for a continuous addition of a co-solvent directly into SC-CO₂. Hence, to study the effect of a polar entrainer on EPA recovery, ethanol was mixed with the seaweed (10%, w/w) in an air-tight container and allowed to equilibrate overnight. This pretreated seaweed was extracted with SC-CO₂ at 20.78 MPa and 45°C.

17.2.4   Analysis of extract

The extracts were transmethylated to fatty acid esters by heating (90°C for 45 min) with 12% BF_3 in methanol (Supelco Co., Bellfonte, PA). The fatty acid methyl esters (FAME) were analyzed by gas chromatography (GC) equipped with a flame ionization detector and an integrator (Perkin-Elmer Ltd., Norwalk, CT). The FAME were separated on a fused silica polar capillary column (30 m × 0.3 mm, Stabilwax, Bellfonte, PA). The GC conditions were: carrier gas, helium; inlet pressure, 64.94 kPa; split ratio, 100:1; injector temperature, 250°C; detector temperature, 270°C; oven temperature program of 160–200°C at 3°C/min, holding at 200°C for 5 min and 200–240°C at 2°C/min.

Lipid classes in the extracts obtained by chloroform/methanol and SC-CO₂ (20.78 MPa and 45°C) with and without ethanol as co-solvent were determined by thin layer chromatography (TLC) (Christae, 1982). The lipid samples were applied on precoated silica gel-G plates (250 μm thickness) and the classes were separated in a solvent system consisting of hexane, diethyl ether and formic acid in the proportions of 80:20:1, respectively. The plates were sprayed with sulfuric acid (50%) and heated in an oven maintained at 200°C for 40 min. Individual classes were identified by comparison with the characteristic R_f values for standards.

17.3   Results and discussion

The solubility of oil from *P. palmata* under different extraction conditions is shown in Table 17.1. The results show that the solubility varied from a

Table 17.1 Solubility of seaweed oil and EPA content of the lipid extracts obtained under different conditions

Pressure (MPa)	Temperature (°C)	Solubility (mg/l)	EPA (% lipid)
20.78	35	3.47	0.59
	45	3.27	1.04
	55	3.37	0.78
41.47	35	3.26	0.27
	45	2.88	0.23
	55	3.63	0.16
62.15	35	3.37	0.19
	45	3.28	0.06
	55	3.33	0.18

minimum of 2.88 mg/l at 41.47 MPa/45°C to 3.63 mg/l at 41.47 MPa/55°C. However, these slight differences in the solubilities of total lipids under these conditions were not statistically significant. This can be explained on the basis of the composition of the extracts obtained at different extraction pressures and temperatures. The lipids of *P. palmata* were found to be a mixture of triglycerides, free fatty acids, sterols, hydrocarbons and phospholipids. These components differ in their molar mass, vapor pressure and polarity, which determine solubility behavior in SC-CO_2. For example, triglycerides are preferentially more soluble in SC-CO_2 than phospholipids.

Due to the small amounts of the extracts, it was not possible to determine the exact composition of each of the lipid components, which otherwise could have ascertained the effects associated with component selectivity. Even though there may be significant differences in solubilities of individual lipid classes as a function of extraction conditions, differences in total lipid solubility were found to be insignificant within the temperature and pressure ranges studied. The fact that an increase in pressure did not always translate into an increase in solubility was reported in earlier studies (Schmitt and Reid, 1986; Hardardottir and Kinsella, 1988; Polak *et al.*, 1989; Vijayan and Buckley, 1991). Hardardottir and Kinsella (1988) reported that the recovery of lipids from rainbow trout muscle remained nearly constant in SC-CO_2 extraction between 13.78–34.46 MPa and 40–50°C. Polak *et al.* (1989) reported no significant difference in the solubility of lipids at 24 and 31 MPa for microalgae. Vijayan and Buckley (1991) found that the solubility of lipids from potato chips was higher at 41.47 MPa than that at 55.15 MPa. No explanation was extended for the results.

Another contributing factor to this behavior may be the hydrodynamic instability in the extraction column leading to channelling. Poor contact between the solvent and seaweed may have resulted in a low recovery of lipids. In the present setup, it was impossible to determine these effects experimentally.

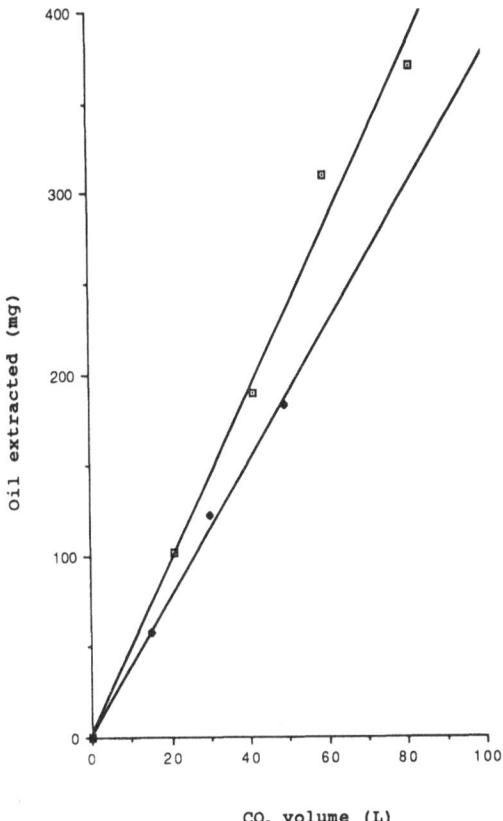

Figure 17.2   Recovery of oil from seaweed (♦) and alumina (□) beads infused with seaweed oil at 62.15 MPa and 45°C.

Although glass beads were mixed with seaweeds, a significant degree of packing of the bed noticed upon opening the cell after each run certainly suggests some channelling.

Fatty acid composition of the extracts was determined and their EPA concentrations are given in Table 17.1. The results indicate higher recovery of EPA at low pressure (20.78 MPa) than at higher pressures. This may be again due to higher selectivity for EPA containing esters at low pressure, as selectivities are inversely related to the density (Prange and Riepe, 1987).

The possibility of absorption of some lipids on the solid matrix also cannot be overruled. Upon extraction of lipid infused alumina beads, we found a difference in the extraction behavior, as depicted in Figure 17.2. The recovery of lipids from beads at 62.15 MPa/45°C was higher (4.70 mg/l) than that from seaweed (3.74 mg/l). This suggests that some of the lipids may be in the bound form, thereby making it difficult to extract with SC-CO$_2$.

The GC analysis of the fatty acid profile of the sequential fractions is

Table 17.2 Fractional recoveries of major fatty acids from alumina beads at 62.15 MPa and 45°C

Fraction	Weight of extract (mg)	14:0 (% lipid)	16:0 (% lipid)	20:5 (% lipid)
1	102	0.09	0.47	1.32
2	87	0.43	1.88	5.18
3	121	0.38	1.79	6.04
4	60	0.33	1.39	6.04
5	71	0.27	0.99	2.98

reported in Table 17.2. The fractions represent the hourly collections of the extract from alumina beads. Since myristic (14:0), palmitic (16:0) and eicosapentaenoic acids (20:5, ω-3) constitute >90% of the total fatty acids, the concentrations of only these fatty acids are reported. In all of the five fractions, the dominant fatty acid was EPA which is consistent with the overall fatty acid profile of the alga. The first fraction showed the least concentration of the three fatty acids, indicating that the majority of the extracted components were hydrocarbons and/or sterols. The concentration of the fatty acids in the extract decreased with time after reaching a maximum at 2 h for 14:0 and 16:0 and 4 h for EPA, possibly due to the depletion of these fatty acids in the sample.

Eicosapentaenoic acid in *P. palmata* is distributed in neutral and polar lipid classes (Mishra *et al.*, 1993). SC-CO$_2$ is essentially a non-polar solvent. Hence, the EPA concentration achieved by SC-CO$_2$ extraction represented only the EPA in the neutral lipids. It was expected that recovery of polar lipids with the non-polar fraction should increase the total concentration of EPA in the extract. When ethanol was added (10% w/w) as an entrainer in the seaweed, the total solubility of lipids increased to 9.35 mg/l CO$_2$ at 20.78 MPa/45°C. Since it was difficult to remove entrained ethanol in the extract, this figure may not represent true solubility. The extract was of light green color due to the presence of chlorophyll pigments. TLC of the extract obtained after addition of ethanol showed the presence of polar lipids (phospholipids) which were absent in the extract obtained by SC-CO$_2$ alone at similar conditions of extraction (20.78 MPa/45°C). This effect was, indeed, expected as ethanol increases the polarity of solvent, thereby increasing the solubility of polar components. The EPA content of the extract increased slightly from 2.29 mg to 2.57 mg by ethanol treatment. Table 17.3 provides a comparison of the fatty acid profiles of lipids extracted with and without addition of ethanol at 20.78 MPa/45°C. By addition of ethanol, there was an increase in the combined concentration of saturated fatty acids, i.e. 14:0, 16:0 and 18:0 (20.8–25.6%), and of 18:1 (ω-7 and ω-9) (5–9.2%). The concentration of EPA in the total fatty acids extracted at 20.78 MPa/45°C/ethanol moderately decreased from 64.1% to 56%.

Table 17.3 Effect of ethanol on the fatty acid profile of the lipids extracted by SC-CO$_2$ at 20.78 MPa and 45°C

Fatty acid	Without ethanol (wt%)	With ethanol (wt%)
14:0	3.3	4.2
16:0	15.7	18.3
16:1 (ω-5, ω-7)	4.5	4.9
18:0	1.8	3.1
18:1 (ω-9, ω-7)	5.0	9.2
18:2 (ω-6)	1.3	1.3
18:4 (ω-3)	1.4	tr
20:4 (ω-6)	2.8	3.0
20:5 (ω-3)	64.1	56.0

17.4 Conclusion

It was technically possible to extract lipids from *P. palmata* using SC-CO$_2$. Based on the solubility results, it is not desirable to use pressures higher than 20.78 MPa, which is economically beneficial. A portion of the lipids in *P. palmata* may be in bound form leading to poor recovery of extract. Addition of ethanol not only leads to an increase in the overall solubility of lipids, co-extraction of polar lipids also increases the amount of EPA in the extract. Further study is needed to determine the effect of high pressure on the solubility.

Even though *P. palmata* contains relatively high initial concentration of EPA (45%), low total lipid content and poor recovery of EPA by SC-CO$_2$ extraction make the economic feasibility of the process based on this raw material questionable. If it is possible to increase the total lipid content by suitable genetic manipulation, this alga can become a potential source of EPA.

Acknowledgements

Cultured *P. palmata* was supplied by Dr. J. S. Craigie of Atlantic Regional Laboratory, National Research Council of Canada, Halifax, NS. Financial support by the Central Research Fund of the University of Alberta, NSERC and Commonwealth Scholarship Plan is gratefully acknowledged.

References

Bligh, E. G. and Dyer, W. J. (1959) A rapid method of total lipid extraction and purification. *Can. J. Biochem. Physiol.* **37**, 911.

Christie, W. (1982) *Lipid Analysis*, 2nd edition, Pergamon Press, New York.

Eisenbach, W. (1984) Supercritical fluid extraction: a film demonstration. *Ber. Bunsenges. Phys. Chem.* **88**, 882.

Hardardottir, I. and Kinsella, J. E. (1988) Extraction of lipid and cholesterol from fish muscle with supercritical fluids. *J. Food Sci.* **53,** 1656.

Kinsella, J. E. (1990) Source of omega-3 fatty acids in human diets, in *Omega-3 Fatty Acids in Health and Diseases,* eds. R. S. Lee and M. Karel, Marcel Dekker, New York.

McHugh, M. A. and Krukonis, V. J. (1986) *Supercritical Fluid Extraction: Principles and Practice.* Butterworths, Stoneham, MA.

Mishra, V. K., Temelli, F. Ooraikul, B., Shacklock, P. F. and Craigie, J. S. (1993) Lipids of the red alga: *Palmaria palmata.* Botanica Marina, **36,** 169.

Morgan, K. C., Shacklock, P. F. and Simpson, F. J. (1980) Some aspects of the culture of *Palmaria palmata* in greenhouse tanks. *Botanica Marina* **23,** 765.

Nilsson, W. B., Gauglitz, E. J., Hudson, J. K., Stout, V. F. and Spinelli, J. (1988) Fractionation of menhaden oil ethyl esters using supercritical fluid CO_2. *J. Am. Oil Chem. Soc.* **65,** 109.

Nilsson, W. B., Gauglitz, E. J. and Hudson, J. K. (1989) Supercritical fluid fractionation of fish oil esters using incremental pressure programming and a temperature gradient. *J. Am. Oil Chem. Soc.* **66,** 1596.

Polak, J. T., Balaban, M., Peplow, A. and Philips, A. J. (1989) Supercritical carbon dioxide extraction of lipids from algae, in *Supercritical Fluid Science and Technology,* eds. K. P. Johnston and J. M. L. Penninger, ACS Symposium Series No. 406, ACS, Washington, DC.

Prange, M. M. and Riepe, W. H. (1987) Studies on phase equilibria of a multicomponent model mixture in supercritical carbon dioxide and trifloromethane. *Chem. Eng. Process.* **22,** 183.

Rizvi, S. S. H., Chao, R. R. and Liew, Y. J. (1988) Concentration of omega-3 fatty acids from fish oil using supercritical carbon dioxide, in *Supercritical Fluid Extraction and Chromatography,* eds. B. A. Charpentier and M. R. Sevenants, ACS Symposium Series No. 366, ACS, Washington, DC.

Schmitt, W. J. and Reid, R. C. (1986) The solubility of twenty-five paraffinic hydrocarbons in supercritical carbon dioxide, presented at the Annual Meeting of AIChE, Miami, FL.

Vijayan, S. and Buckley, L. (1991) Separation of biomaterials by supercritical extraction process: an overview of bench-scale test experience and process economics, presented at the 8th World Congress of Food Science and Technology, Toronto, Canada.

Weaver, B. J. and Holub, B. J. (1988) Health effects and metabolism of dietary eicosapentaenoic acid. *Prog. Foods Nutr. Sci.* **12,** 111.

Yamaguchi, K., Murakami, M., Nakano, H., Konosu, T., Yamamoto, H., Kosaka, M. and Hata, K. (1986) Supercritical carbon dioxide extraction of oils from Antarctic krill. *J. Agric. Food Chem.* **34,** 904.

Yongmanitchai, W. and Ward, O. P. (1989) Omega-3 fatty acids: alternative sources of production. *Process Biochem.* **26,** 117.

18 Commercial feasibility of a supercritical extraction plant for making reduced-calorie peanuts

C. A. PASSEY

Abstract

A process using supercritical extraction with carbon dioxide has been developed at the St. Hyacinthe Food Research and Development Centre for making virtually unbroken calorie-reduced peanuts by removing 27–32% of the oil present in raw blanched peanut kernels. The process is commercially feasible with a good market potential for calorie-reduced peanuts. The demand for this product is estimated at about 70 000 tonnes/year, based on the combined European and North American population of 700 million and a per capita consumption of about 0.1 kg/year or 10% of the current consumption of snack peanuts in North America. A 16 000-l extraction vessel(s) capacity plant, producing about 1000 tonnes/year of calorie-reduced peanuts for the snack food market, and about 200 tonnes/year of 'cold-pressed' quality peanut oil, installed at a capital cost of $6.10 million, is highly profitable, with an after-tax net present value of $3–7 million and corresponding return on equity of some 42–59%, depending upon the location. Moreover, as the production from such a plant would be less than 2% of the market demand for the calorie-reduced peanuts, there should be very little marketing risk involved.

18.1 Introduction

It is estimated that about 70 000 and 700 000 tonnes of shelled peanuts are processed per year in Canada (Clark and Clark, 1984) and United States (USDA, 1984), respectively, for food use. Of the peanuts used as food, about 65% are converted into peanut butter, while the remainder are used in snack foods in one form or another. According to the nutrition recommendations of the Scientific Review Committee (Health and Welfare Canada, 1990) and the US Select Committee on Nutrition and Human Needs (Clydesdale and Francis, 1985), the caloric contribution of fat in a healthy diet should not exceed 30%. In this regard, ordinary peanuts have 70% of the calories contributed by its oil content whereas calorie-reduced peanuts would be more balanced calorie-wise and better satisfy the demand of diet-conscious consumers for snack peanuts ('as is' or in further value-added product forms, e.g. dry roasted, smoked, or chocolate-, honey- or yogurt-coated). They would also have longer shelf-life (Pominski et al., 1971) since they would be less prone to rancidity due to reduced oil content. Calorie-reduced peanuts should help enhance product variety for consumers, production opportunities for farmers, and product and market diversification opportunities for the processing distribution and retailing sectors.

Researchers have tried unsuccessfully to develop an acceptable process for producing unbroken calorie-reduced peanuts, since the early 1960s. Two approaches have been used in the past for partially defatting peanuts: one using extraction with organic solvent(s), and another employing mechanical pressing (Pominski et al., 1964, 1969).

Hexane, a widely used organic solvent for extracting oil from oilseeds, has been tried with peanuts. However, the hexane extracted peanuts must be desolventized almost completely to less than 0.5 ppm residual hexane to satisfy FDA requirements for human consumption. Excessive stripping time and the high temperatures required for such desolventization adversely affect the organoleptic quality of defatted peanuts. Pominski et al. (1964) used hexane to extract oil from the fully or partially roasted peanuts at different solvent/peanut ratios, and concluded that the slow rate of extraction, and difficulties encountered in desolventizing extracted peanuts to acceptable levels without degrading their taste and appearance, seem to negate commercial acceptance of the hexane process.

In a process developed by the Southern Utilization Research and Development Division of the United States Department of Agriculture (USDA) in the 1960s, blanched peanuts (5% moisture) were first mechanically pressed to remove 50–80% of the oil. The pressed, deformed, peanuts were then soaked in hot water in an attempt to regain the original size and shape, and then redried. The USDA process caused considerable splitting (12–43%) and breakage (4–46%) of peanut kernels, and resulted in losses of about 5% of the water soluble solids, mainly sugars and proteins (Pominski et al., 1969). Further, the soaking of pressed defatted peanuts also caused colour and flavour changes.

Passey and Patil (1989a, b) have developed a successful supercritical extraction process for making unbroken calorie-reduced peanut kernels of superior quality, e.g. better retention of kernel appearance (shape and colour), taste, and crunchier texture. Supercritical carbon dioxide ($SC-CO_2$) was selected as the extraction medium because it is non-toxic, non-reactive, non-flammable, physiologically and environmentally safe, and can be easily recovered and recycled. Due to these favourable attributes, extraction with $SC-CO_2$ is likely to be more cost-effective than the organic solvents such as hexane, when the additional costs of explosion-proof construction and safety measures are taken into consideration.

This paper discusses the commercial feasibility of the Passey–Patil process (Passey and Patil, 1989a) of a 1000 tonne/year production capacity plant (16 000 l total extraction vessels' capacity) for the base case scenario (see Table 18.1). This level of production capacity represents less than 2% of the estimated market demand in the combined European and North American markets based on a per capita consumption of about 0.1 kg/year or 10% of the current consumption of snack peanuts in North America (estimate only, not based on rigorous market research). The after-tax discounted cash flow analysis, made over the economic life of the plant (15 years), has been used for assessing the profitability of investment in terms of net present values and rates of return based on equity as well as the total capital cost.

Sensitivity of profitability to various capacity, policy, and revenue and cost assumptions (given in Table 18.1) has also been analysed by varying one or more assumptions at a time, and the results are presented graphically. Profitability is also compared for locating a 16 000 litre capacity plant in three different countries representing a typical mix of economic factors: United States (high labour cost, low interest on debt and high corporate income tax), Canada (medium labour cost, medium interest and medium corporate income tax), and India (low labour cost, high interest and low corporate income tax).

18.2 Market potential for reduced-calorie peanuts

It is estimated that the per capita consumption of snack peanuts is about 1 kg/year. The combined North American and European market potential for snack peanuts is about 700 000 tonnes/year. Although the author has not carried out an exhaustive market research on the latent demand for calorie-reduced peanuts, as it is not within the scope of the mandate of the Food Research & Development Centre, given the consumer preference for low-calorie healthy foods, and the superior quality (e.g. better retention of kernel appearance (shape and colour), taste and crunchier texture) of the calorie-reduced peanuts (Passey and Patil, 1993), it is conceivable that the market penetration by calorie-reduced peanuts could reach 10% over the next 5 years with sound marketing strategy. This represents a realizable market demand of about 70 000 tonnes of calorie-reduced peanuts per year which can support about 70 strategically located plants of 1000 tonnes capacity each. Obviously commercial risk is minimal as the production from a single 16 000-l plant would represent less than 2% of the market demand.

Moreover, the Passey–Patil process can also be used for making other calorie-reduced nut products. Thus, while there is considerable potential for growth, the companies who would be the front runners are likely to reap the most profits from the technology. However, the persons or companies wishing to commercially exploit this technology are advised to satisfy themselves about the market demand through their own market research.

Table 18.1   Assumptions for evaluation of the commercial viability of SFE project for low calorie nuts

Capacity and cost assumptions:

1	CAP–EV	16000	Plant extraction vessels' capacity (l)
2	CCOST	6100000.00	Capital cost of the 16000-l plant ($)

Policy assumptions

3	DER	2.00	Debt/equity ratio for investment
4	%IDBT	15.00	Interest on debt (before corporate income tax) (% year)
5	DISCR	12.00	After tax discount rate (%/year)
6	INFLR	4.00	Inflation rate/year (%/year)
7	PTX&I	1.50	Property tax and insurance rate (% of CCOST)
8	MTNC	1.50	Maintenance cost: % of CCOST
9	AMORTP	15.00	Amortization period (years)
10	DEP		Straight line depreciation for costing
11	SALV	1830000.00	Salvage value ($)
12	CCAF	20.00	Capital cost allowance factor: % of CCA base
13	ITAXR	40.00	Rate of income tax (%)

Revenue and cost assumptions

14	RMCOST	0.85	Raw material: nuts ($/kg)
15	SPRICE– P1	3.00	Low calorie nuts ($/kg)
16	SPRICE– P2	3.00	Extracted oil ($/kg)
17	DEN– P1	0.50	Low calorie nuts (kg/l)
18	DEN – P2	0.90	Extracted oil (kg/l)
19	%OIL	46.00	% oil content of nuts
20	%OILR	40.00	% of oil removed
21	%PLOSS – P1	3.00	Product loss, low calorie nuts (%)
22	%PLOSS – P2	3.00	Product loss, extracted oil (%)
23	WDPY	300.00	Working days/year
24	SPD	3.00	Shifts/day
25	DEPS	4.00	Direct employees/shift
26	HPS	8.00	Paid hours/shift
27	DPC	2.00	Days per cycle
28	DLR	20.00	Direct labour rate ($/h) including benefits
29	GCOST	0.08	Cost of liquid CO_2 ($/kg)
30	ELER	0.05	Cost of electricity ($/kW-h)
31	STMCOST	10.00	Cost of steam ($/1000 kg)
32	CWCOST	0.20	Cost of cooling water ($/kl)
33	SOISCF	1.00	Solubility of extracted oil in SCF (%)
34	%GL	2.00	Gas lost, % of QSCFPC
35	%CWL	10.00	Cooling water lost (%)
36	CEMHPC	20.00	Circulation-extract mode (CEM) (h/cycle)

Calculated values

37	EQUITY	2033333.33	Equity part of CCOST ($)
38	DEBT	4066666.67	Debt part of CCOST ($)
39	PCAP-P1	6332.16	Production capacity, low calorie nuts (kg/cycle)
40	PCAP-P2	1427.84	Production capacity, extracted oil (kg/cycle)
41	QSCFPC	147200.00	SCF circulated (kg/cycle)
42	SCFCR	7360.00	SCF circulation rate during CEM (kg/h)
43	ELEL	460.00	Electrical load (kw)
44	STM	23920.00	Steam used/cycle (kg/cycle)
45	CWC	2024.00	Cooling water circulated/cycle (kl/cycle)

Reference levels

CAP-EVR	8000.00	Ref. CAP – EV (l)

Table 18.1 *continued*

CCOSTR	4000000.00	Ref. CCOST ($)
SIZECSF	0.60 exponent	Capital cost scaling factor (exponent)
SALVF	0.30	Salvage value factor of CCOST
DEPSR	2	For 8000 l (EVR) plant (no. of persons)
SOISCFR	1.00	Solubility of oil in SCF (%)
RSCFCR	4000.00	SCF circulation rate in circulation-extract mode (kg/h)
ELELR	250.00	For EVR plant (kW)
STMR	13000.00	For EVR plant (kg/cycle)
CWCR	1100.00	For EVR plant (kl/cycle)

Analyses for other sets of assumptions can be obtained from St-Hyacinthe Food Research and Development Centre, Quebec.

18.3   Plant and process design criteria

18.3.1   Design considerations

A schematic flow diagram of a supercritical fluid extraction (SFE) plant for carrying out the Passey–Patil process is shown in Figure 18.1. Although the equipment is not materially different from conventional SFE plants, it must be properly designed and controlled for optimally executing the Passey–Patil process. While only one each of the extraction and separation vessels, liquid CO_2 pump, and various heat exchangers, are shown in Figure 18.1, the design calculations for a 16 000-l (total capacity of extraction vessels) plant indicate that multiple units of these components (especially the extraction vessels) would be more cost effective than a single large unit, and would also provide considerable operational flexibility (Passey, unpublished design information) as each of the multiple extraction vessels could be loaded and operations sequenced in order to carry out the extraction of several batches (at different stages of extraction) simultaneously to achieve semi-continuous production.

The high pressure parts (e.g. the extraction vessels, liquid CO_2 pumps, solvent and extract-phase heat exchangers, etc.) should be rated at a working pressure of 500 MPa at 100°C, and designed for 50 000–100 000 pressurization–depressurization cycles.

Although not shown in Figure 18.1, the equipment for gradual decompression of the extraction vessel during or at the end of the extraction cycle, and for efficient recovery of carbon dioxide removed from the extraction vessel is included in the capital cost estimates (see Table 18.2). This equipment includes a gas compressor, condenser/heat exchanger, and the associated piping and control valves.

18.3.2   The extraction process

Passey and Patil (1989a) have shown that to prevent kernel breakage during

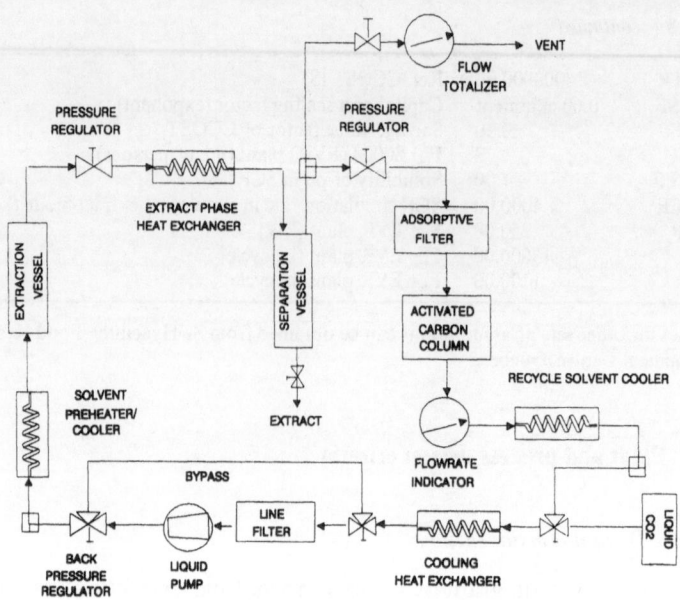

Figure 18.1   Schematic diagram of the SFE system.

extraction, the raw blanched peanut kernels should be humidified to a moisture content of 8–14% prior to loading them in the extraction vessel. The extraction vessel is loaded with suitably pretreated unbroken peanut kernels, charged with liquid (or supercritical) CO_2, and pressurized. The peanuts are extracted to the required degree, or until further extraction becomes uneconomical due to diminishing rate of extraction, by passing supercritical carbon dioxide through the extraction vessel under optimum conditions (see below). The extraction vessel is then gradually depressurized to atmospheric pressure, while recovering most of the carbon dioxide removed from the ('source') extraction vessel at the end of the extraction cycle. This recovery process includes the steps of (i) transferring the carbon dioxide (as liquid or supercritical CO_2) removed from the source extraction vessel to another extraction vessel(s) which is ready for charging, until pressure equilibration is achieved; (ii) condensing and diverting the condensed CO_2 into the make-up liquid CO_2 vessel until the pressure in the source extraction vessel drops to about 60 bar; and (iii) operating the gas compressor to continue depressurization of the source extraction vessel, condensing the carbon dioxide gas and sending condensed CO_2 into the liquid CO_2 vessel until the pressure in the source extraction vessel drops to about 2 bar at which time the remaining gas may be vented for depressurizing the source extraction vessel to atmospheric pressure so that it can be opened for unloading (and reloading). The caloriferous matter consisting of 'cold-pressed' quality oil collected in the separation vessel can be withdrawn as required or automatically when it reaches a certain level in the separation vessel.

Table 18.2 Cost estimate of a reference plant of 8000-l capacity

Quantity	Item	Pressure(MPa)		Cost
		Working	Design	
Vessels				
4	2000-l extraction vessels	500	725	1000000
2	2000-l separation vessels	125	200	230000
1	10000-l liquid CO_2 vessel	125	200	400000
2	1000-l oil receiving vessels	7		20000
25	Extraction baskets			50000
Compressors				
1	5000 kg/h, 60–500 MPa	500	590	502500
1	250 m³/h, 2–60 MPa			100000
Heat Exchangers (HX)				
1	Solvent HX, 4000 kg/h (steam 300 kg/h)	500		25000
1	Extract phase HX (steam 400 kg/h)	500		30000
1	CO_2 water-cooled condenser 210000 kcal/h (450 l/min)	125		30000
1	Liquid CO_2 water-cooled subcooler 16000 kcal/h (70 l/min)	125		6500
1	Refrigerated liquid CO_2 subcooler to − 20°C, 100000 kcal/h	125		20000
1	Refrigeration system 35 tons, − 20°C			70000
1	Nut conditioning equipment			50000
Subtotal 1				2534000
Installation, piping and controls at 40% of Subtotal 1				1013600
Subtotal 2				3547600
Project engineering, fee at 10% of Subtotal 2				354760
Subtotal 3				3902360
Contingencies, at 10% of Subtotal 3				390236
Total				3937836
Estimated capital cost used in analysis (8000-l plant)				4000000

While suitably pretreated unbroken peanut kernels may be extracted with supercritical CO_2 in the continuous mode, three new extraction modes have been developed for enhancing the extraction rate and process effectiveness. These hold-and-extract modes, shown schematically in Figures 18.2 (a–c), are more fully described elsewhere (Passey and Patil, 1989a). Briefly, the extraction is carried out in a single stage comprising several hold-and-extract steps in the case of the hold-and extract mode of Figure 18.2(a). On the other hand, extraction is carried out in two or more stages with intervening partial decompression in case of the hold-and-extract mode of Figure 18.2(b), with the first

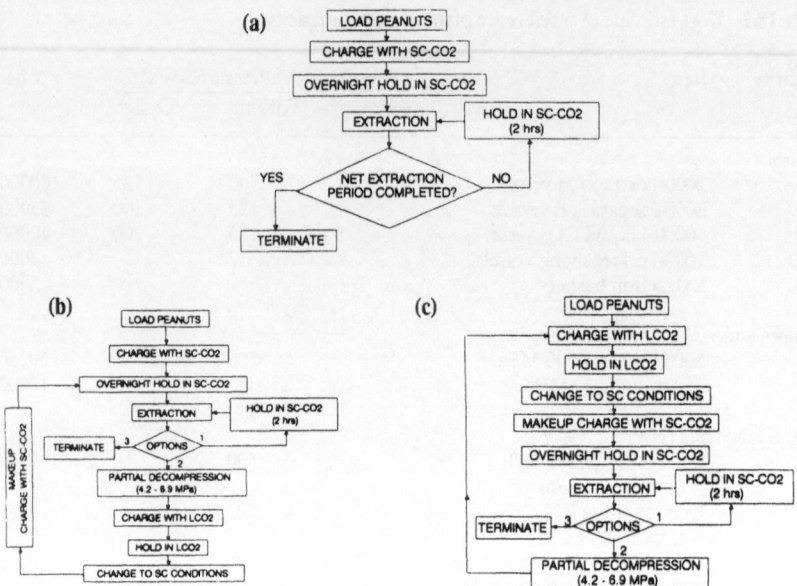

Figure 18.2   Schematic process flow chart for the hold-and-extract extraction modes. **(a)** Initial charge with SC-CO$_2$ with no intervening decompression step during extraction. **(b)** Initial charge with SC-CO$_2$ with an intervening decompression step during extraction. **(c)** Initial charge with liquid CO$_2$ with an intervening decompression step during extraction.

extraction stage starting with hold in SC-CO$_2$. The hold-and-extract mode of Figure 18.2(c) differs from that of Figure 18.2(b) in that the initial hold is done in liquid carbon dioxide. As compared with the continuous extraction mode, Passey and Patil (unpublished) were able to remove 2–3 times more oil using the hold-and-extract modes of Figures 18.2(b) and 18.2(c).

18.4   Commercial viability

A set of assumptions for evaluating the commercial viability of a 16 000-l SFE plant (comprising four extraction vessels of 4000 l each) for making 1000 tonnes/year of calorie-reduced peanuts and 200 tonnes/year of its joint-product, the 'cold-pressed' quality peanut oil, are given in Table 18.1. All the costs and revenues mentioned in this paper are in US dollars.

18.4.1   Basic assumptions

It has been assumed that the SFE plant is an addition to an ongoing peanut processing operation, would occupy its existing building/facility, and would use its existing packaging machinery, and warehousing space (i.e. no investment in land, building, and packaging machinery is included in the capital cost estimates). Assumed utility costs are adequate to pay for installing any additional capacity required, so no provision needs to be made for it explicitly in the capital cost estimates.

18.4.2   Capacity and cost assumptions

Detailed capital cost breakdown is given in Table 18.2 for a plant of 8000-l total capacity of extraction vessels. However, computer simulation of viability of a plant of this size indicated that although still viable, it will be only marginal if the capital cost were to increase by 20%. Therefore, the plant capacity was increased to 16 000 l for the base case scenario. The capital cost of such a plant is estimated at $6.10 million, using cost scaling exponent factor (SIZECSF) of 0.60.

18.4.3   Policy assumptions

Usually, most countries offer certain capital grants, interest subsidies, loan guarantees, and income tax based incentives as a part of their socio-economic development policies to attract industries to locate in certain regions within the country. Because such policies can vary from country to country, and from region to region within a country, and moreover may change with the government or policy changes, it would be too risky to include these incentives in viability assessment. Therefore such benefits, which could add significantly to profitability, have been ignored in the viability analysis presented in this paper. However, if deemed necessary, the present value of the net after-tax effect of these benefits can be separately calculated and added to the net present values reported in this paper.

The corporate income tax and the capital cost allowance rates can have significant effect on the after-tax cost of interest on debt and the tax-shield provided by the depreciation. For the base case scenario, they are assumed to be 40% and 20%, respectively. The interest on debt and the leverage (debt/equity ratio) can also vary from situation to situation, and are included in the analysis at 15%/year and 2, respectively. The interest on debt is assumed to be compounded annually, and the debt is assumed to be retired by 15 equal instalments (consisting of principal and interest) made at the end of each year. Sensitivity of profitability to these four variables in three representative countries has also been evaluated and the results are presented in Table 18.4

Inflation rate is assumed to be 4%/year. While most of the capital equipment (vessels, heat exchangers, piping, etc.) is stationary and hence would have considerable economic life, a project life of 15 years has been assumed with a salvage value (30%) realized in the 16th year for purposes of analysis. Further, the straight line depreciation policy is assumed for costing purposes.

An after-tax discount rate of 12% has been assumed for making the discounted cash flow analysis.

Table 18.3 Estimate of operating costs and profitability (for the year 1) for a plant based on assumptions given in Table 18.1

	$/year	$/day	Lowcal nuts ($/kg)	Oil ($/kg)	Conversion cost $/day	Conversion cost Percent
Revenue						
Lowcal nuts	2849472.00	9498.24	3.00			
Oil	642528.00	2141.76		3.00		
Total revenue (TR)	3492000.00	11640.00	3.00	3.00		
Variable cost						
Raw material	1020000.00	3400.00	0.88	0.88	1920.00	33.67
Direct labour (VCDL)	576000.00	1920.00	0.49	0.49		
Utilities (VCU):						
CO$_2$	35328.00	117.76	0.03	0.03	117.76	2.06
Electricity	165600.00	552.00	0.14	0.14	552.00	9.68
Steam	35880.00	119.60	0.03	0.03	119.60	2.10
Cooling water	6072.00	20.24	0.01	0.01	20.24	0.35
Total variable cost (TVC)	1838880.00	6129.60	1.58	1.58	2729.60	47.86
Contribution margin (TR − TVC)	1653120.00	5510.40	1.42	1.42		
Fixed cost						
Depreciation	284666.67	948.89	0.24	0.24	948.89	16.64
Average interest cost	424358.24	1414.53	0.36	0.36	1414.53	24.80
Property tax and insurance	91500.00	305.00	0.08	0.08	305.00	5.35
Maintenance	91500.00	305.00	0.08	0.08	305.00	5.35
Total fixed cost (TFC)	892024.91	2973.42	0.76	0.76	2973.42	52.14
Total cost (TC = TVC + TFC)	2730904.91	9103.02	2.34	2.34	5703.02	100.00
Net income before tax (NIBT = TR − TC)	761095.09	2536.98	0.66	0.66		
Add back: Depreciation cost Average interest cost	284666.67 year 1 424358.24					
Income before depreciation, interest and income tax	1470120.00 year 1					

Table 18.4 Profitability of a 16 000-l SFE project affected by plant location (assumptions same as in Table 18.1, except for the country-specific assumptions below)[a]

Country specific assumptions	Base case scenario	Country 1 USA	Country 2 Canada	Country 3 India
Policy assumptions				
Income Tax Rate (ITAXR) (%)	40.00	40.00	30.00	20.00
Capital Cost Allowance Rate (CCAF) (%)	20.00	20.00	35.00	50.00
Debt Equity Ratio (DER)	2.00	3.00	2.00	1.00
Interest on Debt (%IDBT) (%/year)	15.00	9.00	12.00	15.00
Revenue and cost assumptions				
Raw Material Cost (RMCOST) ($/kg)	0.85	0.85	0.85	0.75
SPRICE-P1 (low calorie peanuts) ($/kg)	3.00	3.00	3.00	2.75
SPRICE-P2 (extracted peanut oil) ($/kg)	3.00	3.00	3.00	2.75
Working days per year (WDPY) (days/year)	300.00	250.00	250.00	300.00
Direct labor rate (DLR) ($/h)	20.00	20.00	15.00	3.00
Cost of electricity (ELER) $/kW-h)	0.05	0.08	0.05	0.08
Conversion costs (% of total conversion cost)				
Direct labor (%)	33.67	32.10	26.55	6.54
Utilities (%)	14.19	19.07	14.92	25.92
Fixed (%)	52.14	48.84	58.53	67.54
Profitability				
NPV based on equity (NPVEB) (million $)	3.76	2.78	4.03	6.62
NPV based on assets (NPVTCCB) (million $)	4.43	2.07	4.03	7.29
Return on equity (%ROE) (%)	40.90	42.42	44.47	59.00
Return on assets (%IRR) (%)	24.05	17.98	23.16	30.38
Payback, equity-based (years)	2.93	2.71	2.54	1.88
Payback, assets-based (years)	5.82	8.57	6.17	4.47

[a]Capacity and cost assumptions: CAP–EV 16 000-l (plant extraction vessels' capacity); CCOST $6 100 000.00 (capital cost of the 16 000-l plant).

18.4.4   Cost and revenue assumptions

As any viability study, of necessity, must be based on estimates of future cash flows, it is necessary to have long-run cost and reasonable revenue projections into the future. However, it would be a mistake to estimate future long-run raw material costs from the 'current' prices which are often very much affected by the vagaries of the weather and at times can be quite deceptive if used as estimator of the future cost of raw material. For example, in August 1990 (when this study was commenced), the price of raw blanched peanuts in North America was about twice as high as the year before due to the drought in the United States. A year later, in August 1991, the prices were back to the level in August 1989.

18.4.4.1   Raw material costs.   For reasons given above, raw material costs used in this study were based on regression (Passey, unpublished data) of the available historic price data for the 14-year period from 1970 to 1983 (USDA,

1984) and discussions with industry sources. Based on this regression ($R^2 =$ 0.9118, $P < 0.01$),* the cost of raw blanched peanuts in July 1991 was estimated at $0.77/kg with a standard error of ± $0.02/kg based on 5-year moving average centred around 1991. To be on the safe side, the cost of raw blanched peanuts was therefore assumed as $0.85/kg for the base case scenario, 10% higher than the long-run regressed price. Similarly, the raw material cost inflation rate of 4%/year was assumed for the base case scenario, instead of 3.8%/year based on the 5-year moving average of regressed prices over the 15-year period from 1974 to 1993.

18.4.4.2 Selling prices of products. Although, based on industry estimates (October 1990), the calorie-reduced peanuts can be marketed at an ex-factory selling price of $3.39/kg, an ex-factory selling price of $3.00/kg has been assumed for the base case scenario. Similarly, although the 'cold-pressed' quality peanut oil obtained as a joint product could command an ex-factory price of $3.49/kg estimated from the prevailing wholesale prices (August, 1991), a lower ex-factory selling price of $3.00/kg has been assumed for the base case scenario.

Further, it should be noted that the analysis takes into account product losses at a rate of 3%.

Direct labour rate (DLR), number of working days per year (WDPY) and the cost of electricity (ELER) can vary widely from country to country, and the effect of these costs on profitability is compared in the country scenarios presented in Table 18.4.

18.4.5   Measures of profitability

An excellent treatment of the capital expenditure decision-making process is presented by Gordon and Shillinglaw (1974). The basic concept in assessing profitability is to compute the present value of the future cash flows (positive when received, and negative when spent or paid out) on an after-tax basis by discounting, and set it off against the present value of the investment. The terms net present value, internal rate of return and the payback period, used in this paper are briefly explained in the Appendix.

18.5   Results of feasibility analysis

18.5.1   Base case scenario

A 16 000-l SFE plant comprising four extraction vessels (4 000 l each), four

---

*That is, it can be said with 99% confidence that 91% of the variance in price is explained by the year of production/sale, and only 9% is due to other economic and marketing factors.

separation vessels (2000 l each), and at least one each of a liquid CO_2 pump (60–500 bar), a CO_2 evacuation compressor (2–60 bar) and a liquid CO_2 storage vessel (10 000 l), as well as the necessary heat exchangers and circulation pumps, refrigeration system, and peanut conditioning equipment is estimated to cost $6.10 million (mid-June 1991 prices). It would require about 200 m^2 of floor space for plant installation.

18.5.1.1 Capacity, sales and revenue. Under the base case scenario, a 16 000-l plant is expected to produce about 1 000 tonnes of calorie-reduced peanuts and 200 tonnes of peanut oil per year, with $3.49 million in sales and a net income of $0.76 million or 22% of sales (see Table 18.3). Corresponding annual income before depreciation, interest and corporate income tax is about $1.47 million.

18.5.1.2 Conversion costs. Following accepted accounting practice, the fixed and variable costs have been allocated to the two joint products in proportion to their contribution to sales revenue. On this basis, and under the base case scenario, the conversion cost (sum of the total fixed cost and the total variable cost except the cost of raw material) is about $1.46/kg for each product. This is about evenly split between the total variable cost (48%) and total fixed cost (52%). Direct labour accounts for about 34% of the conversion cost.

18.5.1.3 Profitability. Equity-based after-tax cash flow (EBCF) and discounted cash flow (EBDCF) as well as accumulated discounted cash flow (EBADCF) are plotted in Figure 18.3 with respect to the elapsed time measured in years from start-up, as is the total capital cost or assets-based accumulated discounted cash flow (TCCBADCF). From these graphs (or Table 18.4), it can be seen that the equity- and assets-based payback periods are about 2.93 and 5.82 years, respectively. Also the equity- and assets-based after-tax net present values of the project are $3.76 and $4.43 million, and the corresponding after-tax internal rates of return are about 41% and 24%, respectively.

It should be noted that, had the actual raw material cost of $1.15/kg (instead of $0.85/kg) and the actual product selling prices of $3.39/kg (instead of $3.00/kg) for reduced-calorie peanuts, and $3.49/kg (instead of $3.00/kg) for peanut oil been used in the base case scenario, other things being equal, the sales revenue would be $3.97 million (instead of $3.49 million), a net income of $0.88 million (instead of $0.76 million) or 22% of sales, and corresponding annual income before depreciation, interest and corporate income tax of about $1.59 million (instead of $1.47 million). The after-tax profitability of the project would also improve considerably: net present value of the project will increase by about $0.60 million (over $3.76 million), the return on equity by 4 percentage points (over 41%), and return on assets by 1.5 percentage points (over 24%).

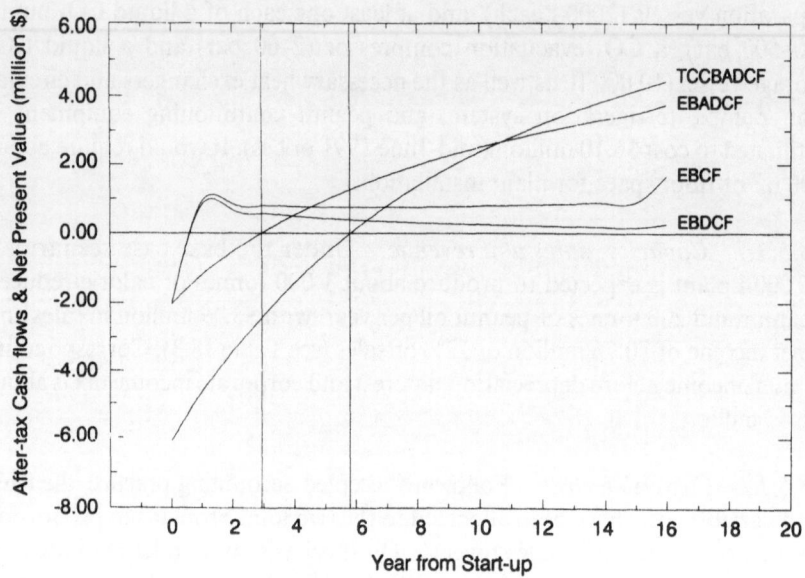

Figure 18.3   After-tax equality-based cash flow (EBCF), discounted cash flow (EBDCF) and net present values based on equity (EBADCF) and total capital cost (TCCBADCF).

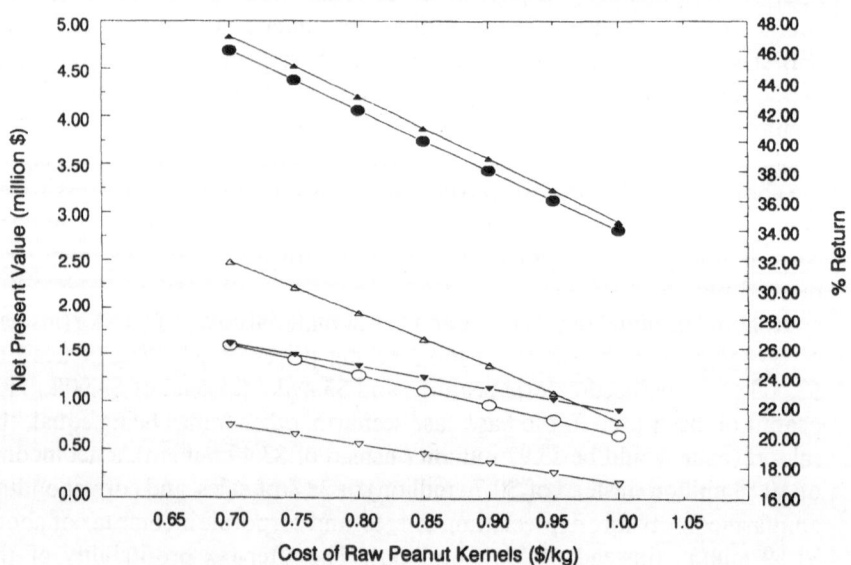

Figure 18.4   Effect of raw material cost on the after-tax net present value (NPV ●, 16 kl,O, 8 kl), return on equity (%ROE ▲, 16 kl; △, 8kl) and return on total capital cost (%IRR ▼, 16 kl; ▽, 8 kl) of 8000 and 16 000-l extraction vessels' capacity SCFE plants.

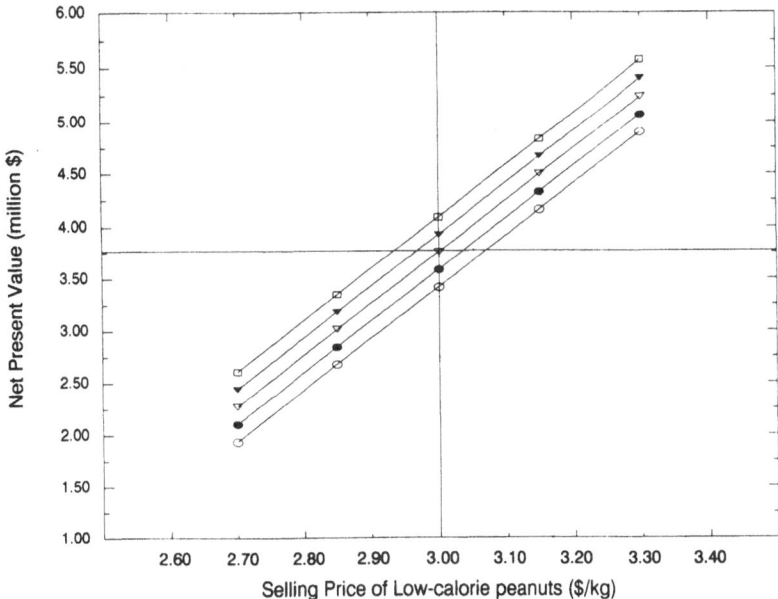

Figure 18.5 Effect of selling prices of low-calorie peanuts and extracted oil on the net present value of a 16 000-l extraction vessels' capacity SCFE plant. □, $3.30/kg; ▼, 3.15; ▽, 3.00; ●, 2.85; ○, 2.70.

18.5.2 Sensitivity analysis

In what follows, the sensitivity of profitability to various cost and other economic factors is presented. With the exception of results presented in Table 18.4, comparing the profitability in three different representative countries, only one or two factors are varied at a time for a particular analysis, other factors being held constant at the levels indicated in Table 18.1 for the base case scenario. Further, unless indicated otherwise, the net present values and returns given in the sensitivity analysis are after-tax and based on equity.

Effect of raw material cost and plant size on the after-tax net present value (NPV) and return on equity (ROE), and return on assets (IRR) is shown in Figure 18.4. The NPV of the 16 000-l plant is more than three times that of the smaller 8000-l plant, and is only about half as sensitive to change in raw material cost ($\approx - 1.4\%/\%$). The ROE and IRR of the larger plant are some 15 and 5 percentage points better than those of the smaller plant, also being some 30% and 8% less sensitive to change in raw material cost. The small NPV ($0.65 million) at higher raw material cost ($1.00/kg) indicates room for only a small margin of error in cost estimates beyond which the smaller, 8000-l plant would become uneconomical due to raw material cost. This, however, is not the case with the larger, 16 000-l plant with an NPV of about $2.82 million.

Figure 18.5 illustrates that the larger, 16 000-l plant, is still quite viable even if the selling prices were to decline from $3.00/kg to $2.70/kg for both the

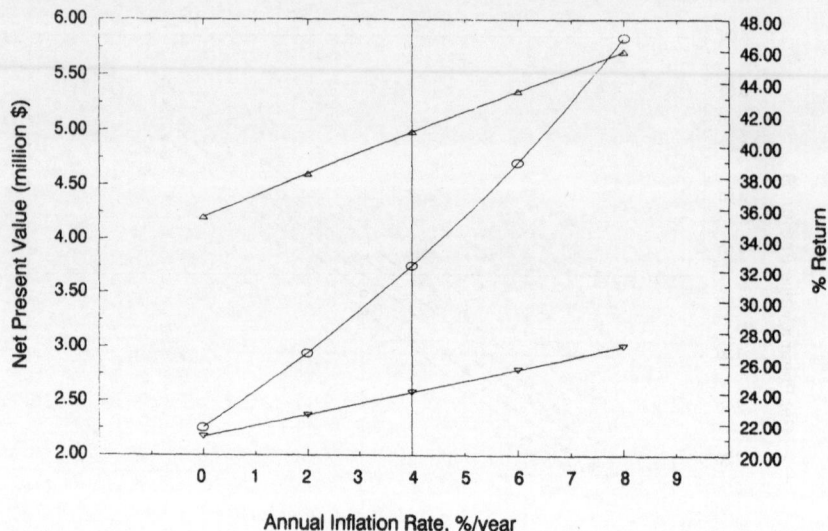

Figure 18.6 Effect of inflation on the after-tax equity-based net present value (NPV, O), return on equity (%ROE, △) and return on total capital (%IRR, ▽) for a 16 000-l extraction vessels' capacity SCFE plant.

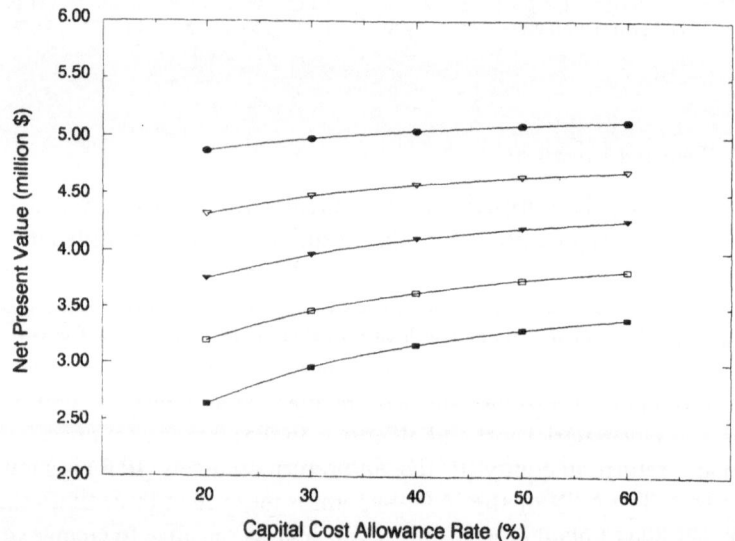

Figure 18.7 Effect of capital cost allowance rates and income tax rates on the after-tax net present value of a 16 000-l extraction vessels' capacity SCFE plant. ●, 20%; ▽, 30%; ▼, 40%; □, 50%; ■, 60%.

calorie-reduced peanuts and peanut oil, with an NPV of about $2.00 million. Sensitivity of NPV to changes in selling price of calorie-reduced peanuts is about 4%/% compared with about 1%/% in the case of peanut oil.

Results presented in Figure 18.6 show that the project's profitability will

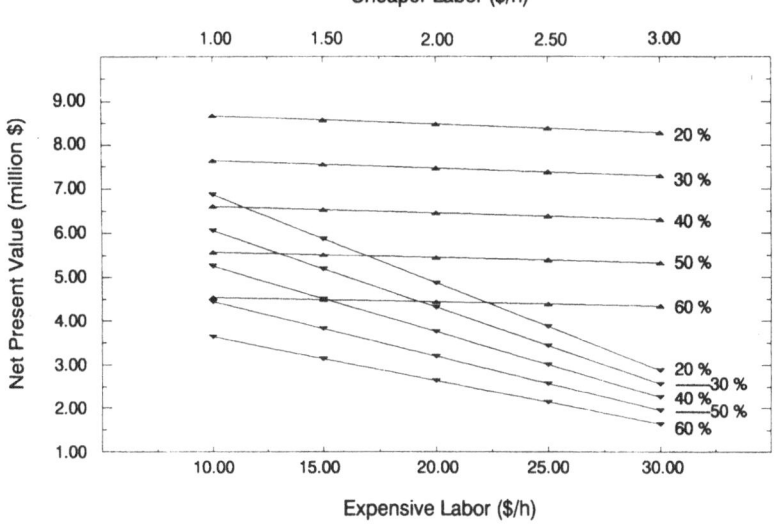

Figure 18.8 Effect of labour cost and the income tax rate (ITAXR) on the after-tax net present value of a 16 000-l extraction vessels' capacity SCFE plant. Cheaper labour, ▲; expensive labour,▼.

improve with inflation in the costs and revenue. An improvement of about $0.50 million in NPV, and about 1.3 and 0.7 percentage points, respectively, in ROE and IRR, is indicated for every percentage point increase in inflation.

A very interesting, and intuitively expected, effect of taxation policy (i.e. the corporate income tax (ITAXR) and capital cost allowance (CCAF) rates) on the NPV, shown in Figure 18.7, clearly demonstrates the beneficial effect of locating the plant in countries with (i) lower income tax rate and (ii) faster write-offs (higher capital cost allowance rates). The loss of NPV is about 5–7 times more sensitive to an increase in income tax rate than to a drop in the capital cost allowance rate. For example, even the highest NPV at 60% ITAXR ($3.39 million, at 60% CCAF) is still lower than the lowest NPV at 40% ITAXR ($3.76 million, at 20% CCAF).

Another very interesting aspect of profitability of the 16 000-l plant, as influenced by the labour cost and corporate income tax rates, is examined in Figure 18.8. As can be seen, the project will be most profitable (NPV ≥ $7.5 million) if cheap labour (up to $3.00/h) and low corporate income taxes (≤ 30%) are both available, as is the case in many developing countries. With the wage rates prevailing in Canada and the United States ($15–20/h), income tax rates will have to be less than 30%, and then too the plant will be able to compete only with low wage rate countries which have corporate income tax rates of 50% or higher (which is not likely in most low wage countries). Other analytical results (not presented here) indicate that the capacity of a plant located in countries like Canada and the United States would have to be

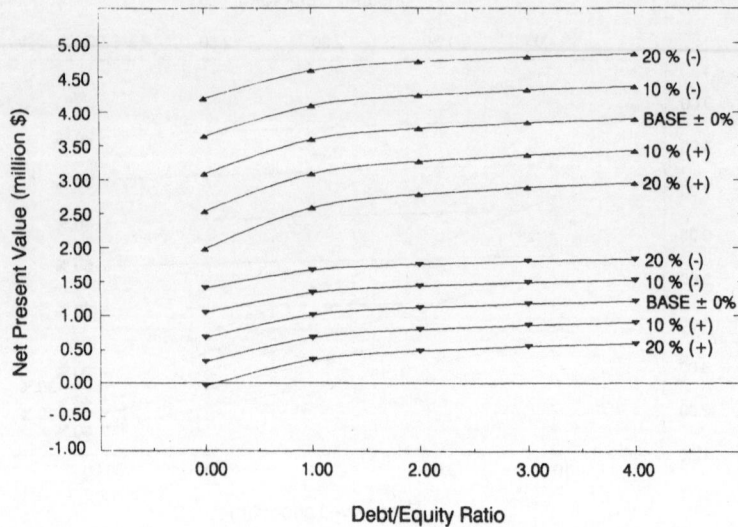

Figure 18.9   Effect of the debt/equity ratio and variation in plant costs (CCOST) on the after-tax net present value of SCFE plants of 8000 and 16 000 l capacity. Size 16 kl, ▲; 8 kl,▼.

20 000 l or more to be able to compete with a 16-000-l plant located in low wage countries.

Further, the steeper slopes of the curves for expensive labour countries, compared with relatively flat curves for cheaper labour countries, clearly indicate that a plant located in expensive labour countries will entail greater risk of reduced profitability due to wage escalation by the same percentage increment.

The effect of the size of the plant is re-examined in Figure 18.9 in the context of the debt/equity ratio and variation in plant cost. As can be seen, the smaller, 8000-l plant, will become unprofitable (NPV less than zero) if it were totally financed by equity and plant cost overrun were to exceed 20%. With debt financing, the plant still remains only marginally profitable. This, however, is not the case with the larger, 16 000-l plant which has an impressive NPV of about $3.00 million even with the plant cost overrun by 20%. This analysis further confirms the conclusion drawn earlier from Figure 18.4 that a plant of 16 000-l (or greater) capacity is to be preferred due to its considerably better profitability.

18.5.3   Considerations for plant location

Of necessity, the sensitivity analysis above has been limited to examining the effect of one, two, or at the most three, economic factors at a time. However, many more economic factors may be different simultaneously among the sites being considered for plant location. Therefore, in what follows, project profitability is compared for locating a 16 000-l capacity plant in three different countries, representing typical mixes of economic factors: the United States (high labour cost, low interest on debt and high corporate income tax); Canada (medium labour cost, medium interest and medium corporate income tax); India (low labour cost, high interest and low corporate income tax).

As can be seen from results presented in Table 18.4, a 16 000-l plant located in a country like India would be the one most profitable with an NPV of $7 million, followed by Canada ($4 million) and United States ($3 million), with corresponding returns on equity of 59, 44, and 42%, respectively. This relative positioning is due to cheaper labour and relatively better taxation policies of India, compared with Canada and United States, despite the highest interest rate on debt and least leverage in financing the capital cost.

Under the circumstances examined, the payback period is also much shorter for the plant located in India, reducing the period during which the invested capital may be regarded as being at risk.

18.6 Economic considerations

Having considered the commercial viability of supercritical extraction plants, it is instructive to also assess the overall economic merit of such an opportunity in a macro sense at the global level. What would it mean to countries supporting such a new industry? Based on the revenue and cost assumptions given in Table 18.1, the two joint products are expected to return a revenue of about $2910.00/tonne of peanut kernels used at a raw material cost of about $850.00. This represents an added value of some $2060.00/tonne of peanut kernels. On the other hand, the alternative traditional crushing of peanuts for oil and meal, even if this oil were valued at the price of the high quality 'cold-pressed' oil from the SFE plant ($3.00/kg) and the residual meal were valued at about $250.00/tonne, would provide an added value of only about $620.00/tonne of peanut kernels used. Thus the proposed new product mix is likely to provide an economic rent of some $1440.00/tonne of the raw material used. This indicates enormous economic benefit that a peanut producing nation can derive by diverting peanuts from the traditional crushing for oil to the production of calorie-reduced peanuts.

18.7 Conclusions

A 16 000-l plant, producing about 1000 tonnes/year of calorie-reduced peanuts for the snack food market, and about 200 tonnes/year of 'cold-pressed' quality peanut oil, installed at a capital cost of $6.10 million, is highly profitable, with an NPV of $3–7 million, depending upon the location, and corresponding return on equity of some 42–59%.

As the production from such a plant is less than 2% of the market demand for the calorie-reduced peanuts, there is very little marketing risk involved.

Appendix: Measures of profitability

The net present value, internal rate of return, and the payback period, based on the discounted cash flow analysis provide a good measure of profitability of capital project. These terms are explained below.

In computing the cash flows for discounted cash flow analysis of a capital project partly financed with debt (total capital cost = debt + equity), it should be remembered that the proceeds of the debt reduce the net cash flow for initial investment to the value of equity. Further, besides other operating costs, the corporate taxes payable, and the interest paid out on debt, the principal paid back to discharge the debt must also be deducted from the future cash receipts to calculate the future cash flows. Moreover, the capital cost allowance (although not by itself a cash flow) and the interest cost reduce the taxable income, thereby shielding some of the profits from corporate tax. With the cash flows calculated in this manner, the net present value, internal rate of return, and the payback period are indeed based on equity. These values can also be calculated, based on assets, by considering the debt to be internalized, i.e. while the interest paid out is regarded as a cost for calculating the taxable income for estimating the income taxes payable, the proceeds of the debt as well as the principal plus interest paid out are not taken into account in computing the cash flows.

The net present value (NPV) of an investment is the sum of the present values of future after-tax cash flows (positive if cash receipt, negative if cash payout) and the present value of the cash flow (negative) at the time of initial investment. The net present value is thus the value of the accumulated discounted cash flow at the end of the project. A net present value of 'zero' means that the project will be earning an after-tax rate of return on its investment equal to the discount rate used in the computations. If there were no other projects requiring capital investment, then making the investment in a project with 'zero' net present value is still justified. Further, the greater the net present value, the greater are the profits measured in present value money that the project would earn over and above recovering its investment also measured in the present value money. However, if there were more than one project competing for the capital, then the project(s) with the potential to maximize the net present value as a whole should be included in the capital budget.

It should be noted that the equity- and assets-based net present values would be equal when the interest on debt and the discount rate are equal because the internalized principal amount of the debt is exactly offset by the present value of future debt payments. However, when the interest on debt is lower, then the assets-based net present value will also be lower and vice versa (see scenarios of United States and India with respect to Canada in Table 18.4).

The internal rate of return is another useful way of looking at the profitability. This is the after-tax discount rate that will make the net present value of the investment and future cash flows equal to 'zero'. It may also be based on equity or assets, and referred to as return on equity, or return on assets, respectively.

Payback period is yet another way of assessing the profitability and more importantly the period during which the investment is at risk, i.e. has not yet been fully recovered. This is determined by plotting the accumulated discounted cash flow with respect to time, and noting the time when the net present value crosses over from negative (i.e. there is still some unrecovered investment) to positive. The payback period can also be based on equity or assets.

References

Clark, J. H. and Clark, J. S. (1984) The economic potential of peanuts for Southern Ontario, a report prepared by the Ontario Regional Development Branch, Agriculture Canada.

Clydesdale, F. M. and Francis, F. J. (1985) Dietary goals, in *Food, Nutrition and Health*, AVI, Westport, CT, pp. 231–238.

Gordon, M. J. and Shillinglaw, G. (1974) The capital expenditure decision, in *Accounting: A Management Approach*, 5th edition, Richard D. Irvin, Homewood, IL 60430, pp. 613–645.

Health and Welfare Canada (1990) Nutrition Recommendations, The Report of the Scientific Review Committee, Dept. of Supply and Services, Cat. no. H49-42/1990E.

Passey, C. A. and Patil, N. D. (1989a) Process for preparing low-calorie nuts. Canadian Patent Application No. 615,387-9.

Passey, C. A. and Patil, N. D. (1989b) Supercritical carbon dioxide extraction process for making low calorie peanuts, in: *Proc. 2nd Int. Conf. on Separations Sci. Technol*, eds. M. H. I. Baird and S. Vijayan, Canadian Society for Chemical Engineering, Ottawa.

Passey, C. A. and Patil, N. D. (1993) Effect of pretreatments on the extraction and quality of calorie-reduced peanuts prepared by the Passey–Patil supercritical extraction process, unpublished.

Passey, C. A. (1993) unpublished data.

Passey, C. A. (1993) unpublished design information.

Pominski, J., Patton, E. L. and Spadaro, J. J. (1964) Pilot plant preparation of defatted peanuts. *J. Am. Oil Chem. Soc.* **41**, 66.

Pominski, J., Pearce Jr., H. M. and Spadaro, J. J. (1969) Factors affecting water solubles of partially defatted peanuts. *Food Technol.* **24**, 76.

Pominski, J., Pearce Jr., H. M. and Spadaro, J. J. (1971) Storage of raw pressed peanuts and roasted partially defatted peanuts. *Am. Peanut Res. Educ. Assoc. J.* **3**, 133.

USDA (1984) *Peanuts: Background for 1985 Farm Legislation*, United States Department of Agriculture, Economic Research Service, Agricultural Information Bulletin No. 469.

19 *In situ* monitoring of selective extraction of a mixture of higher fatty acids with supercritical carbon dioxide

Y. IKUSHIMA, N. SAITO, K. HATAKEDA and S. ITO

Abstract

On-line supercritical fluid extraction (SFE) supercritical fluid chromatography (SFC)/Fourier transform infrared spectrometry (FTIR) with a high pressure flow cell withstanding pressures as high as 50 MPa was built up to make an *in situ* observation at the supercritical state. The SFE SFC/FTIR system enabled the rapid and precise identification of higher fatty acid esters extracted by supercritical carbon dioxide and *in situ* observation of the separation behavior of a mixture of higher fatty acid esters. The amount of solute extracted by unit volume of carbon dioxide was well correlated with a parameter derived from the solubility parameter concept. Furthermore, the separation efficiency of a mixture of higher fatty acids could be represented by the solvation power of supercritical carbon dioxide as well as interaction forces of solute with stationary and mobile (supercritical carbon dioxide) phases.

19.1 Introduction

In the field of food science, supercritical carbon dioxide (SC-CO_2) extraction is known to be a useful method for extraction without denaturation of valuable materials such as docosahexaenoic acid (DHA), eicosapentaenoic acid (EPA), linoleic acid, linolenic acid, and others contained in natural resources (Arai and Saito, 1986; Yamaguchi and Murakami, 1986). When we use SC-CO_2 for selectively extracting a desired component from a mixture of higher fatty acids, the preferred extraction and separation can be attained at around ambient temperatures by introducing entrainers and/or using proper packing for the separation (Ikushima *et al.*, 1988, 1989a). This method enabled a large portion of DHA contained in squid oil to be extracted selectively at a concentration over 90wt% (Ikushima *et al.*, 1989b). However, it is difficult to make an accurate determination of the solubility and separation efficiency of these compounds in supercritical fluids (SCF) because the composition of the equilibrium SCF phase is easily shifted by both pressure and temperature on sampling. Much attention has, therefore, been drawn to the measurement *in situ* at the supercritical state.

We report the development of an on-line SFE SFC/FTIR system which allows rapid and accurate identification of solutes extracted with SC-CO$_2$ and *in situ* observations of the separation behavior of mixtures in SC-CO$_2$. Furthermore, the solubilities determined by this system are correlated by a model based on a regular mixing rule. The separation efficiency defined from the retention time of each component is represented by taking into account the physico-chemical properties of SC-CO$_2$ determined spectroscopically as well as computed from the interaction forces of solutes with the mobile (SC-CO$_2$) and stationary phases.

19.2 Materials and methods

19.2.1   SFE SFC/FTIR system

A schematic diagram of the system developed is shown in Figure 19.1(a). Liquid carbon dioxide was charged into a syringe pump through a 1.5-mm tube and a check valve and compressed to the desired pressure. Pressure control was achieved by introducing high-speed flow switching and controlling the outlet pressure irrespective of the mass flow rate of the fluid (Saito *et al.*, 1988). Samples were injected into the SFE SFC system with an injection valve or extracted in a stainless steel extractor of 10 cm^3. A conventional high performance liquid chromatography (HPLC) column (150 mm long and 4.6 mm i.d.) was used for the SFC. The infrared spectrometer was equipped with a narrow bandwidth mercury–cadmium–telluride (MCT) detector cooled by liquid nitrogen. The low frequency cut-off was approximately 800 cm^{-1}. Spectra were acquired at 1 cm^{-1} nominal resolution and interferograms were accumulated at a rate of 1 scan per 1.2 s. A heated stainless steel tube of 0.25 mm i.d. was employed as the transfer tube from the SFE or SFC to an IR flow cell.

19.2.2   IR flow cell

A schematic diagram of the IR flow cell is shown in Figure 19.1(b). The mobile phase was introduced into the bottom of the cell and emerged from the top. The infrared radiation was impinged on the cell at 90° to the flow. The selection of window material was very important for the achievement of optimum sensitivity and adequate resolution of IR measurements at the supercritical state. In this work, an IRTRAN 2 window which is formed by crystalline ZnS was used because the material possesses a high modulus of rupture and a wide, useable mid-IR region (800–5000 cm^{-1}) (Optical Society of America, 1978).

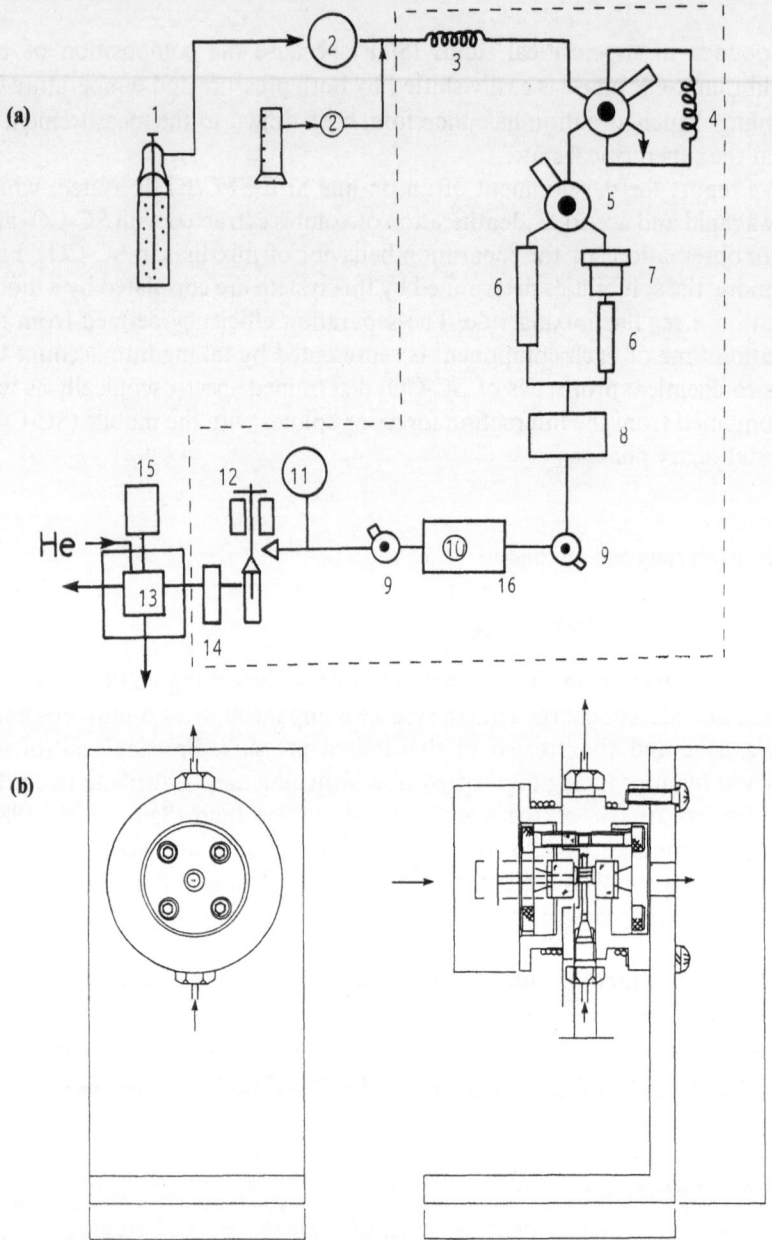

Figure 19.1   (a) Schematic diagram of apparatus used. 1, CO_2 cylinder; 2, CO_2 pump; 3, preheater; 4, injector; 5, three-way valve; 6, separation column; 7, extractor; 8, UV/vis detector; 9, on-off valve; 10, flow cell; 11, pressure transducer; 12, pressure regulator; 13, FID detector; 14, gas sampler; 15, recorder; 16, FTIR. (b) Schematic diagram of IR flow cell.

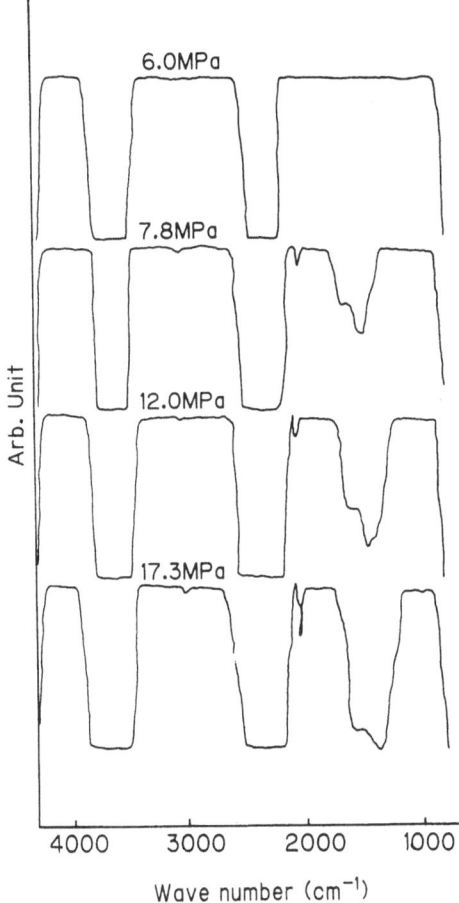

Figure 19.2   FTIR spectra of SC-CO$_2$ at 323 K and pressures given.

This IR flow cell can withstand a pressure of 30 MPa and a temperature of 373 K by using 5-mm thick windows. By comparison, to maintain a pressure of 20.7 MPa, NaCl windows would need to be 10 mm thick (Optical Society of America, 1978). The nominal volume of the cell was 16 μl (optical path length of 5 mm, and cross-sectional area of 3 mm^2). This cell which was placed in a cell heating unit was maintained at temperatures of 318–353 (±0.5) K. An ultraviolet/visible (UV/vis) spectrophotometer and a gas chromatograph with FID detector were mounted in series to enhance the analytical capabilities.

19.2.3   *Chemicals*

Lipids such as stearic, linolenic acid methyl esters, DL-α-tocopherol, and their mixtures, which were the analytical standards in the experiments, were purchased from Gasukuro Kogyo Inc. A commercial-grade carbon dioxide

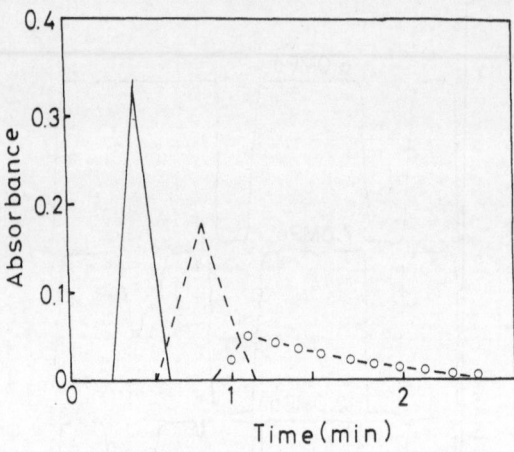

Figure 19.3 The change of absorbance at 1053 cm⁻¹ with time on the extraction of DL-α-tocopherol at 323 K. ——, 18.6 MPa; – –, 11.7 MPa; O, 8.3 MPa.

(above 99.995% purity) was used. All chemicals were used without further purification.

19.3   Results and discussion

19.3.1   IR spectra for SC-CO₂

In order to analyse higher fatty acid esters and DL-α-tocopherol, it is absolutely necessary to know the IR absorption spectra of SC-CO₂. Figure 19.2 shows the IR absorption spectra of SC-CO₂ at 6.0~17.3 MPa and 323 K. The strong absorption between 1200 and 1400 cm⁻¹ due to the Fermi resonance increased with an increase in pressure above the critical pressure. The absorption band, however, does not influence the precision in the IR analysis since a spectrum of SC-CO₂ alone at each pressure and temperature can be measured as the background spectrum and be subtracted from the spectrum of solutes measured under the same conditions. Although the asymmetric stretching band of CO₂ around 2350 cm⁻¹ obscures the spectrum in this region, few compounds possess characteristic absorption bands near this wave number. In addition, the IR transparency of SC-CO₂ is nearly ideal, and this SC-CO₂ medium is suitable for IR measurements.

19.3.2   Solubilities determined spectroscopically

The amount of DL-α-tocopherol extracted by unit volume of CO₂ during the whole course of extraction was examined as a function of pressure with the on-line SFE/FTIR. Figure 19.3 shows the time-course of the absorption intensity on the extraction at 323 K. The absorption band at 1054 cm⁻¹ is due

Table 19.1 Amount of DL-α-tocopherol extracted with supercritical CO_2 at 323 K determined by on-line FTIR

Pressure (MPa)	Weight (mg)	Flow rate (Nl/min)	Amount of extract (mg/min)
18.6	7.8	0.55	22.02
11.7	5.8	0.56	14.44
8.3	5.3	0.46	2.73

to C–O–C stretching of DL-α-tocopherol. The time at the strongest absorption intensity, which is hereafter called the 'retention time', became longer with decrease in pressure. The variation in intensity as a function of time is considered to correspond to the change in concentration of solute. Therefore, the amount of solute extracted per unit volume of CO_2 can be easily determined from the velocity of CO_2 and the weight of solute charged into the extractor. Table 19.1 provides the pressure-dependence of the amount of DL-α-tocopherol extracted per unit volume, which is defined as the solubility. It is noted that the solubility of DL-α-tocopherol in SC-CO_2 is significantly enhanced with pressure above the critical pressure. These values are comparable to those reported elsewhere (Ohgaki *et al.*, 1987). In addition, the solubilities of fatty acid esters in SC-CO_2 can also be determined by the above-mentioned method. Thus, with the SFE/FTIR system, the solubilities of non-volatile and heavy compounds in SC-CO_2 can be determined with high accuracy using a very small amount of solute.

19.3.3 In situ *observation of separation behavior*

The SFC/FTIR system was used to investigate the influence of pressure on separation performance of a mixture of fatty acid methyl esters. A mixture of 0.3 mg stearic and 0.3 mg linolenic acid methyl esters in 10 μl chloroform was injected into the system. The flow rate of CO_2 was 0.5 Nl/min (1 Nl = 1 l of gas at 293 K and 101.3 kPa). $AgNO_3$ (0.1wt%) supported on silica gel was employed as a packing for separation because a mixture of fatty acids with different degree of unsaturation could not be separated well using only silica gel as the packing (Ikushima *et al.*, 1988, 1989b). Figure 19.4 shows the change in IR spectra with time at 14.7 MPa and 323 K. The separation behavior could be confirmed in real time by FTIR. The absorption band at 1744 cm^{-1} due to >C=O stretching of fatty acid methyl ester showed the major two peaks at 12.2 and 21.0 min, respectively. This result indicates that the mixture of stearic and linolenic acids can be separated well, which may be due to the difference in interaction forces between stearic and linolenic acids with $AgNO_3$ supported on silica gel. The latter peak can be assigned to linolenic acid because $AgNO_3$ has a stronger interaction for fatty acids with higher degree of unsaturation (Ikushima *et al.*, 1988). This was also confirmed by gas chromatography with

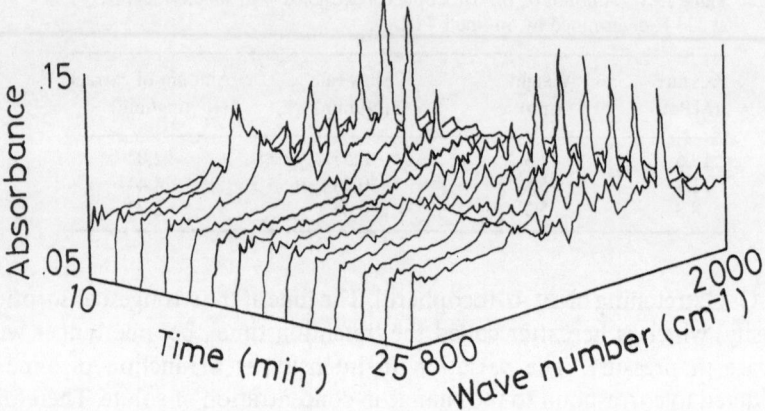

Figure 19.4 The change of FTIR spectra with time on the separation of a mixture of stearic and linolenic acid methyl esters with SC-CO_2 at 14.7 MPa using $AgNO_3$ supported on silica gel column.

a FID detector connected to the SFC/FTIR system. The SFC/FTIR system can monitor *in situ* separation of a mixture of lipids at the supercritical state.

Table 19.2 gives the pressure-dependence of the retention time of chloroform, stearic, and linolenic acid methyl esters on separation at 323 K. The identification of chloroform was made with C–Cl stretching of the absorption band at 1219 cm^{-1}. The retention time is influenced by the linear velocity of CO_2, and for comparison it must be normalized by the retention time of chloroform as reference in such a way that $T_j = (t_j - t_{ref})/t_{ref}$, where t_j is the retention time of component j and subscript ref indicates chloroform. Thus, the separation efficiency which is abbreviated as SE between components i and j is defined as follows:

$$SE = T_i/T_j, \quad T_i > T_j \tag{19.1}$$

As shown in Table 19.2, the retention time is longer for a decrease in pressure, but the separation efficiency was improved. This is in agreement with the results reported previously (Ikushima *et al.*, 1988, 1989b).

19.3.4 Correlation by model

We model the extraction and separation behavior of solutes through the $AgNO_3$-supported silica gel column as a chromatographic mechanism. That is, the amount of each solute eluted from the column strongly depends upon the difference in interactions with mobile (SC-CO_2) and stationary phases. We attempt to represent the interactions of solutes with the two phases by a model that includes the solubility parameter on the assumption of the regular mixing

Table 19.2 Retention time of strongest intensity of each component on separation of a mixture of linolenic and stearic acid methyl esters in chloroform using 0.1wt% AgNO₃ supported on silica gel column with SC-CO₂ at 323 K and linear velocity of 50 cm/s

Pressure (MPa)	Retention time (min)			
	Linolenic ($1744\ cm^{-1}$)	Stearic ($1744\ cm^{-1}$)	Chloroform ($1219\ cm^{-1}$)	$T_{li}/T_{st}{}^a$
18.6	15.4	9.2	1.4	1.79
14.7	21.0	12.2	1.4	1.82
12.7	30.1	16.7	1.3	1.87

$^aT_i = (t_i - t_{ch})/t_{ch}$.

rule. If we take the pure liquid to be the standard state, the regular mixing rule describes the activity coefficient (γ) of a solute (i) in a phase (l) as

$$\ln \gamma_i = V_i(\delta_i - \delta_l)^2/RT \qquad (19.2)$$

where δ is the solubility parameter, R is the gas constant, and V is the molar volume. The capacity factor (Schoermakers *et al.*, 1978) k, used as the retention parameter represents the chromatographic separation power. It can be expressed as the ratio of the activity coefficients of solute in the two phases as follows:

$$k = (\gamma_m/n_m)/(\gamma_s/n_s) \qquad (19.3)$$

where n is the number of moles in the column and the subscripts m and s indicate the mobile and stationary phases, respectively. The capacity factor can be expressed from equations (19.2) and (19.3) as follows:

$$\ln k_i = V_i(\delta_{CO_2} + \delta_s - 2\delta_i)(\delta_{CO_2} - \delta_s)/RT + \ln(n_s/n_{CO_2}) \qquad (19.4)$$

where the subscript CO_2 indicates SC-CO₂. The solubility parameter of SC-CO₂ was previously determined elsewhere (Ikushima *et al.*, 1987).

19.3.4.1 Correlation of solubility. Substituting zero for δ_s in equation (19.4) gives

$$\ln k_i = V_i(\delta_{CO_2} - 2\delta_i)\delta_{CO_2}/RT \qquad (19.5)$$

Equation (19.5) expresses the capacity factor of solute (i) in SC-CO₂ without using packing, and can be used to express the interaction force between solute (i) and SC-CO₂ with no packing. Figure 19.5 shows the relationship between the solubility of DL-α-tocopherol and the capacity factor determined by equation (19.5) which indicates a good linear relationship. The solubilities of higher fatty acid esters in SC-CO₂ can also be correlated with equation (19.5).

19.3.4.2 Correlation of separation efficiency. We assume that the above-

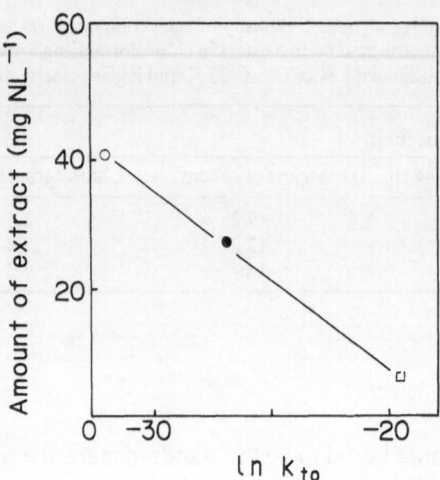

Figure 19.5 Relationship between the extracted amount and $\ln k_{to}$ in the extraction of DL-α-tocopherol with SC-CO2 at 323 K. O, 18.6 MPa; ●, 11.7 MPa; ◻, 8.7 MPa.

mentioned separation efficiency is dependent upon several aspects of solvation of SC-CO$_2$ as well as the difference in interaction forces of solute with the mobile and stationary phases and hence the following multiparameter approach is examined:

$$\ln T_i/T_j = \ln(k_i/k_j) + s\pi^* + a\alpha + bB_{MeOD} + d\delta^2 \qquad (19.6)$$

where π^* is an index of solvent dipolarity/polarizability (Kamlet *et al.*, 1977), α is a measure of the solvent hydrogen-bond donor acidity (Kamlet *et al.*, 1977), B_{MeOD} is a measure of the solvent hydrogen-bond acceptor basicity (Burder *et al.*, 1976), and the coefficients s, a, b, and d describe the sensitivity to the separation efficiency. Ikushima *et al.* (1991) have determined the values of B_{MeOD} and π^* of SC-CO$_2$ by FTIR spectroscopy methods and the α-value of SC-CO$_2$ by UV/vis spectroscopy, respectively.

Equation (19.6) is applied to estimate the pressure-dependence of the separation efficiency of a mixture of stearic and linolenic acid methyl esters in chloroform with SC-CO$_2$ using a AgNO$_3$-supported silica gel column. The separation efficiency ($\ln T_{li}/T_{st}$) calculated by equation (19.6) is indicated in Figure 19.6. The subscripts li and st indicate linolenic and stearic acid methyl esters, respectively. The $\ln(T_{li}/T_{st})$ values shown in Table 19.2 are also plotted in Figure 19.6. The coefficients in equation (19.6) are determined by least squares. It was reported that the solubility parameter of the uncoated silica gel is close to that of alumina and its value is about 16 cal$^{1/2}$cm$^{-3/2}$ (Karger *et al.*, 1978). We presume that the solubility parameter of 0.1wt% AgNO$_3$ supported on silica gel is 16.5 cal$^{1/2}$cm$^{-3/2}$ with the assumption that the δ_s is nearly

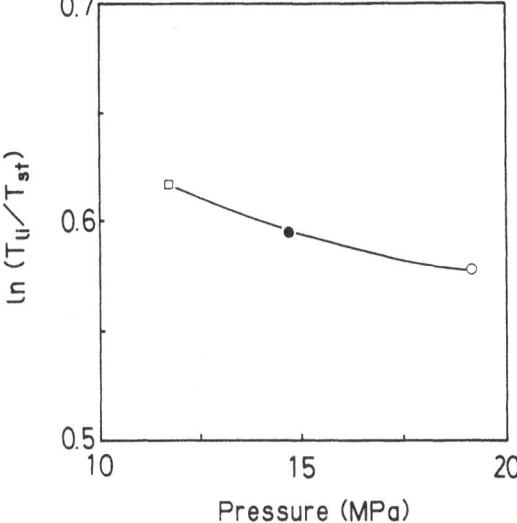

Figure 19.6 Correlation of ln (T_{li}/T_{st}) by equation (19.6) on the separation of a mixture of linolenic and stearic acid methyl esters in chloroform using a silica gel column with SC-CO_2 at 323 K and pressures given. ——, Calculated; O, 19.1 MPa; ●, 14.7 MPa; □, 11.7 MPa.

proportional to the mole fractions of uncoated silica gel and Ag. As shown in Figure 19.6, the experimental data can be correlated by equation (19.6) very well. Thus, this model is suitable for estimating the pressure-dependence of the separation efficiency of a mixture of higher fatty acids with different degree of unsaturation in SC-CO_2 using packing.

References

Arai, K. and Saito, S. (1986) Application of supercritical fluid extraction method to extraction of physiologically active substances from natural products. *J. Jpn. Oil Chem. Soc.* **35**, 267.

Burder, A. G., Collier, G. and Shorter, J. (1976) Influence of aprotic solvent on the O–D stretching bond of methan[^{2}H]ol. *J. Chem. Soc. Perkin Trans. 2* 1627.

Ikushima, Y. Goto, T. and Arai, M. (1987) Modified solubility parameter as an index to correlate the solubility in supercritical fluids. *Bull. Chem. Soc. Jpn.* **60**, 4145.

Ikushima, Y., Hatakeda, K., Saito, N., Ito, S., Asano, T. and Goto, T. (1988) A supercritical carbon dioxide extraction from mixtures of triglycerides and higher fatty acid methyl esters using a gas-effusion-type system. *Ind. Eng. Chem. Res.* **27**, 818.

Ikushima, Y., Saito, N. and Goto, T. (1989a) Selective extraction of oleic, linoleic, and linolenic acid methyl esters from their mixtures with supercritical carbon dioxide-entrainer systems and a correlation of the extraction efficiency with a solubility parameter. *Ind. Eng. Chem. Res.* **28**, 1364.

Ikushima, Y., Hatakeda, K., Saito, N., Ito, S. and Goto, T. (1989b) Selective extraction of docosahexaenoic acid (DHA) from entrails of squid with supercritical carbon dioxide. *Kagaku Kogaku Ronbunshu* **15**, 511.

Ikushima, Y., Saito, N., Arai, M. and Arai, K. (1991) Solvent polarity parameters of supercritical carbon dioxide measured by infrared spectroscopy. *Bull. Chem. Soc. Jpn.* **64**, 2224.

Kamlet, M. J., Abboud, J. L. and Taft, R. W. (1977) The solvatochromic comparison method. The π^* scale of solvent polarities. *J. Am. Chem. Soc.* **99**, 6027.

Karger, B. L., Eon, C. and Synder, L. R. (1978) An expanded solubility parameter treatment for classification and use of chromatographic solvents and absorbents. Parameters for dispersion, dipole and hydrogen-bonding interactions. *J. Chromatogr.* **125**, 71.

Ohgaki, K., Tsukahara, K., Semba, K. and Katayama, T. (1987) A fundamental study of supercritical fluid extraction—solubilities of α-tocopherol, palmitic, and tripalmitic in compressed carbon dioxide at 298 and 313 K. *Kagaku Kogaku Ronbunshu* **13**, 298.

Optical Society of America (1978) *Handbook of Optics*, ed H. B. Crawford and B. Carson, McGraw-Hill, New York.

Saito, M., Yamaguchi, Y., Kashiwazaki, H. and Sugiyama, M. (1988) New pressure regulating system for constant mass flow supercritical fluid chromatography and physico-chemical analysis of mass-flow reduction in pressure programming by analogous circuit model. *Chromatographia* **25**, 801.

Schoermakers, P. J., Billiet, H. A. H., Tijissen, R. and de Galan, L. J. (1978) Gradient selection in reversed-phase liquid chromatography. *J. Chromatogr.* **149**, 519.

Yamaguchi, K. and Murakami, M. (1986) Application of supercritical fluid extraction to aquatic organisms. *J. Jpn. Oil Chem. Soc.* **35**, 260.

Index